T0214044

# Communications
# in Computer and Information Science 1125

*Commenced Publication in 2007*
Founding and Former Series Editors:
Phoebe Chen, Alfredo Cuzzocrea, Xiaoyong Du, Orhun Kara, Ting Liu,
Krishna M. Sivalingam, Dominik Ślęzak, Takashi Washio, Xiaokang Yang,
and Junsong Yuan

More information about this series at http://www.springer.com/series/7899

Jürgen Gerhard · Ilias Kotsireas (Eds.)

# Maple in Mathematics Education and Research

Third Maple Conference, MC 2019
Waterloo, Ontario, Canada, October 15–17, 2019
Proceedings

 Springer

*Editors*
Jürgen Gerhard
Maplesoft
Waterloo, ON, Canada

Ilias Kotsireas (ID)
Wilfrid Laurier University
Waterloo, ON, Canada

ISSN 1865-0929          ISSN 1865-0937   (electronic)
Communications in Computer and Information Science
ISBN 978-3-030-41257-9          ISBN 978-3-030-41258-6   (eBook)
https://doi.org/10.1007/978-3-030-41258-6

This Springer imprint is published by the registered company Springer Nature Switzerland AG
The registered company address is: Gewerbestrasse 11, 6330 Cham, Switzerland

# Preface

The Maple Conference is a meeting of Maple enthusiasts and experts from around the world, covering topics in education, algorithms, and applications of the mathematical software Maple. This conference has existed in various forms on and off for over two decades, permitting participants to benefit from the experiences and insights of both leading researchers in mathematics and computer science, and of passionate educators. It was a great pleasure to be part of its revival this year, and to welcome participants from around the world, including Canada, the USA, France, China, Spain, Belgium, and the UK.

In three decades, Maple has grown from a research project in symbolic computation to a complete environment for mathematical problem solving and exploration that is used by mathematicians, engineers, scientists, educators, and students around the globe. Thousands of sophisticated algorithms are implemented in Maple today, and this tremendous growth is due to both Maple visionaries and developers, but also to Maple users. Maple has been used as a crucial ingredient in countless Master and PhD theses in academia worldwide, as well as in commercial and government projects in robotics, medicine, green energy, space exploration, and more.

The Maple Conference 2019 was held at the University of Waterloo in Waterloo, Ontario, Canada, during October 15–17, 2019. This conference featured two keynote speakers:

- Dr. Marvin Weinstein, physicist at the Stanford Linear Accelerator Center for 42 years and now CSO of Quantum Insights
- Dr. Laurent Bernardin, President and CEO of Maplesoft, the developers of Maple

This volume contains the contributed papers from the conference, as well as extended abstracts of contributed talks. All submissions were reviewed by an international committee, and the quality of the accepted papers was very impressive.

We would like to thank the authors and presenters for making the conference both interesting and useful. We thank the members of the Maple Conference 2019 Program Committee and additional referees for their insightful comments and suggestions for the authors. We also offer thanks to Kathleen McNichol, Eithne Murray, Stacey Nichols, and Jennifer Iorgulescu from Maplesoft, for their tireless support of this conference and the proceedings. Additionally, we thank the staff at Springer for their help in completing this Maple Conference 2019 book of proceedings.

We are proud to recognize the invaluable patronage of the Maple Conference 2019 partners and sponsors, namely the Perimeter Institute for Theoretical Physics, the Fields Institute for Mathematical Sciences, Springer, the University of Waterloo, and Wilfrid Laurier University.

Finally, we are delighted to serve as Proceedings Editors for the present volume and hope Maple users will refer to it in the future, to learn the latest algorithmic developments, be inspired to conduct further research in interesting topics using Maple, and enlarge their outlook on how to use Maple effectively in educational contexts in high school and university.

October 2019                                                          Jürgen Gerhard
                                                                     Ilias Kotsireas

# Organization

## Scientific Program Committee

| | |
|---|---|
| Alin Bostan | Inria Saclay, France |
| Jacques Carette | McMaster University, Canada |
| Bruce Char | Drexel University, USA |
| Shaoshi Chen | Chinese Academy of Sciences, China |
| Paulina Chin | Maplesoft, Canada |
| Frédéric Chyzak | Inria Saclay, France |
| Robert Corless | Western University, Canada |
| Thierry Dana-Picard | Jerusalem College of Technology, Israel |
| James Davenport | University of Bath, UK |
| Paul DeMarco | Maplesoft, Canada |
| Jürgen Gerhard | Maplesoft, Canada |
| Mark Giesbrecht | University of Waterloo, Canada |
| Laureano Gonzalez-Vega | Universidad de Cantabria, Spain |
| Jonathan Hauenstein | University of Notre Dame, USA |
| David Jeffrey | Western University, Canada |
| Manuel Kauers | Johannes Kepler University, Austria |
| Alexander Kobel | Max Planck Institut für Informatik, Germany |
| George Labahn | University of Waterloo, Canada |
| Victor Levandovskyy | RWTH Aachen University, Germany |
| Robert Martin | University of Manitoba, Canada |
| John May | Maplesoft, USA |
| Michael Monagan | Simon Fraser University, Canada |
| Guillaume Moroz | Inria Nancy, France |
| Eithne Murray | Maplesoft, Canada |
| Erik Postma | Maplesoft, Canada |
| Clemens Raab | Johannes Kepler University, Austria |
| Fabrice Rouillier | Inria Paris, France |
| Sivabal Sivaloganathan | University of Waterloo, Canada |
| Arne Storjohann | University of Waterloo, Canada |
| Mark van Hoeij | Florida State University, USA |
| Gilles Villard | Université de Lyon, France |
| Stephen Watt | University of Waterloo, Canada |
| Thomas Wolf | Brock University, Canada |

## Organizing Committee

| | |
|---|---|
| Ilias S. Kotsireas (Chair) | Wilfrid Laurier University, Canada |
| Kathleen McNichol | Maplesoft, Canada |
| Jürgen Gerhard | Maplesoft, Canada |
| Eithne Murray | Maplesoft, Canada |

# Contents

## Extended Abstracts – Research Stream

## Extended Abstracts – Education/Applications Stream

# Keynote

# Your Data Wants You to Ask Better Questions. Do It!

Marvin Weinstein[✉]

Quantum Insights Inc., Menlo Park, CA, USA
Mweinstein@quantuminsights.io

"Correlation is not causation", or so the old trope goes.

This statement is true if we are talking about correlations revealed by familiar methods that are used to analyze data. In this talk, I want to put a new spin on this statement. I will argue that a new technology, quantum generated density maps, changes the way we look at complex data. The main point is that visualizing *all the correlations* hidden in a set of data, their shapes and *their relationship* to one another, can force us to ask questions that strongly suggest causal relationships. To make this point I will discuss the analysis of melanoma data provided to Quantum Insights Inc. by Genentech.

**The Data and What Our Client Wished to Know**

The information studied consists of RNA-sequence data. Essentially, it is a spreadsheet where each of the 550 rows of the spreadsheet contain RNA data for a single tumor. The columns give the individual tumor expression levels for each of the 30,684 RNA-transcripts.

The aim at the outset was to answer two questions. First, characterize the range of variation in the data and how many clusters – or distinct types - of tumor behavior would be revealed by our Dynamic Quantum Clustering (DQC) technology. Second, would these clusters correlate with the presence or absence of a BRAF mutation and the use of two drugs that had been administered to the patients. The simple answers to these questions are: the data shows significant clustering; these clusters do not show a positive relation between BRAF and drug administered.

**The DQC Analysis**

The results of a DQC analysis is usually shown as an animation. The first frame in the animation displays the original data plotted in the space of features. The frames that follow show the data moving towards the regions of higher density. The animation is constructed for all data dimensions. But, since we can only display three dimensions at a time, we use PCA (Principal Component Analysis) to rotate the data. This guarantees that the first three dimensions show a view of the data that has the most variation. Furthermore, in the next three dimensions the data will have somewhat less variation, etc.

The initial plot of the data in the first three PCA components is shown in Fig. 1a. The important points to know about this plot are:

1. Each point represents a single data entry in the 30,684-dimensional data space. This plot shows the first 3 components of that entry in the coordinate system chosen by the PCA transformation. As a result of applying DQC to this data we were see

© Springer Nature Switzerland AG 2020
J. Gerhard and I. Kotsireas (Eds.): MC 2019, CCIS 1125, pp. 3–10, 2020.
https://doi.org/10.1007/978-3-030-41258-6_1

structure that can be shown to be almost entirely due to a subset of 69 transcripts. Figures 1b, c and d show the results of the DQC analysis done for the $550 \times 69$ data matrix restricted to just these 69 RNA-transcripts.

2. It is important to point out distance matters in DQC. In fact, the distance between two datapoints directly measures the correlation between the corresponding rows in the data matrix.

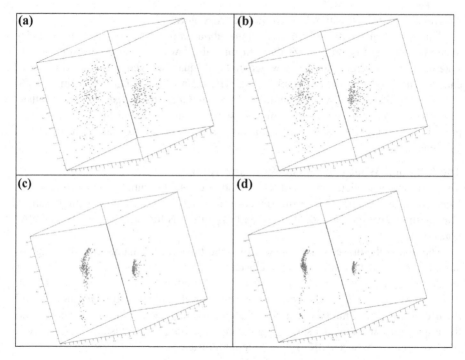

**Fig. 1.**

Figure 1a shows what are apparently two largish regions that are separated from one another, but at this point, since this only shows what is going on in three of sixty-nine dimensions, we can't trust this conclusion. The next three images show how the data evolves due to DQC evolution. What we see is that the rightmost region contacts to a small, tight cluster (eventually it contracts to a point) and an extended curved structure that never contracts to a point. It also shows outliers that never join either of the two dominant clusters. Clearly, the picture shown in the first three PCA dimensions does not explain why the outliers do not coalesce into the major structures, since they don't seem particularly far from the main clusters. The fact is that that they are further away in the other PCA dimensions. The message is that DQC evolution in high dimension is important, and using PCA to do strong dimensional reduction is dangerous.

**Answering Genentech's Question**

The fully evolved data in Fig. 1d reveals the full set of clusters and outliers that appear in all 69 dimensions of the melanoma data. The next step is to see if the clusters can predict whether the melanomas have a BRAF mutation, or reveal the drug used to treat the tumors. The simplest way to answer these questions is to simply re-color the points. This coloring is shown below, in Figs. 2a to d. The patients that have a BRAF mutation and took drug 1 are colored blue; patients who don't have a BRAF mutation and took drug 1 orange; the ones that took drug 2 are colored red.

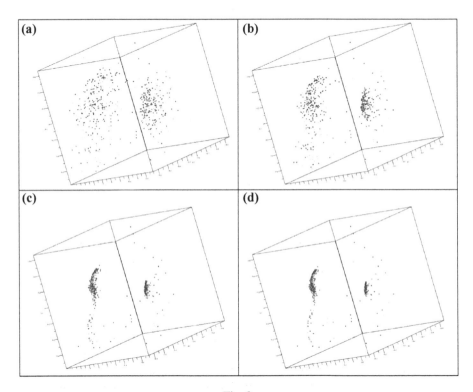

**Fig. 2.**

Clearly these plots do not show separation of the colors between the clusters and outliers. What we see is essentially "confetti". This says that whatever the clustering of tumors by RNA-expression levels is telling us, it has nothing to do with the BRAF status of the tumors; nor, does it strongly reflect the drug given to the patient. This cannot be the end of the story. Strong clusters do exist, and they must tell us something about the melanomas.

The answer to the question of what the clusters tell us is answered when we color by tumor stage.

## Tumor Stage

Pathologists assign tumors a stage that reflects tumor's morphology, whether it has invaded distant tissues, etc. The table below tells how we color the data points according to the known stage classification. The classifications range from IIB to IIC. As we move from black to magenta the common wisdom is that we move from less advanced tumors to more advanced – harder to treat – tumors. Figures 3a and b show the original data and the DQC evolved data re-colored according to this information.

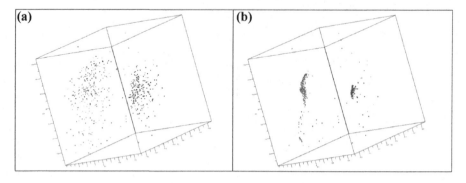

**Fig. 3.**

| Stage | Color |
|---|---|
| IIB | Black |
| IIC | **Blue** |
| IIIA | Brown |
| IIIB | Orange |
| IIIc | Magenta |

Figures 3a and b show a strong separation between the less advanced IIB, IIC and IIIA tumors and the tumors classified as IIIB and above. The dominant clusters are at some distance from one another, and this shows that the tumors in one cluster should differ significantly in their RNA-expression levels from those in the other cluster. At this point we have succeeded in showing how the RNA-data is distributed and understand that this difference in expression levels is correlated with the stage of the tumor. This answers our original questions. But, the shape and relationship of the two main clusters to one another, raises a new question a question that is begging to be asked.

## Better Questions

In Fig. 3b we see an extended cluster on the left-hand side of the plot, and a much tighter cluster that eventually collapses to a point in succeeding frames. Since the left-hand cluster - or "hook" - never collapses, there must be some reason for this shape. The fact that nearby datapoints have very similar RNA-expression levels implies that for this subset of the data something about the RNA-expression levels varies continuously along the hook. So better questions are "What is varying?" and "What is it telling us?".

To answer these questions, zoom in on the "hook" and divide it up into 22 contiguous regions -as shown in Fig. 4. We will define a new feature - "severity" – as the location along the hook, running from left to right.

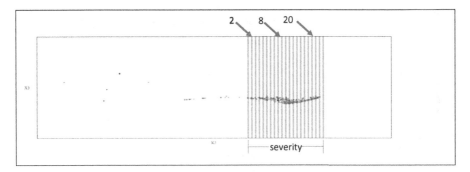

**Fig. 4.**

**NOTE:** Severity is not correlated with stage, since the coloring by stage as we move along the line shows a "confetti"-like behavior. Severity is something new and unexpected! The reason for this name will become clear in a moment.

To understand what the "hook" is trying to tell us, we averaged the expression levels for each of these 24 segments of the hook. We then compared these averages to the average of the expression levels for the tight cluster of less advanced tumors. The result of this process is shown for segments 2, 8 and 20 in Fig. 5.

The three plots show the expression level of each of the 69 RNA-transcripts in a way that reveals how close the tumors in each segment of the hook are to the less advanced tumors. The horizontal axis in each plot is a number - from 1 to 69 – indicating the labels of the RNA-transcripts. The vertical axis in each plot runs from -3 to 3. If the value of a transcript lies on the horizontal axis, that indicates that the expression value is the same as the average of the less advanced cluster for that transcript. On the other hand, if it takes a negative value (many of the transcripts in segment 2), it is less expressed than the average of the less advance cluster.

A value of 1 means that the expression level has the same value as the average of the transcripts for all points in the "hook". Finally, a value greater than 1 means that transcript is more expressed than the average value for the more advanced tumors.

Going from left to right along the line we see that in segment 2, 19 out of the 69 transcripts are less expressed than the average of the more advanced tumors. In fact, 11 of these are even less expressed than the average of the less advanced tumors. By the time we get to segment 8 the situation has changed dramatically. Now all of the transcripts are being expressed in the same way as the average of the more advanced tumors. Finally, by segment 20 we see that 15 transcripts have become much more expressed than the average for the more advanced tumors. This strongly suggests that

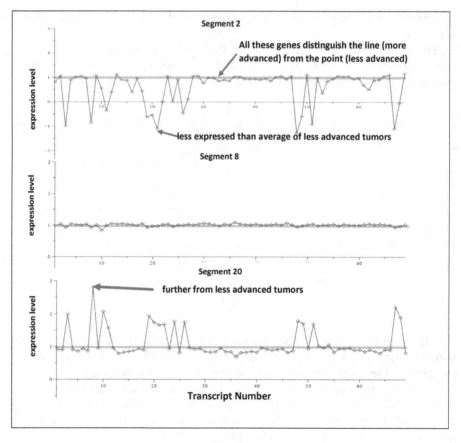

**Fig. 5.**

as we move from left to right along the line the tumors go from less serious tumors to more serious tumors. This is why I invented the name "severity" for the distance from the left hand side of the hook to the right hand side of the hook.

### What Medical Outcome Is Related to Severity?

The only information we had that related to a medical outcome is whether a patient had the tumor recur within a specified period of time. Since we didn't have a lot of patients, we divided the "severity" line into three equal segments and computed the number of patients that had a recurrence in each segment divided by the total number of patients in the same segment. The resulting values for each of the three segments is shown in Fig. 6. There is a clear progression in the probability of recurrence as we go from left to right along the "hook" from slightly more than 15%, to almost 50%.

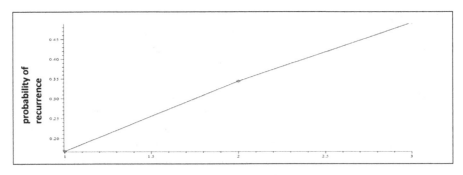

**Fig. 6.**

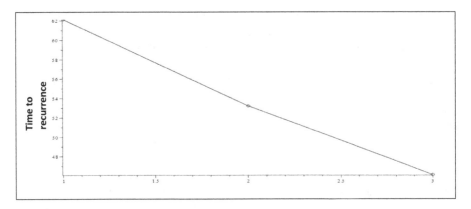

**Fig. 7.**

This is, of course, entirely consistent with our interpretation of the variation in RNA-expression values as we go, segment-by-segment, from left to right. Another piece of evidence that implies that "severity" is tied to recurrence is shown in the plot of the time to recurrence. Figure 7 shows that the time to recurrence decreases linearly with severity, going from 62 months to less than 48 months.

**Why Do These Transcripts Move in Lockstep?**
The DQC analysis of the melanoma data led us to identify a subset of 16 out of 69 transcripts that separate stage II from stage III melanoma that are prognostic for a recurrence. This forces us to ask why the expression levels of these transcripts change in lockstep as we move along the "severity" axis. This behavior strongly suggests that they are all acting under the influence of some promoter that is controlling this behavior. In effect, it suggests that these transcripts belong to a known biological pathway, or perhaps something new? This suggests a new line of research.

There is more that can be extracted from this data; however I choose not to pursue this further in this talk. I think I have said enough to make my initial point about how a comprehensive density map of data changes how we look at the statement *"Correlation is not causation"*.

**Summing Up**

I began with the quote *"Correlation is not causation"*. I then claimed that I would show how knowing all the correlations in a dataset, the shape of these correlations and their spatial relationship to one another could lead us to an understanding of causal relationships hidden in the data.

In the case of the melanoma dataset we revealed the existence of two clusters, or correlated subsets of RNA-expression data, that characterized the information contained in the dataset. We then were able to show that these two clusters strongly correlate with the stage of the tumors. Finally, we saw that the less advanced tumors form a tight (eventually point-like) cluster, whereas the more advanced tumors formed a one-dimensional "hook". It was the surprising existence of the hook that suggested there was more to learn.

Trying to extract the meaning of the hook led us to identify a biological fingerprint, involving a small number of transcripts, that explained the shape of the hook. The correlation of the changes of the transcripts along the hook led to the notion of severity, which – contrary to expectations - was not correlated with the stage of the tumor. However, it turned out that severity is directly correlated with the likelihood of – and time to - recurrence. Finally, the fact that 16 transcripts vary in lockstep with changes in severity led us to ask if these transcripts - and an unknown promoter – could define a new biological pathway. Thus, we went from trying to simply characterize the data to asking about a potential causal relationship. Obviously, we cannot, without doing further experiments, nail down the validity of this conclusion. However, our discussion has revealed what the next steps should be.

My takeaway from all of this is that complex data often contains a lot of information. It is the appearance of unexpected structure in the data, the analysis of the shape of these structures, and their relationship to one another, that allows us to build a move complete picture of what the data has to say.

**Maple's Role in All of This**

Since this is a Maple conference, I would be remiss if I didn't point out that Maple played an important role in the birth of DQC. The first implementation of the algorithm upon which DQC is based was done in Maple. The virtue of using Maple to create this software was that it provided an excellent GUI, an easy way to call special purpose compiled libraries that implement the algorithm and good support for multi-threading. We couldn't have gotten up and running nearly as quickly without Maple.

# Full Papers – Research Stream

# The LegendreSobolev Package and Its Applications in Handwriting Recognition

Parisa Alvandi$^{(\boxtimes)}$ and Stephen M. Watt$^{(\boxtimes)}$

David R. Cheriton School of Computer Science, University of Waterloo,
Waterloo, Canada
{palvandi,smwatt}@uwaterloo.ca

**Abstract.** The present work is motivated by the problem of mathematical handwriting recognition where symbols are represented as parametric plane curves in a Legendre-Sobolev basis. An early work showed that approximating the coordinate functions as truncated series in a Legendre-Sobolev basis yields fast and effective recognition rates. Furthermore, this representation allows one to study the geometrical features of handwritten characters as a whole. These geometrical features are equivalent to baselines, bounding boxes, loops, and cusps appearing in handwritten characters. The study of these features becomes a crucial task when dealing with two-dimensional math formulas and the large set of math characters with different variations in style and size.

In an early paper, we proposed methods for computing the derivatives, roots, and gcds of polynomials in Legendre-Sobolev bases to find such features without needing to convert the approximations to the monomial basis. Furthermore, in this paper, we propose a new formulation for the conversion matrix for constructing Legendre-Sobolev representation of the coordinate functions from their moment integrals.

Our findings in employing parametrized Legendre-Sobolev approximations for representing handwritten characters and studying the geometrical features of such representation has led us to develop two MAPLE packages called LegendreSobolev and HandwritingRecognitionTesting. The methods in these packages rely on MAPLE's linear algebra routines.

## 1 Introduction

Orthogonal polynomials have many applications in different recognition problems such as face [15], speech [5], speech emotion [13], and gesture [14] recognition. We are particularly interested in using orthogonal polynomials to represent handwritten characters for the purpose of mathematical handwriting recognition. In fact, modelling handwritten characters as parametrized curves $X(\lambda)$ and $Y(\lambda)$ on an orthogonal basis accurately captures the shape of handwritten mathematical characters using few parameters. This new representation of handwritten characters turns the recognition into a writer and device independent problem as one does not need to deal with the factors such as different device resolutions

© Springer Nature Switzerland AG 2020
J. Gerhard and I. Kotsireas (Eds.): MC 2019, CCIS 1125, pp. 13–29, 2020.
https://doi.org/10.1007/978-3-030-41258-6_2

and the number of points in the sampling process of a handwritten character. The authors of [6] have found that even without further similarity-processing, the polynomial coefficients from the writing samples form clusters which often contain the same characters written by different test users. This work uses the Chebyshev basis to represent handwritten characters as parametrized curves.

The work [7] is inspired by a similar idea as [6] to represent handwritten data, but uses a different functional basis. The basic idea of [7] is to compute moments of the coordinate curves in real time as the character is being written and then to construct the coefficients of the coordinate curves in the Legendre basis from moments at the time pen is lifted. As a matter of fact, the Legendre representation is just as suitable in practice for representation and analysis of ink traces as the Chebyshev representation, but has the benefit that it can be computed in a small, fixed number of arithmetic operations at the end when a stroke is written, completely. This approach works for any inner product with a linear weight function.

Representing handwritten characters with parametrized coordinate curves in the Chebyshev and Legendre series yields low RMS error rates. However, low RMS error rates do not guarantee the shape similarity of two characters, see Fig. 1. In fact, the problem with the blue and red curves on the right side of Fig. 1 is that the corners of these two curves are not in the right places, despite the fact that they demonstrate a lower RMS error rate compared to the curves on the left. One solution to this obstacle might be to work in a jet space to force coordinate and derivative functions of the letters from the same class to have a similar form.

**Fig. 1.** Despite the fact that the two curves on the right have lower RMS error compared to the ones on the left, they do not have a similar shape. (Color figure online)

The work [7,9] use a special case of Legendre-Sobolev polynomials to represent handwritten characters based on the computation of moment integrals. These special polynomials, which are also called Althammer polynomials, enables us to work in a first jet space. The authors have reported experimental results that demonstrate that representing coordinate curves in a Legendre-Sobolev basis has higher detection rates compared to when these curves are represented in the Legendre basis.

In [2], we have proposed an online method for computing Legendre-Sobolev representations of handwritten characters from their moments by a matrix multiplication. Furthermore, we have presented methods in [3] for computing the derivatives, roots, and gcd of polynomials in Legendre-Sobolev bases by relying on linear algebra arithmetic operations. The goal of the latter work is to study the geometry and features of handwritten curves by relying on the Legendre-Sobolev coefficients of the approximated parametrized curves corresponding to handwritten characters. In [12], the authors proposed an algorithm to compute some of the important features of the Legendre-Sobolev approximations of hand-written characters by relying on the Newton method. Thus, conversion from a Legendre-Sobolev basis to the monomial basis is required. But the work [3] avoids this conversion because such conversion is known to be ill-conditioned.

In the present work, we give a new formulation for the matrix $C$ in Proposition 1, which is inspired by the work [2], to compute the Legendre-Sobolev coefficients of the truncated parametrized curves of handwritten characters from their moments. The results in [2,3] and Proposition 1 have led to the development of a MAPLE package called LegendreSobolev[1]. This package has all the necessary tools for representing handwritten characters as parametrized curves in a Legendre-Sobolev basis for the purpose of handwriting recognition.

After giving the preliminaries and our new result on the construction of the Legendre-Sobolev coefficients from moments, we explain the structure of the package LegendreSobolev with demonstrative examples in Sect. 4. Finally, we illustrate how to compute Legendre-Sobolev representations of handwritten curves from their corresponding digital inks, by relying on LegendreSobolev package, in Sect. 5.

## 2   Preliminaries

Digital ink is generated by sampling points from handwritten characters and is a collection of points $(x, y, t)$ with position $(x, y)$ and timestamp $t$. We use these points to compute moment integrals and from them, we approximate the coefficients of the coordinate curves $X(\lambda)$ and $Y(\lambda)$ on an orthogonal basis, where $\lambda$ is either time or length of handwritten curves. In this paper, we assume that sample values of $X(\lambda)$ and $Y(\lambda)$ are received as a real time signal and $\lambda$ corresponds to the length of handwritten curves. We assume the sample points are equally spaced with $\Delta t = 1$.

The works [2,6,7] showed that the coordinate curves $X(\lambda)$ and $Y(\lambda)$ for handwritten characters can be modelled by truncated Chebyshev, Legendre, and Legendre-Sobolev series, respectively. The coefficients of such series can be used for classification and recognition of characters. In this paper, we are interested in Legendre-Sobolev approximations for representing handwritten characters.

For two functions $f, g : [-1, 1] \rightarrow \mathbb{R}$, consider the following inner product which is given as

$$\langle f(\lambda), g(\lambda) \rangle = \int_{-1}^{1} f(\lambda)\, g(\lambda) d\lambda + \mu \int_{-1}^{1} f'(\lambda) g'(\lambda) d\lambda, \tag{1}$$

---

[1] This package is publicly available at www.maplesoft.com/applications/view.aspx? SID=154553.

where $\mu \in \mathbb{R}_{\geq 0}$. This inner product is a special case of the Legendre-Sobolev inner product. In fact, the Legendre-Sobolev inner product may involve terms corresponding to higher order derivatives, but for the purpose of this paper we restrict ourselves to the first order derivative. Systems of orthogonal polynomials corresponding to the above inner product can be computed by applying the Gram-Schmidt orthogonalization process to the monomial basis. We denote the Legendre-Sobolev polynomials corresponding to the inner product given by Eq. 1 (which are also called Althammer polynomials, see [1]) of degree $n$ by $S_n^\mu(\lambda)$. When $\mu = 0$ in Eq. 1, we denote the polynomial $S_n^0(\lambda)$ by $P_n(\lambda)$, as well. In fact, the polynomial $P_n(\lambda)$ is the Legendre polynomial of degree $n$.

A function $f : [-1, 1] \to \mathbb{R}$, when the integrals involved in the inner product are well-defined for $f(\lambda)$, can be represented by an infinite linear combination of orthogonal polynomials $\{S_0^\mu(\lambda), S_1^\mu(\lambda), \ldots\}$ as

$$f(\lambda) = \sum_{i=0}^{\infty} \alpha_i S_i^\mu(\lambda).$$

The coefficients of the series in the new basis can be computed by the formula

$$\alpha_i = \frac{\langle f(\lambda), S_i^\mu(\lambda)\rangle}{\langle S_i^\mu(\lambda), S_i^\mu(\lambda)\rangle}, \, i = 0, 1, \ldots,$$

where $\langle ., . \rangle$ stands for the Legendre-Sobolev inner product given by Eq. (1). For representing handwritten characters, we use truncated linear combinations of Legendre-Sobolev polynomials to represent the function $f(\lambda)$. In fact, the closest polynomial of degree $d$ to function $f(\lambda)$ with respect to Euclidean norm induced by the given inner product is the following series

$$f(\lambda) \simeq \sum_{i=0}^{d} \alpha_i S_i^\mu(\lambda).$$

Such approximation allows to think of functions as points $(\alpha_0, \ldots, \alpha_d)$ in $(d+1)$-dimensional vector space. That means that one can establish a method for measuring how close two functions are to each other in such $(d+1)$-dimensional vector space. In other words, for two functions $f, g : [-1, 1] \to \mathbb{R}$, if we approximate $f(\lambda)$ and $g(\lambda)$ as following

$$f(\lambda) \simeq \sum_{i=0}^{d} \alpha_i S_i^\mu(\lambda), g(\lambda) \simeq \sum_{i=0}^{d} \beta_i S_i^\mu(\lambda),$$

then one can measure how close $f(\lambda)$ and $g(\lambda)$ are by computing the quantity

$$\|f(\lambda) - g(\lambda)\| \simeq \sqrt{\sum_{i=0}^{d} (\alpha_i - \beta_i)^2}.$$

This method of measuring the distance of two functions is the basic and important rule in the handwriting recognition method used in [8].

The moments of a function $f(\lambda)$ defined on the interval $[a, b]$ are the integrals:

$$\int_a^b \lambda^k f(\lambda) d\lambda.$$

A key aspect of the approach used in [6] for the purpose of interpolating the coordinate curves $X(\lambda)$ and $Y(\lambda)$ corresponding to handwritten strokes is to recover these curves from their moments. This is the Hausdorff moment problem [10,11], known to be ill-conditioned. For the purpose of this paper, the moments of a function $f$ are defined over an unbounded half-line since the curve may be traced over an arbitrary length:

$$m_k(f(\lambda), \ell) = \int_0^\ell \lambda^k f(\lambda) d\lambda.$$

In our application, we assume that discrete sample values of $f(\lambda)$ are received as a real-time signal. We use these values to compute approximate values for the moment integrals. After a curve is traced out, we will have computed its moments over some length $L$, with $L$ known only at the time the pen is lifted. The problem is now to scale $L$ to a standard interval and compute the truncated Legendre-Sobolev series coefficients for the scaled function from the moments of the unscaled function, $m_k(f(\lambda), L)$.

Having represented handwritten characters as Legendre-Sobolev coefficients of parametrized coordinate curves, we can study the geometrical features of handwritten curves by means of linear algebra calculations, such as matrix multiplication, and solving Diophantine equations. The work [3] gives methods for computing the derivatives, roots, and gcd of polynomials in Legendre-Sobolev bases based on their coefficient matrix.

## 3   Construction of Handwritten Curves from Moments

The goal of this section is to present our new result on computing the representations of handwritten characters as approximated parametrized curves in Legendre-Sobolev bases from their corresponding digital inks, see Proposition 1.

To find such representation, one needs to compute moment integrals as a curve is being traced out, first, and then use Proposition 1 to compute the Legendre-Sobolev coefficients of the parametrized coordinate curves in a Legendre-Sobolev basis. In Section V in [2], we have explained how to compute moment integrals from the input digital ink.

**Proposition 1.** *Suppose that $m_i(f(\lambda), L)$ is defined as $\int_0^L f(\lambda) \lambda^i d\lambda$ where $L$ is the length of a given curve and $f(\lambda)$ is either $X(\lambda)$ or $Y(\lambda)$, for $i = 0, \ldots, d$. Let also $\hat{f}(\lambda) = \sum_{i=0}^d \alpha_i S_i^\mu(\lambda)$ be the corresponding scaled function of $f(\lambda)$ in the interval $[-1, 1]$. Then for $i = 0, \ldots, d$, one may compute $\alpha_i$ as*

$$\alpha_i = \sum_{j=0}^d \frac{1}{L^{j+1}} C_{ij} \, m_j(f(\lambda), L),$$

*where*

$$(-1)^{i+j} C_{ij} = \frac{1}{2i!(i+1)!} \left( B_{ij} \left( \lfloor \tfrac{d-j}{2} \rfloor \right) - B_{ij} \left( \max(0, \lceil \tfrac{i-j}{2} - 1 \rceil) \right) \right) \left( \frac{1}{a_j(\mu)} - \frac{1}{a_{j+2}(\mu)} \right)$$
$$+ (2j+1) \binom{j}{i} \binom{i+j}{j} \frac{1}{a_j(\mu)},$$
$$B_{ij}(k) = \frac{(i+j+2k+2)!}{(j-i+2k)!}, \quad taking \ \frac{1}{(-n)!} = 0, \ for \ k, n \in \mathbb{Z}^+,$$

$$a_0(\mu) = 1, \ a_i(\mu) = \sum_{k=0}^{\lfloor \frac{i-1}{2} \rfloor} \left( \frac{\mu}{4} \right)^k \frac{(i+2k-1)!}{(2k)!(i-2k-1)!}, \ for \ i \geq 1. \tag{2}$$

**Proof.** We have simplified the following summation as

$$\sum_{\ell=k_1}^{k_2} (2j+1+4\ell) \binom{j+2\ell}{i} \binom{i+j+2\ell}{j+2\ell} = \frac{1}{2i!(i+1)!} \left( B_{ij}(k_2) - B_{ij}(k_1 - 1) \right),$$

for $k_1, k_2 \in \mathbb{Z}_{\geq 0}$, while taking $\frac{1}{(-n)!} = 0$, for $n \in \mathbb{Z}^+$. Then, the proof of this proposition is followed by substituting the above simplified form in the counter-part representation of matrix $C$ in [2]. □

Note that the coefficients $C_{ij}$ are independent of the problem and may be computed as constants, in advance.

We have developed the MAPLE's package HandwritingRecognition Testing[2] which implements the idea of Proposition 1 for computing Legendre-Sobolev approximations of handwritten curves. Section 5 explains how to use this package to compute such representations.

## 4   LegendreSobolev Package

We have developed a MAPLE package for applying different mathematical operations in Legendre-Sobolev bases called LegendreSobolev (see Fig. 2). The work [2,3] and Proposition 1 support the theory behind the commands in this package. The operations in this package rely on linear algebra arithmetic operations such as matrix multiplication, and solving Diophantine equations.

In this section, we explain how one can use the commands in LegendreSobolev package to compute Legendre-Sobolev polynomials of a given degree and parameter $\mu$, change the representation of polynomials with respect to different bases, and find roots and gcds of polynomials in Legendre-Sobolev bases.

---

[2] This package is publicly available at www.maplesoft.com/applications/view.aspx?
SID=154553.

> **read** "LS.mpl" :
> with(LegendreSobolev);
>   [ ComradeMatrix, DerivativeInLS, DerivativeMatrixInLS, GcdInLS,
>   LSToLegendreMatrix, LSToMonomialMatrix, LegendreToLSMatrix,
>   MomentsToLSMatrix, MonomialToLSMatrix, P, S, $\alpha$ ]

**Fig. 2.** The functions of `LegendreSobolev` package.

## 4.1 Legendre-Sobolev Polynomials

As Fig. 3 demonstrates, one can use the command S in `LegendreSobolev` package to compute a Legendre-Sobolev polynomial $S_n^\mu(\lambda)$. Furthermore, we can use both $S_n^0(\lambda)$ and $P_n(\lambda)$ to compute the Legendre polynomial of degree $n$, see Fig. 4.

> n := 20 :
> $\mu$ := 0.125 :
> S[n, $\mu$](x);
> $5.8754582398620977895900768 \, 10^{13} - 1.1177242397192269290374167 \, 10^{16} \, x^2$
> $+3.5243802999731541193038741 \, 10^{17} \, x^4 - 4.3268454823302905511612081 \, 10^{18} \, x^6$
> $+2.7061754002348252015842451 \, 10^{19} \, x^8 - 9.7480590243981247807226921 \, 10^{19} \, x^{10}$
> $+2.1427243648177985254531152 \, 10^{20} \, x^{12} - 2.9210697557111092015824162 \, 10^{20} \, x^{14}$
> $+2.4108358540453217625442430 \, 10^{20} \, x^{16} - 1.1033834468388575891747713 \, 10^{20} \, x^{18}$
> $+2.1493662705437245105787467 \, 10^{19} \, x^{20}$

**Fig. 3.** Legendre-Sobolev polynomial computation with `LegendreSobolev` package.

> S[5, 0](x);
> $\frac{15}{8}x - \frac{35}{4}x^3 + \frac{63}{8}x^5$

> P[5](x);
> $\frac{15}{8}x - \frac{35}{4}x^3 + \frac{63}{8}x^5$

**Fig. 4.** The Legendre polynomial computation with `LegendreSobolev` package with two different functions.

Legendre-Sobolev polynomials of a given degree $n$ can also be computed as a function in $\mu$ (see Fig. 5). The function $\alpha$ in `LegendreSobolev` package gives a polynomial with respect to degree $n$ and parameter $\mu$. In fact, the command $\alpha$ is used to compute Legendre-Sobolev polynomials with respect to the Legendre polynomials and implements $a_n(\mu)$ given by Eq. (2). The formula for computing Legendre-Sobolev polynomials from $a_n(\mu)$ and the Legendre polynomials is given by the relation: $S_n^\mu(\lambda) = S_{n-2}^\mu(\lambda) + a_n(\mu) \left( P_n(\lambda) - P_{n-2}(\lambda) \right)$.

$$> S[5, \mu](x);$$
$$x + (1 + 3\mu)(-\tfrac{5}{2}x + \tfrac{5}{2}x^3) + (105\mu^2 + 45\mu + 1)(\tfrac{27}{8}x - \tfrac{45}{4}x^3 + \tfrac{63}{8}x^5)$$
$$> \text{alpha}[5, \mu];$$
$$105\mu^2 + 45\mu + 1$$

**Fig. 5.** Legendre-Sobolev polynomial computation as a function in $\mu$.

## 4.2   Changing a Polynomial Representation w.r.t Different Bases: Legendre-Sobolev, Legendre, and Monomial

When a polynomial $f(\lambda)$ is given in the monomial basis, one can compute the coefficients of the polynomial in any Legendre-Sobolev basis, with parameter $\mu$, by using `MonomialToLSMatrix` command of `LegendreSobolev` package (see Fig. 6). In fact, one needs to multiply the coefficient matrix of the given polynomial in the monomial basis and the matrix in the output of the command `MonomialToLSMatrix`$(\text{degree}(f), \mu)$.

```
> f := (x − 1)²(x + 2)³ :
  fcoeffs := Matrix([seq(coeff(f, x, i), i = 0..degree(f))]) :
  μ := 0.125 :
  fcoeffsInLS := Multiply(fcoeffs, MonomialToLSMatrix(degree(f), μ));
    fcoeffsInLS := [ 5.46666666666667  − 2.70649350661521  − 3.78467908902692
                      0.691130587508162 0.318012422360248 0.0153629189534213 ]
```

**Fig. 6.** Changing the representation of $f$ from the monomial to Legendre-Sobolev basis with $\mu = 0.125$.

When a polynomial $f(\lambda)$ is given in the monomial basis, it is possible to convert this polynomial to the Legendre basis, as well, by a matrix multiplication. In fact, to compute the Legendre representation of $f(\lambda)$, one needs to multiply the coefficient matrix of $f(\lambda)$ in the monomial basis and the matrix in the output of the command `MonomialToLSMatrix`$(\text{degree}(f), \mu)$, where $\mu = 0$ (see Fig. 7).

```
> f := (x − 1)²(x + 2)³ :
  fcoeffs := Matrix([seq(coeff(f, x, i), i = 0..degree(f))]) :
  fcoeffsInL := Multiply(fcoeffs, MonomialToLSMatrix(degree(f), 0));
    fcoeffsInL := [ 82/15  − 104/35  − 92/21  38/45  32/35  8/63 ]
```

**Fig. 7.** Changing the representation of $f$ from the monomial to Legendre basis.

Furthermore, when a polynomial $f(\lambda)$ is given in the Legendre basis, one can change its representation to a Legendre-Sobolev one by multiplying the coefficient matrix of $f(\lambda)$ in the Legendre basis and the matrix from the output of the command **LegendreToLSMatrix**(degree($f$), $\mu$), where $\mu$ is given (see Fig. 8a).

> fcoeffsInL := Matrix([[1, 1, 1, 1, 1, 1]]) :
$\mu := \frac{1}{5}$ :
fcoeffsInLS := Multiply(fcoeffsInL,
LegendreToLSMatrix(5, $\mu$))
fcoeffsInLS := $[\,1 \;\; \frac{7}{4} \;\; \frac{7}{4} \;\; \frac{335}{284} \;\; \frac{1}{4} \;\; \frac{5}{71}\,]$

(a)

> fcoeffsInLS := Matrix([[1, 1, 1, 1, 1, 1]]) :
$\mu := \frac{1}{5}$ :
fcoeffsInL := Multiply(fcoeffsInLS,
LSToLegendreMatrix(5, $\mu$))
fcoeffsInLS := $[\,1 \;\; -\frac{1}{5} \;\; -2 \;\; -11 \;\; 4 \;\; \frac{71}{5}\,]$

(b)

**Fig. 8.** Changing the representation of $f$ from (a) the Legendre to Legendre-Sobolev and (b) Legendre-Sobolev to Legendre basis with $\mu = \frac{1}{5}$.

Finally, for a polynomial $f(\lambda)$ in a Legendre-Sobolev basis, we can compute the Legendre representation of this polynomial by multiplying the corresponding coefficients in the Legendre-Sobolev basis and the matrix in the output of the command **LSToLegendreMatrix**(degree($f$), $\mu$), where $\mu$ is the same parameter as the one in the Legendre-Sobolev representation of $f(\lambda)$, see Fig. 8b.

### 4.3 Computing the Derivates of Polynomials in Legendre-Sobolev Bases

When a polynomial is given in a Legendre-Sobolev basis, one can compute the derivative of such polynomial by a matrix multiplication (see [3]). Figure 9 shows how to compute the derivative of a polynomial by using the command **DerivativeInLS** of **LegendreSobolev** package.

> f := $(x - 1)^2 \cdot (x + 2)^3$;
fcoeffs := Matrix([seq(coeff(f, x, i), i = 0..degree(f))]) :
$\mu$ := .125 :
fcoeffsInLS := fcoeffsInL · LegendreToLSMatrix(degree(f), $\mu$);
der := DerivativeInLS($\mu$, convert(fcoeffsInLS, list))
fcoeffsInLS := [ 5.46666666666667  − 2.70649350661521
−3.78467908902692  .691130587508162  .318012422360248
0.0153629189534213 ]
der := [ −2.00000000015363, −8.65454545474783, 5.60248447177297
4.65454545570186, .397515527919776]

**Fig. 9.** Computing the derivative of the polynomial $f$ by a matrix multiplication.

```
> f := randpoly(x, degree = 12, terms = 10);
  fcoeffs := Matrix([seq(coeff(f, x, i), i = 0..degree(f))]) :
  μ := 0.125 :
  fcoeffsInLS := Multiply(fcoeffs, MonomialToLSMatrix(degree(f), μ)) :
  C := ComradeMatrix(degree(f), μ, convert(fcoeffsInLS, list)) :
  LinearAlgebra : −Eigenvalues(C) :
  rootList := [seq(%[i], i = 1..degree(f))]
```
$f := 75\,x^{12} - 92\,x^{10} + 6\,x^9 + 74\,x^8 + 72\,x^7 + 37\,x^6 - 23\,x^5 + 87\,x^4 + 44\,x^3 + 29\,x$
```
  rootList := [ 1.141014116 + 0.529676826 I, 1.141014116 − 0.529676826 I,
   −1.471711477 10^−8 + 0 I,  0.563935206 + 0.629828559 I, 0.563935206 − 0.629828559 I,
   −0.4044380673 + 0.900856959 I, −0.404438067 − 0.900856959 I,
   0.1365209364 + 0.583947922 I, 0.136520936 − 0.583947922 I,
   −1.0327465955 + 0.373007812 I,  −1.0327465955 + 0.373007812 I,
   −0.808571018 + 0 I ]
> ResidualError := norm(eval(f, x = rootList[1]), 2);
  ResidualError := 0.0129507410089049
```

**Fig. 10.** Computation of the roots of the polynomial $f$ by finding the eigenvalues of the corresponding comrade matrix.

### 4.4  Computing the Roots of Polynomials in Legendre-Sobolev Bases

Here, we have computed the roots of a polynomial $f(\lambda)$ of degree 12, which is randomly created by using the command **randpoly**. We have first computed the coefficient matrix corresponding to $f(\lambda)$ in the Legendre-Sobolev basis corresponding to $\mu = 0.125$; then we have used this matrix to compute the comrade matrix $C$ of $f(\lambda)$ in the Legendre-Sobolev basis (see [4]), and the roots of $f(\lambda)$ by computing the eigenvalues of matrix $C$ (see Fig. 10).

### 4.5  Computing Gcds of Polynomials in Legendre-Sobolev Bases

When two polynomials $f(\lambda)$ and $h(\lambda)$ are given in a Legendre-Sobolev basis, then one can find the monic gcd of these polynomials in the same basis by using the command **GcdInLS** as illustrated in Fig. 11. The work [3] explains the theory behind **GcdInLS** command which is essentially based on solving Diophantine equations arising from a comrade matrix.

## 5    Handwriting Recognition with LegendreSobolev Package

In this section, we explain how to compute the parametrized approximations of handwritten curves in a Legendre-Sobolev basis using **LegendreSobolev** package. To compute such approximations, we have implemented a package called **HandwritingRecognitionTesting** (see Fig. 12). The functions in the latter package are implemented based on Sect. 3 and [2].

> Digits := 17 :
$\mu$ := 0.125 :
f1 := randpoly(x, degree = 10);
g := randpoly(x, degree = 9);
h1 := randpoly(x, degree = 8);
f1 := $40\,x^9 - 81\,x^7 + 91\,x^3 + 68\,x^2 - 10\,x + 31$
g := $55\,x^8 - 28\,x^6 + 16\,x^4 + 30\,x^3 - 27\,x^2 - 15\,x$
h1 := $72\,x^8 - 87\,x^7 + 47\,x^6 - 90\,x^4 + 43\,x^3 + 92\,x$

> f := f1 g :
fcoeffs := Matrix([seq(coeff(f, x, i), i = 0..degree(f))]) :
fcoeffsInLS := Multiply(fcoeffs, MonomialToLSMatrix(degree(f), mu)) :
h := h1 g :
hcoeffs := Matrix([seq(coeff(h, x, i), i = 0..degree(h))]) :
hcoeffsInLS := Multiply(hcoeffs, MonomialToLSMatrix(degree(h), mu)) :
gcdLS := GcdInLS(convert(fcoeffsInLS, list), convert(hcoeffsInLS, list), $\mu$);
gcdLS := [ $-0.67070707067873429, 0.11404958676877247, 0.23366585106067264$,
$0.15867768593779621, 0.12037071477726489, -7.6571048185969832\,10^{-13}$,
$0.0039091068982669860, -4.9471370550988164\,10^{-17}, 0.000034125246982815868$ ]

> gcoeffs := Matrix([seq(gcdLS[i], i = 1..nops(gcdLS))]) :
gcoeffsInMonomial := Multiply(gcoeffs, LSToMonomialMatrix(nops(gcdLS) $- 1, \mu$)) :
computedGcd := add(gcoeffsInMonomial[1][i] $\cdot x^{i-1}$, i = 1..nops(gcdLS)) :
RelativePolynomialError := evalf($\frac{\text{norm}(\text{simplify}(\%-\frac{g}{\text{lcoeff}(g)},2))}{\text{norm}(\frac{g}{\text{lcoeff}(g)},2)}$)
RelativePolynomialError := $4.8855925361391707\,10^{-11}$

**Fig. 11.** Monic gcd computation using `LegendreSobolev` package.

Handwritten curves are presented as points $(x, y, t)$ with coordinates $(x, y)$ and timestamp $t$. To compute the parametrized approximation of a handwritten curve in a Legendre-Sobolev basis, we first compute the arc-lengths at each time for which a data point $(x, y)$ is collected. To do so, we use the command `NormalizeArcLength`$(x, y, m)$, where $x$ and $y$ are the tables containing all $x_i$ and $y_i$, respectively, which are sampled at $t_i$, for $i = 0, \ldots, m$, where $m$ is the number of times a data point is collected (see Fig. 13).

The next step is to compute the matrix of moment integrals corresponding to the points which are given by $x$ and $y$. To do so, we use the command `MomentIntegrals`$(x, y, ArcLength, L, numSteps)$, where $ArcLength$ is the table

> **read** "HandwrittingRectesting.mpl" :
with(HandwritingRecognitionTesting);
[ ApproximateCurveFromCurves, ApproximateCurveFromPoints,
BoundingBox, MomentIntegrals, NormalizeArcLength ]

**Fig. 12.** The functions of `HandwritingRecognitionTesting` package.

$$
\begin{bmatrix}
> \textbf{read} \text{``m\_inkml''} : \\
\text{ArcLength} := \text{NormalizeArcLength(xValues,} \\
\quad \text{yValues, nVals);} \\
\text{ArcLength} := \text{Ltable}
\end{bmatrix}
$$

(a)

$$
\begin{bmatrix}
> \text{L} := \text{ArcLength[nVals];} \\
\quad \text{L} := 748.28709093397816
\end{bmatrix}
$$

(b)

**Fig. 13.** (a) Arc-lengths computation of handwritten characters at each time a point is collected; (b) computing the total arc-length of a handwritten character.

$$
\begin{bmatrix}
> \text{numSteps} := 400 : \\
\quad \text{d} := 18 : \\
\quad \text{xmoments} := \text{MomentIntegrals(} \\
\quad \text{xValues, ArcLength, d, L, numSteps) :} \\
\quad \text{xmomentsVec} := \text{Matrix(} \\
\quad \text{[seq(xmoments[i], i = 0..d)]);} \\
\quad \text{xmomentsVec} := \begin{bmatrix} 1 \times 19 \text{ Matrix} \\ \text{DataType : anything} \\ \text{Storage : rectangular} \\ \text{Order : Frotran\_order} \end{bmatrix}
\end{bmatrix}
$$

$$
\begin{bmatrix}
> \text{ymoments} := \text{MomentIntegrals(} \\
\quad \text{yValues, ArcLength, d, L, numSteps) :} \\
\quad \text{ymomentsVec} := \text{Matrix(} \\
\quad \text{[seq(ymoments[i], i = 0..d)]);} \\
\quad \text{ymomentsVec} := \begin{bmatrix} 1 \times 19 \text{ Matrix} \\ \text{DataType : anything} \\ \text{Storage : rectangular} \\ \text{Order : Frotran\_order} \end{bmatrix}
\end{bmatrix}
$$

**Fig. 14.** Moments computation with `HandwritingRecognitionTesting` package.

of arc lengths corresponding to the data points given by $x$ and $y$ at a time, $L$ the total arc length, and *numSteps* the number of steps in numerical integration for computing moment integrals (see Fig. 14).

Figure 15a illustrates how to compute coefficients of an approximated curve in the Legendre-Sobolev basis with $\mu = 0.125$ from moments. The matrix which is computed by the command `MomentsToLSMatrix` is given by Proposition 1.

$$
\begin{bmatrix}
> \mu := 0.125 : \\
\quad \text{C} := \text{MomentsToLSMatrix(d, } \mu\text{);} \\
\quad \text{C} := \begin{bmatrix} \textbf{19 x 19} \quad (Matrix) \\ \textbf{Data Type:} \; anything \\ \textbf{Storage:} \; rectangular \\ \textbf{Order:} \; Fortran\_order \end{bmatrix}
\end{bmatrix}
$$

(a)

$$
\begin{bmatrix}
> \text{N} := \text{Matrix([seq([seq(0, i = 1..j - 1), 1/L}^j, \\
\quad \text{seq(0, i = j + 1..d + 1)], j = 1..d + 1)])} \\
\quad \text{M} := \text{N} \cdot \text{C;} \\
\quad \text{N} := \begin{bmatrix} \textbf{19 x 19} \quad (Matrix) \\ \textbf{Data Type:} \; anything \\ \textbf{Storage:} \; rectangular \\ \textbf{Order:} \; Fortran\_order \end{bmatrix}
\end{bmatrix}
$$

(b)

**Fig. 15.** (a) Matrix $C$ computation with `LegendreSobolev` package. (b) Computation of matrix $N$ to rescale the parametrized curves from $[0, L]$ to $[-1, 1]$.

Figure 15b shows how to compute matrix $N$ which scales a handwritten curve to be defined over $[-1, 1]$ instead of $[0, L]$. To scale the approximations to be defined over the interval $[-1, 1]$, we multiply two matrices $N$ and $C$ and compute the conversion matrix $M$. Note that the matrix $C$ is independent of the problem and can be computed, in advance, but the matrix $N$ is only known at the time a handwritten curve is completely written and pen is lifted up.

$$
\left[
\begin{array}{l}
> \text{xCoeffsVec} := \text{xmomentsVec} \cdot \text{M} \\
\text{xCoeffsVec} := \left[
\begin{array}{l}
\text{1 x 19 Matrix} \\
\text{DataType : anything} \\
\text{Storage : rectangular} \\
\text{Order : Frotran\_order}
\end{array}
\right]
\end{array}
\right]
\qquad
\left[
\begin{array}{l}
> \text{yCoeffsVec} := \text{ymomentsVec} \cdot \text{M} \\
\text{yCoeffsVec} := \left[
\begin{array}{l}
\text{1 x 19 Matrix} \\
\text{DataType : anything} \\
\text{Storage : rectangular} \\
\text{Order : Frotran\_order}
\end{array}
\right]
\end{array}
\right]
$$

**Fig. 16.** Computing the Legendre-Sobolev coefficients of the coordinate curves from moment integrals.

Now one can compute the coefficients of parametrized approximations of handwritten characters in the Legendre-Sobolev basis, with $\mu = 0.125$, by multiplying the moment integral matrices and conversion matrix $M$, see Fig. 16.

After the coefficient matrices corresponding to Legendre-Sobolev approximations of the given handwritten curve are computed, we can recover the corresponding monomial representation and then plot the handwritten curve, see Figs. 17 and 18, respectively.

```
> xCoeffs := convert(xCoeffsVec, list);
  yCoeffs := convert(yCoeffsVec, list);
  XLS := add(xCoeffs[kk] · S[kk − 1, mu](x), kk = 1..d + 1);
  YLS := add(yCoeffs[kk] · S[kk − 1, mu](x), kk = 1..d + 1);
```
$$XLS := 189.73347007344971 - 43174.6127062900\, x^{17} + 195134.26543686008\, x^{15}$$
$$-372294.97257625631\, x^{13} + 392139.51651262814\, x^{11} - 253289.82060430780\, x^{9}$$
$$+107985.69508828471\, x^{7} + 259085.21825307995\, x^{18}$$
$$-1124537.0937011107000000\, x^{16} + 2009145.0748245059000000\, x^{14}$$
$$-1899073.9014249837000000\, x^{12} + 1016510.0846958727000000\, x^{10}$$
$$-309032.67958172588\, x^{8} - 32230.336065891532\, x^{5} + 54268.035000655102\, x^{6}$$
$$+6111.3437078230404\, x^{3} - 7124.1885409608675\, x^{4} + 788.36849480813250\, x^{2}$$
$$-440.31476515699463\, x$$
$$YLS := 287.64746656757988 - 41062.137598260\, x^{17} + 188484.7429259304\, x^{15}$$
$$-356524.0808428460\, x^{13} + 357612.16332286471\, x^{11} - 204629.82886696718\, x^{9}$$
$$+67770.765572894417\, x^{7} + 285020.31716937217\, x^{18} - 1270274.4758773500000000\, x^{16}$$
$$+2379548.2831858383000000\, x^{14} - 2430565.7566198024000000\, x^{12}$$
$$+1463461.9788810968000000\, x^{10} - 519221.25154222754\, x^{8} - 12983.274033705085\, x^{5}$$
$$+99749.532330325308\, x^{6} + 1403.1811689782580\, x^{3} - 7549.097513537004\, x^{4}$$
$$-151.99583336566333\, x^{2} + 39.1334243192727\, x$$

**Fig. 17.** Recovering the monomial representation of the coordinate curves.

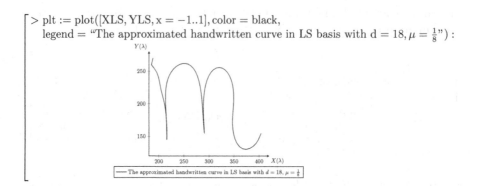

> plt := plot([XLS, YLS, x = −1..1], color = black,
>     legend = "The approximated handwritten curve in LS basis with d = 18, $\mu = \frac{1}{8}$") :

**Fig. 18.** The approximated curve which is constructed from moment integrals.

> Digits := 35;
>     yder := DerivativeInLS($\mu$, yCoeffs)
>     yder := [−57.80172489055115070, 241.95681501221907573147241310639816,
>     −72.00758972940440400672553544459, 492.00134475978062028773451382 44,
>     63.75423428512149041797064603 85, 58.99816519068351673720139527266,
>     7.11192325070875765571005103 8617, 1.59685483656958251903351 2096538,
>     −0.90261179600677726273432558825 12, 0.157592748228409828215 8739019031,
>     0.00493824811802938845646294027 403, 0.0013812349976958496416 992079236,
>     −0.0001105497163631583841037056 8755, 0.00002317647414312603663 8876816426,
>     −2.034462810502421113042576637 × 10⁻⁸, 4.29727128179245603791119 2695 × 10⁻⁷,
>     −8.2086385062481384968179940 88 × 10⁻⁹, 2.150421156817197978999224 152 × 10⁻⁹]

**Fig. 19.** Derivative computation using `LegendreSobolev` package.

## 5.1   Baselines and Cusps

One can compute the cusps and baselines of handwritten characters by computing the critical points of the parametrized approximations of the handwritten curves in a Legendre-Sobolev basis. The experimental results in [3], suggests to use quadruple precision for the calculations. To do so, one can compute the points corresponding to values of $\lambda$ for which either $X'(\lambda) = 0$ or $Y'(\lambda) = 0$, but here we restrict ourselves to the latter case. We apply our calculations for the example which is given in Sect. 5 for the handwritten letter "m", where the degree of the approximation is $d = 18$ and $\mu = 0.125$. To do so, we first compute the coefficients of $Y'(\lambda)$ in the given Legendre-Sobolev basis. The command `DerivativeInLS` implements this functionality, see Fig. 19.

Then, we need to compute the comrade matrix corresponding to $Y'(\lambda)$. In fact, the roots of the system $\{Y'(\lambda) = 0\}$ are equivalent to the eigenvalues of the comrade matrix corresponding to $Y'(\lambda)$, Fig. 20. Now, we can compute the critical points on the handwritten curve as given in Fig. 21.

> C := ComradeMatrix(d − 1, μ, yder);
  evalf(LinearAlgebra : −Eigenvalues(C));
  yroots := [seq(%[i], i = 1..d − 1)];
  realyroots := [];
  **for** i **to** d − 1 **do**
    **if** Im(yroots[i]) = 0 **and** Re(yroots[i]) > −1 **and** Re(yroots[i]) < 1 **then**
      realyroots := [op(realyroots), Re(yroots[i])]
    **end if**
  **end do**;
  realyroots
  [0.8972284851500363926929570653160, 0.4794661941107668225997865145204,
  0.1356368739064478593469426001940, −0.2265522866205856397528476347307,
  −0.6290657031696375888060499900325]

**Fig. 20.** Critical point computation using **LegendreSobolev** package.

> pnts := [seq(eval([XLS, YLS], x = realyroots[i]), i = 1..nops(realyroots))]) :
  b := plots : −pointplot(pnts, symbol = solidbox, color = red,
  legend = "Points on the approximated curve in LS basis with $Y'(\lambda) = 0$") :
  a1 := plot([XLS, YLS, x = −1..1], color = black,
  legend = "The approximated handwritten curve in LS basis with d = 18, $\mu = \frac{1}{8}$") :
  plots : −display([b, a1]);

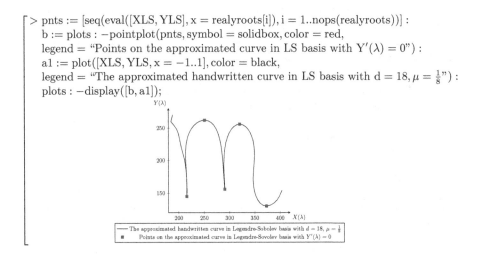

**Fig. 21.** Critical points corresponding to the approximated curve in Legendre-Sobolev basis constructed from moment integrals, with $\mu = 0.125$.

## 5.2 Regions of the Characters in a Handwritten Math Expression

Using the coefficients of the Legendre-Sobolev approximations of handwritten characters, one can find the corresponding regions of individual characters, automatically, by relying on computation of critical points of the corresponding parametrized approximations. In Fig. 22, we have approximated the individual characters in a handwritten math expression by degree 10 polynomials in the Legendre-Sobolev basis with $\mu = \frac{1}{5}$. The command BoundingBox($Cx, Cy, \mu$) implements this method, where $Cx$ and $Cy$ are the $X(\lambda)$ and $Y(\lambda)$ coordinate approximation coefficients in the Legendre-Sobolev basis.

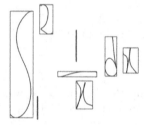

**Fig. 22.** Finding the regions of the characters in a math expression.

## 6    Concluding Remarks

The new MAPLE package `LegendreSobolev` performs various operations on polynomials in Legendre-Sobolev bases. All these operations rely on linear algebra arithmetic operations.

The package `LegendreSobolev` offers a command for recovering the coefficients of polynomials in Legendre-Sobolev bases from their moment integrals. This functionality is very useful in the problem of on-line handwriting recognition, when having the ability of real-time encoding of handwritten characters from their digital inks is crucial. It is also possible to study the geometrical features of handwritten characters by relying on computations of critical and singular points in Legendre-Sobolev bases.

Investigation of how these features might improve the mathematical handwriting recognition rates is a work in progress.

## References

1. Althammer, P.: Eine Erweiterung des Orthogonalitätsbegriffes bei Polynomen und deren Anwendung auf die beste approximation. J. Reine Ang. Math. **211**, 192–204 (1962)
2. Alvandi, P., Watt, S.M.: Real-time computation of Legendre-Sobolev approximations. In: SYNASC, pp. 67–74 (2018)
3. Alvandi, P., Watt, S.M.: Handwriting feature extraction via Legendre-Sobolev matrix representation (2019, preprint)
4. Barnett, S.: A companion matrix analogue for orthogonal polynomials. Linear Algebr. Appl. **12**(3), 197–202 (1975)
5. Carballo, G., Álvarez Nodarse, R., Dehesa, J.S.: Chebychev polynomials in a speech recognition model. Appl. Math. Lett. **14**(5), 581–585 (2001)
6. Char, B.W., Watt, S.M.: Representing and characterizing handwritten mathematical symbols through succinct functional approximation. In: ICDAR, vol. 2, pp. 1198–1202 (2007)
7. Golubitsky, O., Watt, S.M.: Online stroke modeling for handwriting recognition. In: CASCON, pp. 72–80 (2008)
8. Golubitsky, O., Watt, S.M.: Online computation of similarity between handwritten characters. In: Document Recognition and Retrieval XVI, Part of the IS&T-SPIE Electronic Imaging Symposium, pp. C1–C10 (2009)

9. Golubitsky, O., Watt, S.M.: Online recognition of multi-stroke symbols with orthogonal series. In: ICDAR, pp. 1265–1269 (2009)

10. Hausdorff, F.: Summationsmethoden und Momentfolgen. I. Math. Z. **9**, 74–109 (1921)

11. Hausdorff, F.: Summationsmethoden und Momentfolgen. II. Math. Z. **9**, 280–299 (1921)

12. Hu, R., Watt, S.M.: Identifying features via homotopy on handwritten mathematical symbols. In: SYNASC, pp. 61–67 (2013)

13. Wang, K., An, N., Li, B.N., Zhang, Y., Li, L.: Speech emotion recognition using Fourier parameters. IEEE Trans. Affect. Comput. **6**(1), 69–75 (2015)

14. Zhiqi, Y.: Gesture learning and recognition based on the Chebyshev polynomial neural network. In: 2016 IEEE Information Technology, Networking, Electronic and Automation Control Conference, pp. 931–934 (2016)

15. Zhu, L., Zhu, S.: Face recognition based on orthogonal discriminant locality preserving projections. Neurocomputing **70**(7), 1543–1546 (2007). Advances in Computational Intelligence and Learning

# On the Effective Computation
# of Stabilizing Controllers of 2D Systems

Yacine Bouzidi[1], Thomas Cluzeau[2], Alban Quadrat[3($\boxtimes$)], and Fabrice Rouillier[3]

[1] Inria Lille - Nord Europe, 40 Avenue Halley, 59650 Villeneuve d'Ascq, France
yacine.bouzidi@inria.fr
[2] CNRS, XLIM UMR 7252, 87060 Limoges Cedex, France
thomas.cluzeau@unilim.fr
[3] Inria Paris, Institut de Mathématiques de Jussieu Paris-Rive Gauche,
Sorbonne Université, Paris Université, Paris, France
{alban.quadrat,fabrice.rouillier}@inria.fr

**Abstract.** In this paper, we show how stabilizing controllers for 2D systems can effectively be computed based on computer algebra methods dedicated to polynomial systems, module theory and homological algebra. The complete chain of algorithms for the computation of stabilizing controllers, implemented in Maple, is illustrated with an explicit example.

**Keywords:** Multidimensional systems theory · 2D systems · Stability analysis · Stabilization · Computation of stabilizing controllers · Polynomial systems · Module theory · Homological algebra

## 1 Introduction

In the eighties, *the fractional representation approach to analysis and synthesis problems* was introduced by Vidyasagar, Desoer, etc., to unify different problems studied in the control theory community (e.g., internal/strong/simultaneous/optimal/robust stabilizability) within a unique mathematical framework [12, 27]. Within this approach, different classes of linear systems (e.g., discrete, continuous, finite-dimensional systems, infinite-dimensional systems, multidimensional systems) can be studied by means of a common mathematical formulation.

The main idea of this approach is to reformulate the concept of *stability* − central in control theory − as a *membership problem*. More precisely, a *single-input single-output (SISO) linear system*, also called *plant*, is defined as an element $p$ of the *quotient field (field of fractions)* $Q(A) = \left\{ \frac{n}{d} \mid 0 \neq d, n \in A \right\}$ of an integral domain $A$ of SISO stable plants [27]. Hence, if $p \in A$, then $p$ is *A-stable* (simply *stable* when the reference to $A$ is clear) and unstable if $p \in Q(A) \backslash A$. More generally, a *multi-input multi-output (MIMO)* plant can be defined by a matrix $P \in Q(A)^{q \times r}$. Hence, it is stable if $P \in A^{q \times r}$, unstable otherwise.

Different integral domains $A$ of SISO stable plants are considered in the control theory literature depending on the class of systems which is studied. For

J. Gerhard and I. Kotsireas (Eds.): MC 2019, CCIS 1125, pp. 30–49, 2020.
https://doi.org/10.1007/978-3-030-41258-6_3

instance, the *Hardy (Banach) algebra* $H^\infty(\mathbb{C}_+)$ formed by all the holomorphic functions in the open-right half plane $\mathbb{C}_+ = \{s \in \mathbb{C} \mid \Re(s) > 0\}$ which are bounded for the norm $\|f\|_\infty = \sup_{s \in \mathbb{C}_+} |f(s)|$ plays a fundamental role in stabilization problems of infinite-dimensional linear time-invariant systems (e.g., differential time-delay systems, partial differential systems) since its elements can be interpreted as the *Laplace transform* of $L^2(\mathbb{R}_+) - L^2(\mathbb{R}_+)$-stable plant (i.e., any input $u$ of the system in $L^2(\mathbb{R}_+)$ yields an output $y$ in $L^2(\mathbb{R}_+)$) [10]. Similarly, the integral domain $RH_\infty$ of proper and stable rational functions, i.e., the ring of all rational functions in $H^\infty(\mathbb{C}_+)$, corresponds to the ring of *exponentially stable* finite-dimensional linear time-invariant systems (i.e., exponentially stable ordinary differential systems with constant coefficients) [27].

In this paper, we shall focus on the class of *discrete multidimensional systems* which are defined by multivariate recurrence relations with constant coefficients or, using the standard $\mathcal{Z}$-*transform*, by elements of the field $\mathbb{R}(z_1, \ldots, z_n)$ of real rational functions in $z_1, \ldots, z_n$. The latter is the field of fractions $Q(A)$ of the integral domain $A$ of *SISO structurally stable plants* defined by

$$\mathbb{R}(z_1, \ldots, z_n)_S := \left\{ \frac{n}{d} \mid 0 \neq d, \, n \in B, \, \gcd(d, n) = 1, \, V(\langle d \rangle) \cap \mathbb{U}^n = \emptyset \right\},$$

where $B := \mathbb{R}[z_1, \ldots, z_n]$ denotes the polynomial ring in $z_1, \ldots, z_n$ with coefficients in $\mathbb{R}$, $\gcd(d, n)$ the *greatest common divisor* of $d, \, n \in B$, and

$$V(\langle d \rangle) = \{z = (z_1, \ldots, z_n) \in \mathbb{C}^n \mid d(z) = 0\}$$

the *affine algebraic set* defined by $d \in B$, i.e., the complex zeros of $d$, and finally

$$\mathbb{U}^n = \{z = (z_1, \ldots, z_n) \in \mathbb{C}^n \mid |z_i| \leq 1, \, i = 1, \ldots, n\}$$

the *closed unit polydisc* of $\mathbb{C}^n$. It can be shown that $p \in \mathbb{R}(z_1, \ldots, z_n)_S$ implies that the plant $p$ is *bounded-input bounded-output stable* in the sense that an input $u$ in $l^\infty(\mathbb{Z}_+^n)$ yields an output $y$ in $l^\infty(\mathbb{Z}_+^n)$. See, e.g., [15, 16].

Despite the simplicity of the main idea of the fractional representation approach, i.e., to express stability as a membership problem, many problems studied in control theory were reformulated as algebraic (analysis) problems. For instance, internal/strong/simultaneous/optimal/robust stabilizability can be reformulated within this mathematical approach and solved for particular integral domains $A$ such as $RH_\infty$ [27]. But these problems are still open for the class of infinite-dimensional systems [10] and multidimensional systems [15, 16].

The goal of this article is to combine results obtained in [2, 4–6, 20, 21] to obtain a complete algorithmic approach to the computation of stabilizing controllers for 2D stabilizable MIMO systems. In [5], the problem was solved for SISO systems. To handle the class of MIMO systems, we use the module-theoretic approach to the fractional representation approach [20, 21]. More precisely, in [6], the main steps towards an algorithmic computation of stabilizing controllers for general $n$D systems are explained based on computer algebra methods. In this paper, we focus on 2D MIMO systems for which the so-called *Polydisk Nullstellensatz* [7] has received an effective version in [5] (which is not the case for general $n$D systems), which yields a complete algorithmic approach to the computation of stabilizing controllers for 2D stabilizable MIMO systems.

Our algorithms were implemented in the computer algebra system `Maple`, based on both the package `nDStab` – dedicated to stability and stabilizability of $n$D systems – and on the OREMODULES package [9] which aims to study linear systems theory based on effective module theory and homological algebra.

## 2   The Fractional Representation Approach

In what follows, we shall use the following notations. $A$ will denote an integral domain of SISO plants, $K := Q(A) = \left\{ \frac{n}{d} \mid 0 \neq d, n \in A \right\}$ its quotient field, $P \in K^{q \times r}$ a plant, $C \in K^{r \times q}$ a controller, and $p = q + r$.

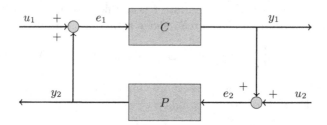

**Fig. 1.** Closed-loop system

With the notations of Fig. 1 defining the closed-loop system formed by the plant $P$ and the controller $C$, we get $(e_1^T \quad e_2^T)^T = H(P, C) \, (u_1^T \quad u_2^T)^T$, where the transfer matrix $H(P,C) \in K^{p \times p}$ is defined by:

$$H(P, C) := \begin{pmatrix} I_q & -P \\ -C & I_r \end{pmatrix}^{-1} = \begin{pmatrix} (I_q - PC)^{-1} & (I_q - PC)^{-1} P \\ C\,(I_q - PC)^{-1} \, I_r + C\,(I_q - PC)^{-1} P \end{pmatrix}$$

$$= \begin{pmatrix} I_q + P\,(I_r - CP)^{-1} C & P\,(I_r - CP)^{-1} \\ (I_r - CP)^{-1} C & (I_r - CP)^{-1} \end{pmatrix}.$$

**Definition 1 ([27]).** *A plant $P \in K^{q \times r}$ is internally stabilizable if there exists $C \in K^{r \times q}$ such that $H(P,C) \in A^{p \times p}$. Then, $C$ is a stabilizing controller of $P$.*

If $H(P,C) \in A^{p \times p}$, then we can easily prove that all the entries of any transfer matrix between two signals appearing in Fig. 1 are $A$-stable (see, e.g., [27]).

A fundamental issue in control theory is to first test if a given plant $P$ is internally stabilizable and if so, to explicitly compute a stabilizing controller of $P$, and by extension the family Stab($P$) of all its stabilizing controllers.

To do that and to study other stabilization problems such as robust control, the fractional approach to systems was introduced in control theory [12,27].

**Definition 2.** A fractional representation *of* $P \in K^{q \times p}$ *is defined by* $P = D^{-1} N = \widetilde{N} \widetilde{D}^{-1}$, *where* $R := (D \quad -N) \in A^{q \times p}$ *and* $\widetilde{R} = (\widetilde{N}^T \quad \widetilde{D}^T)^T \in A^{p \times r}$.

A plant $P \in K^{q \times r}$ always admits a fractional representation since we can always consider $D = d\, I_q$, $\widetilde{D} = d\, I_r$, where $d$ is the product of all the denominators of the entries of $P$, which yields $N = D\, P \in A^{q \times r}$ and $\widetilde{N} = P\, \widetilde{D} \in A^{q \times r}$.

## 3   Testing Stability of Multidimensional Systems

### 3.1   Stability Tests for $n$D Systems

A fundamental issue in the fractional representation approach is to be able to solve the following *membership problem*: let $p \in K := Q(A)$, check whether or not $p \in A$. The answer to this problem depends on $A$.

In this paper, we shall focus on the case $A := \mathbb{Q}(z_1, \ldots, z_n)_S$ defined in the introduction, and mainly on 2D systems for the stabilization issue, i.e., on $A = \mathbb{Q}(z_1, z_2)_S$. Since we shall only consider exact computation methods based on computer algebra techniques, in what follows, we consider the ground field to be $\mathbb{Q}$ instead of $\mathbb{R}$. Tests for stability of multidimensional systems have largely been investigated in both the control theory and signal processing literatures. For more details, see the surveys [15,16] and the references therein.

Let $B := \mathbb{Q}[z_1, \ldots, z_n]$ be the commutative polynomial ring over $\mathbb{Q}$. We first note that $K := Q(A) = \mathbb{Q}(z_1, \ldots, z_n) = Q(B)$. Moreover, $A = \mathbb{Q}(z_1, \ldots, z_n)_S$ is the *localization* $B_S := S^{-1} B = \{b/s \mid b \in B, s \in S\}$ of the polynomial ring $B$ with respect to the *(saturated) multiplicatively closed subset* of $B$ defined by:

$$S := \{b \in B \mid V(\langle b \rangle) \cap \mathbb{U}^n = \emptyset\}.$$

Any element $p \in K$ can be written as $p = n/d$ with $0 \neq d$, $n \in B$. Moreover, we can always assume that the greatest common divisor $\gcd(d, n)$ is reduced to 1. Hence, given an element $p = n/d \in K$, $\gcd(d, n) = 1$, we get that $p \in A = B_S$ iff $d \in S$. The membership problem is reduced to checking whether or not $d \in S$. Let us study how this problem, i.e., the stability test, can be effectively checked.

Setting $z_k = u_k + i\, v_k$, where $u_k, v_k \in \mathbb{R}$, for $k = 1, \ldots, n$, and writing $b(z_1, \ldots, z_n) = c_1(u_1, \ldots, u_n, v_1, \ldots, v_n) + i\, c_2(u_1, \ldots, u_n, v_1, \ldots, v_n)$, where the $c_i$'s are two real polynomials in the real variables $u_k$'s and $v_k$'s, $V(\langle b \rangle) \cap \mathbb{U}^n$ yields the following semi-algebraic set:

$$\begin{cases} c_1(u_1, \ldots, u_n, v_1, \ldots, v_n) = 0, \\ c_2(u_1, \ldots, u_n, v_1, \ldots, v_n) = 0, \\ u_k^2 + v_k^2 \leq 1, \ k = 1, \ldots, n. \end{cases} \tag{1}$$

Real algebraic methods such as CAD [1] can then be used to solve this problem for small $n$. But they quickly become impracticable in practice for $n$D systems, even for $n \geq 2$, since the number of unknowns has been doubled in (1). Hence, an algebraic formulation, more tractable for explicit computation, must be found. The next theorem gives a mathematical characterization of $V(\langle b \rangle) \cap \mathbb{U}^n = \emptyset$.

**Theorem 1** ([11]). *Let* $b \in \mathbb{R}[z_1, \ldots, z_n]$ *and* $\mathbb{T}^n = \prod_{i=1}^{n}\{z_i \in \mathbb{C} \mid |z_i| = 1\}$. *Then, the following assertions are equivalent:*

1. $V(\langle b(z_1, \ldots, z_n) \rangle) \cap \mathbb{U}^n = \emptyset$.
2.
$$
\begin{cases}
V(\langle b(z_1, 1, \ldots, 1) \rangle) \cap \mathbb{U} = \emptyset, \\
V(\langle b(1, z_2, 1, \ldots, 1) \rangle) \cap \mathbb{U} = \emptyset, \\
\quad \vdots \\
V(\langle b(1, \ldots, 1, z_n) \rangle) \cap \mathbb{U} = \emptyset, \\
V(\langle b(z_1, \ldots, z_n) \rangle) \cap \mathbb{T}^n = \emptyset.
\end{cases}
$$

In [2], it is shown how 2 of Theorem 1 can be effectively tested by means of computer algebra methods. Let us shortly state again the main idea. The first $n$ conditions of 2 of Theorem 1 can be efficiently checked by means of standard stability tests for 1D systems such as [26]. The only remaining difficulty is the last condition, which must be transformed into a more tractable algorithmic formulation. To do that, we can introduce *the Möbius transformation*

$$
\varphi : \overline{\mathbb{R}}^n \longrightarrow \mathbb{T}_o^n,
$$
$$
t := (t_1, \ldots, t_n) \longmapsto z := (z_1, \ldots, z_n) = \left( \frac{t_1 - i}{t_1 + i}, \ldots, \frac{t_n - i}{t_n + i} \right),
$$

with the notations $\overline{\mathbb{R}} = \mathbb{R} \cup \{\infty\}$ and $\mathbb{T}_o = \mathbb{T} \backslash \{1\}$. Note that we have:

$$
z_k = u_k + i\, v_k = \frac{t_k - i}{t_k + i} = \frac{t_k^2 - 1}{t_k^2 + 1} - i\, \frac{2\, t_k}{t_k^2 + 1}.
$$

We can easily check that $\varphi$ is bijective and that $t = \varphi^{-1}(z)$ is defined by:

$$
t = \left( i\, \frac{1 + z_1}{1 - z_1}, \ldots, i\, \frac{1 + z_n}{1 - z_n} \right).
$$

Now, substituting $z = \varphi(t)$ into $b(z)$, we get $b(\varphi(t)) = c_1(t) + i\, c_2(t)$, where the $c_i$'s are two real rational functions of the real vector $t$. Writing $c_j = n_j/d_j$, where $n_j, d_j \in \mathbb{Q}[t]$ and $\gcd(d_j, n_j) = 1$, then $b(z) = 0$ is equivalent to $c_1(t) = 0$ and $c_2(t) = 0$, i.e., to $n_1(t) = 0$ and $n_2(t) = 0$. Hence, the problem of computing $V(\langle b \rangle) \cap \mathbb{T}_o^n$ is equivalent to the problem of computing $V(\langle n_1, n_2 \rangle) \cap \mathbb{R}^n$. In particular, we get $V(\langle b \rangle) \cap \mathbb{T}_o^n = \emptyset$ iff $V(\langle n_1, n_2 \rangle) \cap \mathbb{R}^n = \emptyset$. *Critical point methods* (see [1]) can be used to check the last condition. Indeed, they characterize a real point on every connected component of $V(\langle n_1, n_2 \rangle) \cap \mathbb{R}^n$. Finally, while working over $\mathbb{T}_o^n$ and not over $\mathbb{T}^n$, we are missing to test the stability criterion on the particular set of points $\{(1, z_2, \ldots, z_n), \ldots, (z_1, \ldots, z_{n-1}, 1)\}$ of $\mathbb{T}^n \backslash \mathbb{T}_o^n$. Hence, we have to study the stability of the polynomials $b(1, z_2, \ldots, z_n), \ldots, b(z_1, \ldots, z_{n-1}, 1)$ separately based on the same method. This can be studied inductively on the dimension. The corresponding algorithm [2] is given in Algorithm 1 (see below).

---

**Algorithm 1.** IsStable

---

1: **procedure** IsStable($b(z_1, \ldots, z_n)$)             ▷ Return true if $V(\langle b \rangle) \cap \mathbb{U}^n = \emptyset$
2:     **for** $k = 0$ to $n - 1$ **do**
3:         Compute the set $S_k$ of polynomials obtained by substituting $k$ variables by 1 into $b(z_1, \ldots, z_n)$
4:         **for each** element $b$ in $S_k$ **do**
5:             $\{n_1, n_2\} =$ Möbius_transform($b$)
6:             **if** $V(\langle n_1, n_2 \rangle) \cap \mathbb{R}^n \neq \emptyset$ **then**
7:                 **return** False
8:             **end if**
9:         **end for**
10:     **end for**
11:     **return** True
12: **end procedure**

---

### 3.2  An Efficient Stability Test for 2D Systems

For $n = 2$, with the notations of (1), the last condition of 2 of Theorem 1, i.e., checking whether or not $V(\langle b \rangle) \cap \mathbb{T}^2$ is empty, amounts to search for the real solutions $(u_1, v_1, u_2, v_2)$ of the following *zero-dimensional polynomial system*

$$c_1(u_1, u_2, v_1, v_2) = 0, \ c_2(u_1, u_2, v_1, v_2) = 0, \ u_1^2 + v_1^2 = 1, \ u_2^2 + v_2^2 = 1,$$

i.e., for the real solutions of a polynomial system which only has a finite number of complex solutions. Hence, the problem is reduced to deciding if a polynomial system in two variables, which has finitely many complex solutions, admits real ones. Based on this idea, an efficient algorithm for testing stability of 2D systems is given in [3,4]. Let us explain it. According to the end of Sect. 3.1, using the Möbius transformation $\varphi$, $V(\langle b(z_1, z_2) \rangle) \cap \mathbb{T}^2 = \emptyset$ is equivalent to:

$$\begin{cases} V(\langle n_1, n_2 \rangle) \cap \mathbb{R}^2 = \emptyset, \\ b(1, z_2) \neq 0, \ |z_2| = 1, \\ b(z_1, 1) \neq 0, \ |z_1| = 1. \end{cases}$$

Since the last two conditions of the above system can be efficiently checked by means of, e.g., [26], let us concentrate on the first condition, i.e., on the problem of deciding when a zero-dimensional polynomial system $V(\langle n_1, n_2 \rangle)$ in two variables $t_1$ and $t_2$ has real solutions. The main idea is to reduce the problem of checking the existence of real solutions of such a polynomial system to the problem of checking the existence of real roots of a well-chosen univariate polynomial. A convenient method to do that is the so-called *univariate representation* of the solutions [24,25]. Let us recall this concept. Given a zero-dimensional polynomial ideal $I$ in $\mathbb{Q}[t_1, t_2]$, a univariate representation consists in the datum of a linear form $s = a_1 t_1 + a_2 t_2$, with $a_1, a_2 \in \mathbb{Q}$, and three polynomials $h, h_{t_1}, h_{t_2} \in \mathbb{Q}[s]$ such that the applications

$$\phi : V(I) \longrightarrow V(\langle h \rangle) = \{s \in \mathbb{C} \mid h(s) = 0\} \quad \psi : V(\langle h \rangle) \longrightarrow V(I)$$
$$t = (t_1, t_2) \longmapsto s = a_1 t_1 + a_2 t_2, \qquad\qquad s \longmapsto t = (h_{t_1}(s), h_{t_2}(s)),$$

provide a one-to-one correspondence between the zeros of $I$ and the roots of $h$. In that case, the linear form $s = a_1 t_1 + a_2 t_2$ is said to be *separating*.

A key property of the 1–1 correspondence $\phi = \psi^{-1}$ is that it preserves the real zeros of $V(I)$ in the sense that any real zero of $V(I)$ corresponds a real root of $h$ and conversely. As a consequence, deciding if $V(I)$ has real zeros is equivalent to deciding if the univariate polynomial $h$ has real roots.

From a computational viewpoint, to conclude on the existence of real zeros of $V(I)$, it is sufficient to compute a separating linear form for $V(I)$ and the corresponding univariate polynomial $h$. In [3,4], an efficient algorithm based on *resultants* and *subresultant sequences* [1] is used to perform these computations and, then to conclude about the stability of 2D systems.

# 4    Testing Stabilizability of 2D Systems

## 4.1    Module-Theoretic Conditions for Stabilizability

As explained in Sect. 2, a fractional representation of a plant $P \in K^{q \times p}$ is defined by $P = D^{-1} N$, where $R = (D \quad - N) \in A^{q \times (q+r)}$. See Definition 2. Let us set $p = q + r$. Since $A$ is an integral domain and $R$ is a matrix, we are naturally in the realm of *module theory* [13,23], which is the extension of linear algebra for rings. Given $R \in A^{q \times p}$, we shall consider the following $A$-modules:

$$\begin{cases} \ker_A(.R) := \{\mu \in A^{1 \times q} \mid \mu R = 0\}, \\ \operatorname{im}_A(.R) = A^{1 \times q} R := \{\lambda \in A^{1 \times p} \mid \exists \nu \in A^{1 \times q} : \lambda = \nu R\}, \\ \operatorname{coker}_A(.R) = A^{1 \times p}/(A^{1 \times q} R). \end{cases}$$

We recall that factor $A$-module $M := \operatorname{coker}_A(.R) = A^{1 \times p}/(A^{1 \times q} R)$ is defined by the generators $\{y_j = \pi(f_j)\}_{j=1,\dots,p}$, where $\{f_j\}_{j=1,\dots,p}$ denotes the standard basis of the free $A$-module $A^{1 \times p}$, namely, the basis formed by the standard basis vectors $f_j$'s (i.e., the row vectors of length $p$ with 1 at the $j^{\text{th}}$ position and 0 elsewhere), and $\pi : A^{1 \times p} \longrightarrow M$ is the $A$-homomorphism (i.e., $A$-linear map) which sends $\lambda \in A^{1 \times p}$ onto its residue class $\pi(\lambda)$ in $M$ (i.e., $\pi(\lambda') = \pi(\lambda)$ if there exists $\mu \in A^{1 \times q}$ such that $\lambda' = \lambda + \mu R$). One can prove that the generators $y_j$'s satisfy the $A$-linear relations $\sum_{j=1}^{p} R_{ij} y_j = 0$, $i = 1, \dots, q$, where $R_{ij}$ stands for the $(i, j)$ entry of $R$. For more details, see, e.g., [8,20,22]. Hence, if we note by $y = (y_1, \dots, y_p)^T$, then the above relation can formally be rewritten as $Ry = 0$, which explains why the $A$-module $M$ is used to study the linear system $Ry = 0$. Using module theory, a characterization of stabilizability, namely, of the existence of a stabilizing controller $C$ for a given plant $P$, was obtained.

**Theorem 2** ([21]). *Let $P \in K^{q \times r}$, $p = q + r$, $P = D^{-1} N$ be a fractional representation of $P$, where $R = (D \quad - N) \in A^{q \times p}$, and $M = A^{1 \times p}/(A^{1 \times q} R)$ the $A$-module finitely presented by $R$. Then, $P$ is internally stabilizable iff the $A$-module $M/t(M)$ is projective, where $t(M) := \{m \in M \mid \exists\, a \in A \backslash \{0\} : a\, m = 0\}$ is the torsion submodule of $M$. In other words, $P$ is internally stabilizable iff there exist an $A$-module $L$ and $s \in \mathbb{Z}_{\geq 0}$ such that $L \oplus M/t(M) \cong A^{1 \times s}$.*

According to Theorem 2, we have to study the following two problems:

1. Explicitly characterize $t(M)$, and thus $M/t(M)$.
2. Check whether or not $M/t(M)$ is a projective $A$-module.

Both problems can be solved by homological algebra methods [23]. In the rest of the paper, we shall suppose that $A$ is a *coherent ring* [20,23]. For instance, we can consider the coherent but non-noetherian integral domain $H^\infty(\mathbb{C}_+)$ (see Introduction) or any *noetherian ring* such as $\mathbb{R}(z_1,\dots,z_n)_S$ or $RH_\infty$. The category of finitely presented modules over a coherent integral domain is a natural framework for mathematical systems theory [20].

Let us introduce the definition of an *extension module* [13,23]. Let us consider a finitely presented $A$-module $L := A^{1 \times q_0}/(A^{1 \times q_1} S_1)$, where $S_1 \in A^{q_1 \times q_0}$. Since $A$ is coherent, $\ker_A(.S_1)$ is a finitely generated $A$-module, and thus there exists a finite family of generators of $\ker_A(.S_1)$. Stacking these row vectors of $A^{1 \times q_1}$, we obtain $S_2 \in A^{q_2 \times q_1}$ such that $\ker_A(.S_1) = \mathrm{im}_A(.S_2) = A^{1 \times q_2} S_2$. Repeating the same process with $S_2$, etc., we get the following *exact sequence* of $A$-modules

$$\cdots \xrightarrow{.S_3} A^{1 \times q_2} \xrightarrow{.S_2} A^{1 \times q_1} \xrightarrow{.S_1} A^{1 \times q_0} \xrightarrow{\kappa} L \longrightarrow 0, \qquad (2)$$

i.e., $\ker_A(.S_i) = \mathrm{im}_A(.S_{i+1})$ for $i \geq 1$, where $\kappa$ is the epimorphism which maps $\eta \in A^{1 \times q_0}$ onto its residue class $\kappa(\eta)$ in $L$. This exact sequence is called *free resolution* of $L$ [13,23]. "Transposing" (2), i.e., applying the *contravariant left exact functor* $\mathrm{Rhom}_A(\cdot, A)$ to (2), we get the following *complex* of $A$-modules

$$\cdots \xleftarrow{.S_3^T} A^{1 \times q_2} \xleftarrow{.S_2^T} A^{1 \times q_1} \xleftarrow{.S_1^T} A^{1 \times q_0} \xleftarrow{\phantom{.}} 0,$$

i.e., $\mathrm{im}_A\left(.S_i^T\right) \subseteq \ker_A\left(.S_{i+1}^T\right)$ for $i \geq 1$. We introduce the following $A$-modules:

$$\begin{cases} \mathrm{ext}_A^0(L, A) \cong \ker_A\left(.S_1^T\right), \\ \mathrm{ext}_A^i(L, A) \cong \ker_A\left(.S_{i+1}^T\right)/\mathrm{im}_A\left(.S_i^T\right), \ i \geq 1. \end{cases}$$

It can be shown that, up to isomorphism, the $A$-modules $\mathrm{ext}_A^i(L, A)$'s depend only on $L$ and not on the choice of the free resolution (2) of $L$, i.e., on the choice of the matrices $S_i$'s. It explains the notation $\mathrm{ext}_A^i(L, A)$. See [13,23].

**Theorem 3** ([20]). *With the notations of Theorem 2, we have:*

1. $t(M) \cong \mathrm{ext}_A^1(T(M), A)$, *where* $T(M) := A^{1 \times q}/\left(A^{1 \times p} R^T\right)$ *is the so-called Auslander transpose of $M$ (i.e., the $A$-module finitely presented by $R^T$).*
2. *Let us suppose that the* weak global dimension *of $A$ is finite and is equal to $n$. Then, the obstruction for $M/t(M)$ to be projective is defined by*

$$I = \bigcap_{i=2}^{n} \mathrm{ann}_A\left(\mathrm{ext}_A^i(T(M/t(M)), A)\right),$$

*where* $\mathrm{ann}_A(L) := \{a \in A \mid a\, L = 0\}$ *is the annihilator of a $A$-module $L$. Hence, $M$ is projective iff we have $I = A$.*

Let us now show how Theorem 3 can be used to check whether or not a $n$D system $P$ is stabilizable. Let $A = \mathbb{Q}(z_1, \ldots, z_n)_S$, $B = \mathbb{Q}[z_1, \ldots, z_n]$, $K = Q(A) = \mathbb{Q}(z_1, \ldots, z_n)$, and $P \in K^{q \times r}$. Since $K = Q(B)$, we can write each entry $P_{ij}$ of $P$ as $P_{ij} = n_{ij}/d_{ij}$, where $0 \neq d_{ij}$, $n_{ij} \in B$, and $\gcd(d_{ij}, n_{ij}) = 1$. Let us denote by $d$ the least common multiple of all the $d_{ij}$'s and set $D = d\,I_q \in B^{q \times q}$, $N = DP \in B^{q \times p}$, and $R = (D \quad - N) \in B^{q \times p}$. Let us also consider the finitely presented $B$-module $L := B^{1 \times p}/(B^{1 \times q}\,R)$. Since $A = S^{-1}B$ is a localization (see Sect. 3), $A$ is a *flat* $B$-module [13, 23], we then get that

$$A \otimes_B L \cong A^{1 \times p}/(A^{1 \times q}\,R) = M,$$

where $\otimes_B$ stands for the *tensor product* of $B$-modules [13, 23]. Note that $A \otimes_B L$ can be understood as the $A$-module obtained by extending the coefficients of the $B$-module $L$ to $A$. Using elimination theory over $B$ (e.g., Gröbner bases, Janet bases), given the matrix $R^T$, we can compute $\ker_B\left(.R^T\right)$, i.e., the *second syzygy module* of $T(L) = B^{1 \times q}/\left(B^{1 \times p}\,R^T\right)$ [13, 23]. For more details, see [8, 13, 22] and [9] for the OREMODULES package which handles such computations. Hence, we can compute the beginning of a free resolution of the $B$-module $T(L)$:

$$0 \longleftarrow T(L) \xleftarrow{\ \sigma\ } B^{1 \times q} \xleftarrow{\ .R^T\ } B^{1 \times p} \xleftarrow{\ .Q^T\ } B^{1 \times m}.$$

Applying the functor $\mathrm{Rhom}_B(\,\cdot\,, B)$ to it, we obtain the complex of $B$-modules:

$$0 \longrightarrow B^{1 \times q} \xrightarrow{\ .R\ } B^{1 \times p} \xrightarrow{\ .Q\ } B^{1 \times m}.$$

Therefore, we get $t(L) = \mathrm{ext}_B^1(T(L), B) = \ker_B(.Q)/\mathrm{im}_B(.R)$ and since we can also compute a matrix $R' \in B^{q' \times p}$ such that $\ker_B(.Q) = \mathrm{im}_B(.R')$, we obtain:

$$t(L) = (B^{1 \times q'}\,R')/(B^{1 \times q}\,R) \quad \Rightarrow \quad L/t(L) = B^{1 \times p}/(B^{1 \times q'}\,R').$$

The corresponding computations can be handled by the OREMODULES package [9]. Hence, based, e.g., on Gröbner basis techniques, we can find an explicit presentation $R' \in B^{q' \times p}$ of the $B$-module $L/t(L)$. Since $A = S^{-1}B$, we get [23]

$$A \otimes_B t(L) \cong A \otimes_B \mathrm{ext}_B^1(T(L), B) \cong \mathrm{ext}_A^1(A \otimes_B T(L), A) \cong \mathrm{ext}_A^1(T(M), A) \cong t(M),$$

which yields $A \otimes_B (L/t(L)) \cong (A \otimes_B L)/(A \otimes_B t(L)) \cong M/t(M)$. Thus, we have

$$M/t(M) \cong A \otimes_B \left(B^{1 \times p}/(B^{1 \times q'}\,R')\right) \cong A^{1 \times p}/(A^{1 \times q'}\,R'),$$

which shows that $R' \in B^{q' \times p}$ is a presentation matrix of the $A$-module $M/t(M)$, which explicitly solves the first point.

Let us now consider the second one, i.e., the problem of testing whether or not $M/t(M)$ is a projective $A$-module. Clearly, we have:

$$A \otimes_B T(L/t(L)) = A \otimes_B \left(B^{1 \times q'}/\left(B^{1 \times p}\,R'^T\right)\right) \cong A^{1 \times q'}/\left(A^{1 \times p}\,R'^T\right) = T(M/t(M)).$$

Moreover, since the localization $A = S^{-1} B$ commutes with the intersection of ideals and $S^{-1} \operatorname{ann}_B(P) = \operatorname{ann}_{S^{-1}B}(S^{-1} P)$ for a finitely generated $B$-module $P$ (see, e.g., [13,23]), using the fact that the $B$-modules $\operatorname{ext}_B^i(T(L/t(L)), B)$'s are finitely generated, we then obtain:

$$A \otimes_B \left( \bigcap_{i=2}^n \operatorname{ann}_B(\operatorname{ext}_B^i(T(L/t(L)), B) \right) \cong \bigcap_{i=2}^n \operatorname{ann}_A \left( \operatorname{ext}_A^i(T(M/t(M)), A \right).$$

Hence, if we denote by $I = \bigcap_{i=2}^n \operatorname{ann}_A \left( \operatorname{ext}_A^i(T(M/t(M)), A \right)$ the obstruction for the $A$-module $M/t(M)$ to be projective (see 2 of Theorem 3) and similarly by $J = \bigcap_{i=2}^n \operatorname{ann}_B(\operatorname{ext}_B^i(T(L/t(L)), B)$ the obstruction for the $B$-module $L/t(L)$ to be projective, then $I \cong A \otimes_B J$. We note that the ideal $J$ can be explicitly computed by means of elimination theory (e.g., Gröbner/Janet bases) and implemented in a computer algebra system (see the OREMODULES [9]). For more details, see [8]. Hence, $M/t(M)$ is a projective $A$-module iff $I = A$, i.e., iff $S^{-1} J = S^{-1} B$, i.e., iff $J \cap S \neq \emptyset$, i.e., iff there exists $b \in J$ such that $V(\langle b \rangle) \cap \mathbb{U}^n = \emptyset$.

### 4.2   Towards an Effective Version of the Polydisc Nullstellensatz

The condition $J \cap S \neq \emptyset$ yields the so-called *Polydisc Nullstellensatz*.

**Theorem 4 (*Polydisc Nullstellensatz*, [7]).** *Let $J$ be a finitely generated ideal of $B = \mathbb{Q}[z_1, \ldots, z_n]$. Then, the two assertions are equivalent:*

1. $V(J) \cap \mathbb{U}^n = \emptyset$.
2. *There exists $b \in J$ such that $V(\langle b \rangle) \cap \mathbb{U}^n = \emptyset$.*

To our knowledge, there is no effective version of the Polydisc Nullstellensatz for a general ideal $J$. But, in [5], it is shown how the first condition of Theorem 4 can be effectively tested for a zero-dimensional polynomial system $J$, i.e., when $B/J$ is a finite-dimensional $\mathbb{Q}$-vector space, or equivalently when $V(J)$ is defined by a finite number of complex points. Given a zero-dimensional polynomial ideal $J$ in $\mathbb{Q}[z_1, \ldots, z_n]$, a first step consists in computing a univariate representation of $V(J)$. Such a representation characterizes the zeros $(z_1, \ldots, z_n)$ of $V(J)$ as:

$$\begin{cases} h(t) = 0, \\ z_1 = h_1(t), \\ \quad \vdots \\ z_n = h_n(t). \end{cases} \tag{3}$$

Using (3), we now have to check whether or not $|z_1| \leq 1, \ldots, |z_n| \leq 1$. To do, at the solutions $z_k = u_k + i\, v_k$ of (3), where $u_k, v_k \in \mathbb{R}$, we have to study the sign of the $n$ polynomials $u_k^2 + v_k^2 - 1$ for $k = 1, \ldots, n$. From a computational viewpoint, a problem occurs when one of these polynomials vanishes. In that case, numerical computations are not sufficient to conclude. The algorithm, proposed in [5], proceeds by following the following three main steps:

1. Compute a set of hypercubes in $\mathbb{R}^{2n}$ isolating the zeros of $V(J)$. Each coordinate is represented by a box $B$ in $\mathbb{R}^2$ obtained from the intervals containing its real and imaginary parts.
2. For each $z_k$, compute the number $l_k$ of zeros of $J$ satisfying $|z_k| = 1$. This can be obtained by using classical stability test for 1D systems applied to the *elimination polynomial* $p_k$ of $J$ with respect to $z_k$, i.e., $J \cap \mathbb{Q}[z_k] = \langle p_k \rangle$.
3. For each $z_k$, refine the isolating boxes of the solutions until exactly $l_k$ intervals obtained from the evaluation of $u_k^2 + v_k^2 - 1$ at these boxes contains zero. The boxes that yield strictly positive evaluation are discarded.

At the end of these three steps, if all the isolating boxes were discarded, then $V(J) \cap \mathbb{U}^n = \emptyset$, and thus the plant $P$ is stabilizable. Otherwise, the remaining boxes correspond to elements of $V(J) \cap \mathbb{U}^n$, which shows that $V(J) \cap \mathbb{U}^n \neq \emptyset$ and the plant $P$ is not stabilizable. The algorithm testing 1 of Theorem 4 in the case of a zero-dimensional ideal $J$ is given in Algorithm 2. It should be stressed that the symbol $\square f(B)$ used in this algorithm denotes the interval resulting from the evaluation of the polynomial $f$ at the box $B$ using interval arithmetic.

Based on Algorithm 2, we can effectively check whether or not a 2D system is stabilizable since when $n = 2$, $J = \text{ann}_B(\text{ext}_B^2(T(L/t(L)), B))$, and $B/J$ is either $B$, which corresponds to $L/t(L)$ is a projective $B$-module, or zero-dimensional which corresponds to $L/t(L)$ is torsion-free but not projective. More generally, the method developed in [5] can be applied to a $B$-module $L/t(L)$ satisfying $\text{ext}_B^i(T(L/t(L)), B) = 0$ for $i = 1, \ldots, n - 1$. For instance, if $n = 3$, the last conditions mean that $L/t(L)$ is a *reflexive* $B$-module [8].

---

**Algorithm 2.** IsStabilizable

---

**Input:** A set of $r$ polynomials $\{p_1, \ldots, p_r\} \subset \mathbb{Q}[z_1, \ldots, z_n]$ defining a zero-dimensional ideal $J = \langle p_1, \ldots, p_r \rangle$.
**Output:** True if $V(\langle p_1, \ldots, p_r \rangle) \cap \mathbb{U}^n = \emptyset$, and False otherwise.
**Begin**
  $\diamond$ $\{h, h_{z_1}, \ldots, h_{z_n}\} := \text{Univ\_R}(\{p_1, \ldots, p_r\})$;
  $\diamond$ $\{B_1, \ldots, B_d\} := \text{Isolate}(f)$;
  $\diamond$ $L_B := \{B_1, \ldots, B_d\}$ and $\epsilon := \min_{i=1,\ldots,d}\{w(B_i)\}$;
**For** $k$ from 1 to $n$ **do**
  $\diamond$ $l_k := \#\{z \in V(I) \mid |z_k| = 1\}$;
  **While** $\#\{i \mid 0 \in \square(\Re(h_{z_k})^2 + \Im(h_{z_k})^2 - 1)(B_i)\} > l_k$ **do**
    $\diamond$ $\epsilon := \epsilon/2$;
    $\diamond$ For $i = 1, \ldots, d$, set $B_i := \text{Isolate}(f, B_i, \epsilon)$;
  **End While**
  $\diamond$ $L_B := L_B \setminus \{B_i \mid \square(\Re(h_{z_k})^2 + \Im(h_{z_k})^2 - 1)(B_i) \subset \mathbb{R}_+\}$;
  $\diamond$ If $L_B = \{\}$, then **Return True End If**;
**End For**
**Return False.**
**End**

---

## 5   Computing Stabilizing Controllers of 2D Systems

For a zero-dimensional ideal $J = \langle p_1, \ldots, p_r \rangle$, an algorithm is given in [5] for the computation $b \in J$ satisfying 2 of Theorem 4. In the following, we briefly outline the algorithm for $n = 2$. The method is an effective variant of the approach proposed in [14] which consists in considering the elimination polynomials $r_{z_k}$ with respect to the variable $z_k$ defined by $J \cap \mathbb{Q}[z_k] = \langle r_{z_k} \rangle$ and to factorize them into stable and unstable factors, i.e., $r_{z_k} = r_{z_k,s} \, r_{z_k,u}$, where the roots of $r_{z_k,s}$ (resp., $r_{z_k,u}$) are outside (resp., inside) the closed unit disc $\mathbb{U}$ for $k = 1, 2$. Then, we can define a stable polynomial $b = r_{z_1,s}(z_1) \, r_{z_2,s}(z_2)$ and Gröbner basis methods are finally used to get the cofactors $u_i$'s defined by $b = \sum_{k=1}^r u_k \, p_k$.

Since the factorizations of the polynomials $r_{z_1}$ and $r_{z_2}$ are performed in $\mathbb{C}$, the resulting factors $r_{z_k,s}$ do not have usually their coefficients in $\mathbb{Q}$. This prevents the polynomial $b$ (and the $u_i$'s) from being computed exactly in $\mathbb{Q}[z_1, z_2]$. To overcome this issue, the method developed in [5] uses an univariate representation of the solutions of $V(J)$ to compute approximate factorizations of $r_{z_i}$ over $\mathbb{Q}$ and construct a *Nullstelenstaz* relation on the corresponding approximated ideal. For a suitable approximation, the main result in [5] shows that the obtained cofactors $u_i$'s for the approximated ideal hold for the initial ideal $J$, which yields a stable polynomial $b \in J$, i.e., $b \in J \cap S$.

More precisely, given an ideal $J \subset \mathbb{Q}[z_1, z_2]$, the method proceeds by first computing a univariate representation $\{h(t) = 0, z_1 = h_1(t), z_2 = h_2(t)\}$ of the zeros of $J$ with respect to a separating form $t = a_1 z_1 + a_2 z_2$. Then, we consider the ideal $J_r = \langle h(t), z_1 - h_1(t), z_2 - h_2(t) \rangle \subset \mathbb{Q}[t, z_1, z_2]$ which is the intersection of the ideal $J$ and $\langle t - a_1 z_1 - a_2 z_2 \rangle$. Using the univariate representation, approximations of the polynomials $r_{z_1,s}$, $r_{z_2,s}$, and $b$, respectively denoted by $\tilde{r}_{1,s}, \tilde{r}_{2,s}$ and $\tilde{b}$, are then computed as follows:

1. We approximate the complex roots $\gamma_1, \ldots, \gamma_n$ of $h(t)$ so that their real and imaginary parts are given by rational numbers. The resulting approximations are denoted by $\tilde{\gamma}_1, \ldots, \tilde{\gamma}_n$.
2. For each approximation $\tilde{\gamma}_k$, if $|h_{z_i}(\tilde{\gamma}_k)| > 1$, add the factor $z_i - h_{z_i}(\tilde{\gamma}_k)$ to the polynomial $\tilde{r}_{k,s}$.
3. Compute $\tilde{b} = \prod_{k=1}^n \tilde{r}_{k,s}$.

Finally, using the polynomial $\tilde{b}$, a Nullstellensatz relation is computed for the ideal $\tilde{J}_r = \langle \tilde{h}(t), z_1 - h_{z_1}(t), z_2 - g_{z_2}(t) \rangle$, where $\tilde{h}(t) = \prod_{i=1}^d (t - \tilde{\gamma}_i) \in \mathbb{Q}[t]$, which yields three cofactors $u_1, u_2$ and $u_3$ in $\mathbb{Q}[t, z_1, z_2]$. Moreover, in [8], it is shown that for close-enough approximations of the $\gamma_k$'s, using these cofactors with the exact ideal $J_r$ yields a stable polynomial $b$ in $J$, i.e., $b \in J \cap S$.

The main algorithm, which can be generalized for zero-dimensional ideals $J$, is presented in Algorithm 3.

Now, given a 2D plant $P \in \mathbb{Q}(z_1, z_2)^{q \times p}$, we have shown that we can effectively test whether or not $P$ is internally stabilizable. If so, an important issue is then to explicitly compute a stabilizing controller $C$ (see Definition 1).

---

**Algorithm 3.** StabilizingPolynomial

---

**Input:** $J := \langle p_1, \ldots, p_r \rangle$ such that $J$ is zero-dimensional and $V(J) \cap \mathbb{U}^n = \emptyset$.
**Output:** $b \in J$ such that $V(\langle b \rangle) \cap \mathbb{U}^n = \emptyset$.
**Begin**
$\diamond \{h, h_{z_1}, \ldots, g_{z_2}\} := \text{Univ\_R}(\{p_1, \ldots, p_r\});$
$\diamond \{B_1, \ldots, B_d\} := \text{Isolate}(h);$
$\diamond L_B := \{B_1, \ldots, B_d\}$ and $\epsilon := \min_{i=1,\ldots,d}\{w(B_i)\};$
**Do**
$\quad \diamond [r_1, \ldots, r_n] := [1, \ldots, 1]$ and $\tilde{f} := 1;$
$\quad \diamond$ outside := **False**;
$\quad$ **For each** $B$ in $L_B$ **do**
$\quad\quad$ **While** (outside=**False**) **do**
$\quad\quad\quad$ **For** $i$ from 1 to $n$ **do**
$\quad\quad\quad\quad$ **If** $\square(\Re(h_{z_i})^2 + \Im(h_{z_i})^2 - 1)(B) \subset \mathbb{R}_+$ then
$\quad\quad\quad\quad\quad \diamond \gamma := \text{midpoint}(B);$
$\quad\quad\quad\quad\quad \diamond r_i := r_i(z_i - h_{z_i}(\gamma));$
$\quad\quad\quad\quad\quad \diamond$ outside := **True** and **Break For**;
$\quad\quad\quad\quad$ **End If**
$\quad\quad\quad$ **End For**
$\quad\quad\quad \diamond \epsilon := \epsilon/2;$
$\quad\quad\quad \diamond B := \text{Isolate}(h, B, \epsilon);$ (isolate the real roots of $h$ inside $B$ up to a precision $\epsilon$)
$\quad\quad$ **End While**
$\quad\quad \diamond \tilde{h} := \tilde{h}(t - \gamma);$
$\quad\quad \diamond$ outside := **False**;
$\quad$ **End ForEach**
$\quad \diamond \tilde{b} := \prod_{i=1}^n r_i;$
$\quad \diamond \tilde{b}_t := \tilde{b}$ evaluated at $z_i = h_{z_i}(t);$
$\quad \diamond h_0 := \text{quo}(\tilde{b}_t, \tilde{h})$ in $\mathbb{Q}[t];$
$\quad \diamond b := \tilde{b} - h_0(\tilde{h} - h)$ evaluated at $t = \sum_{k=1}^n a_k z_k;$
**While** (IsStable(s)=**False**)
$\diamond$ Return $b$.
**End**

---

**Theorem 5 ([4]).** *Let* $A = \mathbb{Q}(z_1, \ldots, z_n)_S$, $B = \mathbb{Q}[z_1, \ldots, z_n]$, $K = \mathbb{Q}(z_1, \ldots, z_n)$, *and* $P \in K^{q \times p}$ *be a stabilizable plant. Moreover, let* $P = D^{-1} N$ *a fractional representation of* $P$, *where* $R = (D \quad - N) \in B^{q \times p}$, $L = B^{1 \times p}/(B^{1 \times q} R)$ *the* $B$-*module finitely presented by* $R$, *and* $L/t(L) = B^{1 \times p}/(B^{1 \times q'} R')$. *Finally, let* $J = \bigcap_{i=2}^n \text{ann}_B(\text{ext}_B^i(T(L/t(L)), B))$ *and* $\pi \in J \cap S \neq \emptyset$.

*Then, there exists a generalized inverse* $S' \in B_\pi^{p \times q'}$ *of* $R'$, *i.e.,* $R' S' R' = R'$, *where* $B_\pi = S_\pi^{-1} B$ *and* $S_\pi = \{1, \pi, \pi^2, \ldots\}$. *Writing* $R' = (D' \quad - N')$, *where* $D' \in B^{q' \times q'}$ *and* $N' \in B^{q' \times r}$, *and noting* $S = S' D' D^{-1} \in K^{p \times q}$, *then a stabilizing controller* $C$ *of* $P$ *is defined by* $C = Y X^{-1}$, *where:*

$$S = (X^T \quad Y^T)^T, \quad X \in A^{q \times q}, \quad Y \in A^{r \times q}.$$

According to Theorem 5, if $\pi \in J \cap S$ is known, then a stabilizing controller $C$ of $P$ can obtained by means of the computation of a generalized inverse $S'$ of $R'$ over $B_\pi$. Effective methods exist for solving this last point [8, 13, 22].

# 6   A Maple illustrating example

In this section, we demonstrate the results explained in the above sections on an explicit example first considered in [14]. To do that, we first load the nDStab package dedicated to the stability and stabilizability of multidimensional systems, as well as the OREMODULES package [9] dedicated to the study multidimensional linear systems theory based on algebraic analysis methods.

```
>  with(LinearAlgebra):
>  with(nDStab):
>  with(OreModules):
```

We consider the plant $P \in \mathbb{Q}(z_1, z_2)^{2 \times 2}$ defined by the transfer matrix:

```
>  P := Matrix([[-(z[2]-3*z[1])/(2*z[1]-5), (2*z[1]-5)/(3*(2*z[1]-1))],
>  [(2*z[1]-1)/(8*z[2]+6*z[1]-15),z[2]^2/(2*z[1]-1)]]);
```

$$P := \begin{bmatrix} -\dfrac{z_2 - 3 z_1}{2 z_1 - 5} & \dfrac{2 z_1 - 5}{6 z_1 - 3} \\ \dfrac{2 z_1 - 1}{8 z_2 + 6 z_1 - 15} & \dfrac{z_2{}^2}{2 z_1 - 1} \end{bmatrix}$$

We can check that some entries of $P$ are unstable using the command IsStable:

```
>  map(a->IsStable(denom(a)),P);
```

$$\begin{bmatrix} true & false \\ true & false \end{bmatrix}$$

Let us introduce the polynomial ring $B = \mathbb{Q}[z_1, z_2]$:

```
>  B := DefineOreAlgebra(diff=[z[1],s[1]], diff=[z[2],s[2]],
>  polynom=[s[1],s[2]]):
```

Now, we consider the fractional representation of $P$ defined by $R = (d \ -N)$, where $d \in B^{2 \times 2}$ is the diagonal matrix defined by the polynomial den which is the least common multiple of all the denominators of the entries of $P$, i.e.:

```
> den := lcm(op(convert(map(denom,P),set)));
```
$$den := 3\,(2\,z_1 - 5)\,(2\,z_1 - 1)\,(8\,z_2 + 6\,z_1 - 15)$$

```
> d := DiagonalMatrix([den,den]);
```
$$d := \begin{bmatrix} 3\,(2\,z_1 - 5)\,(2\,z_1 - 1)\,(8\,z_2 + 6\,z_1 - 15) & 0 \\ 0 & 3\,(2\,z_1 - 5)\,(2\,z_1 - 1)\,(8\,z_2 + 6\,z_1 - 15) \end{bmatrix}$$

and the matrix $N = d\,P \in B^{2\times2}$ is defined by:

```
> N := simplify(d.P);
```
$$N := \begin{bmatrix} 3\,(2\,z_1 - 1)\,(8\,z_2 + 6\,z_1 - 15)\,(-z_2 + 3\,z_1) & (8\,z_2 + 6\,z_1 - 15)\,(2\,z_1 - 5)^2 \\ 3\,(2\,z_1 - 5)\,(2\,z_1 - 1)^2 & 3\,(2\,z_1 - 5)\,(8\,z_2 + 6\,z_1 - 15)\,z_2{}^2 \end{bmatrix}$$

Since the notation $D$ is prohibited by Maple, we use here $d$ instead of $D$ as it was done in the above sections. Then, we can define the matrix $R = (d \quad -N) \in B^{2\times4}$

```
> R := Matrix([d, -N]):
```

and the finitely presented $B$-module $L = B^{1\times4}/(B^{1\times2}\,R)$. We first have to compute a presentation matrix for the $B$-module $L/t(L)$. Using the OREMODULES package, this can be done as follows:

```
> Ext1 := Exti(Involution(R,B),B,1):
```

The command Exti returns different matrices. Since the first matrix Ext1[1]

```
> Ext1[1];
```
$$\begin{bmatrix} 8\,z_2 + 6\,z_1 - 15 & 0 & 0 \\ 0 & 24\,z_1{}^3 + 32\,z_1{}^2 z_2 - 132\,z_1{}^2 - 96\,z_2 z_1 + 210\,z_1 + 40\,z_2 - 75 & 0 \\ 0 & 0 & 2\,z_1 - 5 \end{bmatrix}$$

is not reduced to an identity matrix, we deduce that the torsion submodule $t(L) = \{l \in L \mid \exists\, 0 \neq b \in B : bl = 0\}$ of $L$ is not reduced to zero, i.e., $t(L) \neq 0$. The second matrix Ext1[2] of Ext1, denoted by Rp in Maple,

```
> Rp := Ext1[2]:
```

is a presentation matrix of $L/t(L)$, i.e., $L/t(L) = B^{1\times4}/(B^{1\times3}\,R')$, where $R' = $ Rp $\in B^{3\times4}$. For an easy display of Rp, we print it by means of its columns:

```
> SubMatrix(Rp,1..3,1..2);
```

$$
\begin{bmatrix}
12\,z_1{}^2 - 36\,z_1 + 15 & 0 \\
36\,z_1z_2{}^3 - 54\,z_1z_2{}^2 - 90\,z_2{}^3 + 135\,z_2{}^2 & -24\,z_1{}^2z_2 - 32\,z_1z_2{}^2 + 36\,z_1{}^2 + 120\,z_2z_1 + 16\,z_2{}^2 - 180\,z_1 - 150\,z_2 + 225 \\
0 & -12\,z_1{}^2 - 16\,z_2z_1 + 36\,z_1 + 8\,z_2 - 15
\end{bmatrix}
$$

```
> SubMatrix(Rp,1..3,3..3);
```

$$
\begin{bmatrix}
-18\,z_1{}^2 + 6\,z_2z_1 + 9\,z_1 - 3\,z_2 \\
-54\,z_1z_2{}^3 + 18\,z_2{}^4 + 8\,z_1{}^2z_2 + 81\,z_1z_2{}^2 - 27\,z_2{}^3 - 12\,z_1{}^2 - 8\,z_2z_1 + 36\,z_1 + 2\,z_2 - 15 \\
4\,z_1{}^2 - 4\,z_1 + 1
\end{bmatrix}
$$

```
> SubMatrix(Rp,1..3,4..4);
```

$$
\begin{bmatrix}
-4\,z_1{}^2 + 20\,z_1 - 25 \\
16\,z_2{}^4 \\
6\,z_2{}^2z_1 + 8\,z_2{}^3 - 15\,z_2{}^2
\end{bmatrix}
$$

If $d_i$ denotes the $i^{\text{th}}$ diagonal element of the matrix $d$, $R'_{i\bullet}$ the $i^{\text{th}}$ row of $R'$, and $l_i$ the residue class of $R'_{i\bullet}$ in the $B$-module $L = B^{1 \times 4}/(B^{1 \times 3}\,R')$, then we have $d_i\,l_i = 0$. Moreover, $\{l_i\}_{i=1,2,3}$ is a generating set of the torsion $B$-submodule $t(L) = (B^{1 \times 3}\,R')/(B^{1 \times 2}\,R)$ of $L$.

By construction, the $B$-module $L/t(L) = B^{1 \times 4}/(B^{1 \times 3}\,R')$ is torsion-free. Since $P$ is a 2D system, i.e., $n = 2$, the obstruction to projectivity for $L/t(L)$ is only defined by the $B$-module $\text{ext}^2_B(N', B)$, where $N' = B^{1 \times 3}/\left(B^{1 \times 4}\,R'^T\right)$ is the so-called *Auslander transpose* of $L/t(L)$. For more details, see [8,22]. More precisely, one can prove that $L/t(L)$ is a projective $B$-module iff $\text{ext}^2_B(N', B) = 0$. If $\text{ext}^2_B(N', B) \neq 0$, then $\text{ext}^2_B(N', B)$ is *0-dimensional B-module*, i.e., it defines a finite-dimensional $\mathbb{Q}$-vector space. In particular, the following ideal $J$ of $B$

$$
J = \text{ann}_B(\text{ext}^2_B(N', B)) = \left\{ b \in B \mid \forall\, e \in \text{ext}^2_B(N', B),\ b\,e = 0 \right\}
$$

is *zero-dimensional*, i.e., $B/J$ is a finite-dimensional $\mathbb{Q}$-vector space. The ideal $J$ can be directly computed by the OREMODULES command PiPolynomial:

```
> pi := map(factor,PiPolynomial(Rp,B));
```

$$
\pi := [4\,z_2{}^2 - 18\,z_1 - 30\,z_2 + 45,\ (2\,z_2 - 3)\,(2\,z_1 - 5),\ (2\,z_1 - 5)\,(2\,z_1 - 1)]
$$

Since $J \neq B$, we obtain that $L/t(L)$ is not a projective $B$-module.

According to our result on stabilizability, $P$ is internally stabilizable iff the $A = \mathbb{Q}(z_1, z_2)_S = S^{-1}\,B$-module $A \otimes_B (L/t(L)) \cong A^{1 \times 4}/(A^{1 \times 3}\,R')$ is projective of rank 2, i.e., iff $S^{-1}\,J = A$, i.e., iff $J \cap S \neq \emptyset$, where

$$
S = \{ b \in B \mid V(\langle b \rangle) \cap \mathbb{U}^2 = \emptyset \}
$$

is the multiplicatively closed subset of $B$ formed by all the stable polynomials of $B$. Equivalently, by the Polydisk Nullstellensatz, $P$ is internally stabilizable iff

$V(J) \cap \mathbb{U}^2 = \emptyset$. Since $V(J)$ is zero-dimensional, i.e., is formed by a finite number of complex points of $\mathbb{C}^2$, we can effectively test the Polydisk Nullstellensatz condition as follows:

```
> IsStabilizable(pi);
```
$$true$$

Hence, we obtain that $P$ is internally stabilizable.

Let us now construct a stabilizing controller $C$ of $P$. To do that, we must find an element $s \in J \cap S$. We can first try to test whether or not one of the generators of $J$ belongs to $S$:

```
> map(IsStable,pi);
```
$$[true, true, false]$$

The first two generators of $J$ belong to $S$. Let us denote them by $\pi_1$, resp. $\pi_2$.

Since the condition $J \cap S \neq \emptyset$ does not necessarily imply that at least one of the generators of $J$ belongs to $S$, the algorithm which computes an element of $J$ in $S$ has to be used. This can be done by the command `StabilizingPolynomial`:

```
> factor(StabilizingPolynomial(pi));
```
$$(2\,z_2 - 3)\,(2\,z_2 - 15)\,(2\,z_1 - 5)$$

Since we have found elements in $J \cap S$, let us compute a stabilizing controller $C$ of $P$. We note that $R'$ has not full row rank since $\ker_B(.R')$ is defined by:

```
> Rp2 := SyzygyModule(Rp,B);
```
$$Rp2 := \begin{bmatrix} 6\,z_2{}^3 - 9\,z_2{}^2 & -2\,z_1 + 1 & 4\,z_2 z_1 - 6\,z_1 - 2\,z_2 + 15 \end{bmatrix}$$

Hence, $R'_2 = $ Rp2 is such that $R'_2\,R' = 0$, which shows that the rows $\{R'_{i\bullet}\}_{i=1,\dots,4}$ of $R'$ satisfy $\sum_{i=1}^{4}$ Rp2$[i]\,R'_{i\bullet} = 0$.

We consider $\pi = \pi_2$, i.e., $\pi = (2\,z_2 - 3)\,(2\,z_1 - 5)$. Similar results can be obtained by choosing $\pi = \pi_1$ instead of $\pi_2$ (but the outputs are larger to display).

We now have to find a generalized inverse $S' \in B_\pi^{4\times3}$ of $R'$, i.e., $R'\,S'\,R' = R'$, where $B_\pi = S_\pi^{-1} B$ is the localization of $B$ with respect to the multiplicatively closed set $S_\pi = \{1, \pi, \pi^2, \dots\}$. This can be done by first computing a right inverse of $R'_2$ over $B_\pi$. Using the OREMODULES package, we obtain that

```
> Sp2:= Transpose(LocalLeftInverse(Transpose(Rp2),[pi[2]],B));
```
$$Sp2 := \begin{bmatrix} 0 \\ -\dfrac{1}{12}\dfrac{\left(-4\,z_2{}^2+18\,z_1+30\,z_2-45\right)(2\,z_2-3)}{4\,z_2{}^2-18\,z_1-30\,z_2+45} \\ \dfrac{1}{4\,z_2{}^2-18\,z_1-30\,z_2+45}\left(\tfrac{1}{3}\,z_2{}^2 - \tfrac{3}{2}\,z_1 - \tfrac{5}{2}\,z_2 + \tfrac{15}{4}\right) \end{bmatrix}$$

is a right inverse of $R'_2$, i.e., $R'_2\,S'_2 = 1$. Then, defining $\Pi = I_3 - S'_2\,R'_2$, we get that $\Pi^2 = \Pi$, and thus there exists $S' \in B_\pi^{4\times3}$ such that $\Pi = R'\,S'$. Using the OREMODULES package, this matrix can be obtained by factorization as follows:

```
>  Proj := Transpose(simplify(1-Sp2.Rp2)):
>  Sp := simplify(Transpose(Factorize(pi[2]*Proj,
>  Transpose(Rp),B))/pi[2]);
```

$$
Sp := \begin{bmatrix}
-\dfrac{243\,z_1 z_2{}^2 - 81\,z_2{}^3 + 84\,z_1 z_2 - 16\,z_2{}^2 + 18\,z_1 - 186\,z_2 + 99}{(1152\,z_2 - 1728)(2\,z_1 - 5)} & -\dfrac{1}{16}\,\dfrac{-z_2 + 3\,z_1}{(2\,z_2 - 3)(2\,z_1 - 5)} & \dfrac{1}{16}\,\dfrac{-z_2 + 3\,z_1}{2\,z_1 - 5} \\[2ex]
-\dfrac{2\,z_1 - 1}{576\,z_2 - 864} & -\dfrac{1}{48}\,\dfrac{1}{2\,z_2 - 3} & \dfrac{1}{48} \\[2ex]
-\dfrac{81\,z_2{}^2 + 12\,z_1 + 16\,z_2 - 30}{1152\,z_2 - 1728} & -\dfrac{1}{16}\,\dfrac{1}{2\,z_2 - 3} & \dfrac{1}{16} \\[2ex]
\dfrac{6\,z_1 - 8\,z_2 + 9}{128\,z_2 - 192} & 0 & 0
\end{bmatrix}
$$

$S' = \mathrm{Sp}$ satisfies $R'\,S'\,R' = R'$, i.e., $S'$ is a generalized inverse of $R'$ over $B_\pi$.

```
>  simplify(Rp.Sp.Rp-Rp);
```

$$
\begin{bmatrix}
0 & 0 & 0 & 0 \\
0 & 0 & 0 & 0 \\
0 & 0 & 0 & 0
\end{bmatrix}
$$

If we set $S := S'\,d'\,d^{-1}$, where $R' = (d' \quad -N')$, $d' \in B^{2\times2}$ and $N' \in B^{2\times2}$, and split $S'$ as $S' = (X^T \quad Y^T)^T$, where $X \in K^{2\times2}$, $Y \in K^{2\times2}$, and $K = \mathbb{Q}(z_1, z_2)$

```
>  dp := SubMatrix(Rp, 1..3,1..2):
>  S := simplify(Sp.dp.MatrixInverse(d)):
>  X := SubMatrix(Sp, 1..2,1..2):
>  SubMatrix(X, 1..2,1..1):
```

$$
\begin{bmatrix}
-\dfrac{486\,z_1{}^2 z_2{}^2 + 486\,z_1 z_2{}^3 - 216\,z_2{}^4 + 168\,z_1{}^2 z_2 - 1247\,z_1 z_2{}^2 + 405\,z_2{}^3 + 36\,z_1{}^2 - 456\,z_1 z_2 + 16\,z_2{}^2 + 180\,z_1 + 186\,z_2 - 99}{(1152\,z_2 - 1728)(2\,z_1 - 5)(2\,z_1 - 1)(8\,z_2 + 6\,z_1 - 15)} \\[2ex]
-\dfrac{36\,z_2{}^3 + 4\,z_1{}^2 - 54\,z_2{}^2 - 4\,z_1 + 1}{(2304\,z_2 + 1728\,z_1 - 4320)(2\,z_1 - 1)(2\,z_2 - 3)}
\end{bmatrix}
$$

```
>  SubMatrix(X, 1..2,2..2):
```

$$
\begin{bmatrix}
\dfrac{1}{4}\,\dfrac{-z_2 + 3\,z_1}{(2\,z_1 - 1)(2\,z_1 - 5)^2(2\,z_2 - 3)} \\[2ex]
\dfrac{1}{12}\,\dfrac{1}{(2\,z_1 - 1)(2\,z_2 - 3)(2\,z_1 - 5)}
\end{bmatrix}
$$

```
>  Y := SubMatrix(S,3..4,1..2);
```

$$
Y := \begin{bmatrix}
-\dfrac{27\,z_2{}^2 + 4\,z_1 - 2}{(1152\,z_1 - 576)(2\,z_2 - 3)} & \dfrac{1}{4}\,\dfrac{1}{(2\,z_1 - 1)(2\,z_2 - 3)(2\,z_1 - 5)} \\[2ex]
\dfrac{6\,z_1 - 8\,z_2 + 9}{(512\,z_2 + 384\,z_1 - 960)(2\,z_2 - 3)} & 0
\end{bmatrix}
$$

then, the controller $C = Y\,X^{-1}$ internally stabilizes $P$. Hence, we obtain the following stabilizing controller of $P$

```
>  C := map(factor,simplify(Y.MatrixInverse(X))):
```

$$
C :=
$$

$$
\begin{bmatrix}
-\dfrac{(81\,z_2{}^2 + 16\,z_2 - 24)(2\,z_1 - 5)}{-243\,z_1 z_2{}^2 + 81\,z_2{}^3 + 36\,z_1{}^2 - 96\,z_1 z_2 + 16\,z_2{}^2 - 36\,z_1 + 192\,z_2 - 99} & 9\,\dfrac{12\,z_1{}^2 - 16\,z_1 z_2 - 36\,z_1 + 72\,z_2 - 33}{-243\,z_1 z_2{}^2 + 81\,z_2{}^3 + 36\,z_1{}^2 - 96\,z_1 z_2 + 16\,z_2{}^2 - 36\,z_1 + 192\,z_2 - 99} \\[2ex]
9\,\dfrac{(6\,z_1 - 8\,z_2 + 9)(2\,z_1 - 5)}{-243\,z_1 z_2{}^2 + 81\,z_2{}^3 + 36\,z_1{}^2 - 96\,z_1 z_2 + 16\,z_2{}^2 - 36\,z_1 + 192\,z_2 - 99} & -27\,\dfrac{(6\,z_1 - 8\,z_2 + 9)(-z_2 + 3\,z_1)}{-243\,z_1 z_2{}^2 + 81\,z_2{}^3 + 36\,z_1{}^2 - 96\,z_1 z_2 + 16\,z_2{}^2 - 36\,z_1 + 192\,z_2 - 99}
\end{bmatrix}
$$

Finally, we check again that $C$ stabilizes $P$. To do that, we can check again that all the entries of the matrix $H(P, C)$ belong to $A = \mathbb{Q}(z_1, z_2)_S$, i.e., are stable:

```
>  H := MatrixInverse(Matrix([[DiagonalMatrix([1,1],2,2),-P],
>  [-C,DiagonalMatrix([1,1],2,2)]])):
>  denomH := convert(map(denom,H),set):
>  map(IsStable,denomH);
```

$$\{true\}$$

# References

1.  Basu, S., Pollack, R., Roy, M.-F.: Algorithms in Real Algebraic Geometry. Springer, Heidelberg (2006). https://doi.org/10.1007/3-540-33099-2
2.  Bouzidi, Y., Quadrat, A., Rouillier, F.: Computer algebra methods for testing the structural stability of multidimensional systems. In: Proceedings of the IEEE 9th International Workshop on Multidimensional (nD) Systems (2015)
3.  Bouzidi, Y., Rouillier, F.: Certified Algorithms for proving the structural stability of two-dimensional systems possibly with parameters. In: Proceedings of the 22nd International Symposium on Mathematical Theory of Networks and Systems (MTNS 2016) (2016)
4.  Bouzidi, Y., Quadrat, A., Rouillier, F.: Certified non-conservative tests for the structural stability of discrete multidimensional systems. Multidimens. Syst. Sig. Process. **30**(3), 1205–1235 (2019)
5.  Bouzidi, Y., Cluzeau, T., Moroz, G., Quadrat, A.: Computing effectively stabilizing controllers for a class of $n$D systems. In: Proceedings of IFAC 2017 Workshop Congress (2017)
6.  Bouzidi, Y., Cluzeau, T., Quadrat, A.: On the computation of stabilizing controllers of multidimensional systems. In: Proceedings of Joint IFAC Conference 7th SSSC 2019 and 15th TDS 2019 (2019)
7.  Bridges, D., Mines, R., Richman, F., Schuster, P.: The polydisk Nullstellensatz. Proc. Am. Math. Soc. **132**(7), 2133–2140 (2004)
8.  Chyzak, F., Quadrat, A., Robertz, D.: Effective algorithms for parametrizing linear control systems over Ore algebras. Appl. Algebra Engrg. Comm. Comput. **16**, 319–376 (2005)
9.  Chyzak, F., Quadrat, A., Robertz, D.: OreModules: a symbolic package for the study of multidimensional linear systems. In: Chiasson, J., Loiseau, J.J. (eds.) Applications of Time Delay Systems. LNCIS, vol. 352, pp. 233–264. Springer, Heidelberg (2007). https://doi.org/10.1007/978-3-540-49556-7_15
10. Curtain, R.F., Zwart, H.J.: An Introduction to Infinite-Dimensional Linear Systems Theory. TAM, vol. 21. Springer, New York (1995). https://doi.org/10.1007/978-1-4612-4224-6
11. Decarlo, R.A., Murray, J., Saeks, R.: Multivariable Nyquist theory. Int. J. Control **25**(5), 657–675 (1977)
12. Desoer, C.A., Liu, R.W., Murray, J., Saeks, R.: Feedback system design: the fractional representation approach to analysis and synthesis. IEEE Trans. Automat. Control **25**, 399–412 (1980)
13. Eisenbud, D.: Commutative Algebra: with a View Toward Algebraic Geometry. GTM, vol. 150. Springer, New York (1995). https://doi.org/10.1007/978-1-4612-5350-1
14. Li, X., Saito, O., Abe, K.: Output feedback stabilizability and stabilization algorithms for 2D systems. Multidimension. Syst. Sig. Process. **5**, 41–60 (1994)

15. Li, L., Lin, Z.: Stability and stabilisation of linear multidimensional discrete systems in the frequency domain. Int. J. Control **86**(11), 1969–1989 (2013)
16. Lin, Z.: Output feedback stabilizability and stabilization of linear $n$D systems. In: Galkowski, K., Wood, J. (eds.) Multidimensional Signals, Circuits and Systems, pp. 59–76. Taylor & Francis
17. Lin, Z., Lam, J., Galkowski, K., Xu, S.: A constructive approach to stabilizability and stabilization of a class of $n$D systems. Multidimension. Syst. Sig. Process. **12**, 329–343 (2001)
18. Lin, Z.: Feedback stabilizability of MIMO $n$D linear systems. Multidimension. Syst. Sig. Process. **9**, 149–172 (1998)
19. Lin, Z.: Feedback stabilization of MIMO $n$D linear systems. IEEE Trans. Autom. Control **45**, 2419–2424 (2000)
20. Quadrat, A.: The fractional representation approach to synthesis problems: an algebraic analysis viewpoint. Part I: (weakly) doubly coprime factorizations. SIAM J. Control Optim. **42**, 266–299 (2003)
21. Quadrat, A.: The fractional representation approach to synthesis problems: an algebraic analysis viewpoint. Part II: internal stabilization. SIAM J. Control. Optim. **42**, 300–320 (2003)
22. Quadrat, A.: An introduction to constructive algebraic analysis and its applications. Les cours du CIRM **1**(2), 281–471 (2010). Journées Nationales de Calcul Formel. INRIA Research Report n. 7354
23. Rotman, J.J.: An Introduction to Homological Algebra, 2nd edn. Springer, New York (2009). https://doi.org/10.1007/b98977
24. Rouillier, F.: Solving zero-dimensional systems through the rational univariate representation. Appl. Algebra Eng. Commun. Comput. **9**, 433–461 (1999)
25. Rouillier, F.: Algorithmes pour l'étude des solutions réelles des systèmes polynomiaux. Habilitation, University of Paris 6 (2007)
26. Strintzis, M.: Tests of stability of multidimensional filters. IEEE Trans. Circ. Syst, **24**, 432–437 (1977)
27. Vidyasagar, M.: Control System Synthesis: A Factorization Approach. MIT Press, Cambridge (1985)

# Using Maple to Analyse Parallel Robots

Damien Chablat[1], Guillaume Moroz[2], Fabrice Rouillier[3(✉)],
and Philippe Wenger[1]

[1] Laboratoire des Sciences du Numérique de Nantes, UMR CNRS 6004,
Nantes, France
[2] INRIA Nancy-Grand Est & LORIA, Nancy, France
[3] INRIA Paris, Institut de Mathématiques de Jussieu Paris-Rive Gauche,
Sorbonne Université, Paris Université, Paris, France
Fabrice.Rouillier@inria.fr

**Abstract.** We present the SIROPA Maple Library which has been designed to study serial and parallel manipulators at the conception level. We show how modern algorithms in Computer Algebra can be used to study the workspace, the joint space but also the existence of some physical capabilities w.r.t. to some design parameters left as degree of freedom for the designer of the robot.

## 1 Introduction

Compared to classical numerical computations, algebraic computations make it possible to work with formal parameters and equalities. In particular, this allows to study singularities exactly. One major drawback when using exact strategies is the cost of the computations which often limit the range of reachable problems.

For illustrating the article, we choose to study 3-PPPS mechanisms because these are 6-degree-of-freedom parallel robots but where projections with respect to 3 of the coordinates can be combined to get the full information (thus we can have explicit plots to give the right intuition).

## 2 Manipulators and Kinematics Problems

A **serial robot** (see [9]) consists of a number of rigid links connected with joints. In general the joints are active revolute or prismatic while the links are passive, parallel or intersecting.

A **parallel robot** (see [12]) consists of several simple serial chains that support a single platform. A well know parallel robot is the Gough-Stewart platform where the serial chains are made of 6 active actuators connected to the ground and to the platform using passive spherical joints.

The (active) **joint** values (positions of the prismatic joints, angles of the revolute joints) and the end effector positions are linked by the so called kinematic equations which can be turned into an equivalent system of algebraic equations.

© Springer Nature Switzerland AG 2020
J. Gerhard and I. Kotsireas (Eds.): MC 2019, CCIS 1125, pp. 50–64, 2020.
https://doi.org/10.1007/978-3-030-41258-6_4

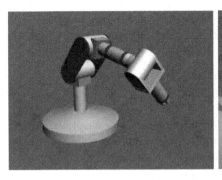

**Fig. 1.** Serial (left) and Parallel (right) manipulators

The solution set of the kinematic equations will be called the kinematic variety in this paper.

The **direct kinematic problem** is the computation of the possible positions of the end effector of the robot knowing the joint values (Fig. 1).

The **inverse kinematic problem** is the computation of the possible joint values knowing the position of the end-effector. It is used, for example, to compute the theoretical instructions to give to the actuators in order to follow a given trajectory.

These problems might have several solutions. For example, for a generic Gough-Stewart platform, the direct kinematics problem might have up to 40 solutions [7].

A **posture** is associated with one solution to the inverse geometric problem of a robot while a **position** is associated with one solution to the direct kinematics problem. The set of points that can be reached by the end-effector is named the **workspace**.

Mathematically, studying the direct or the inverse kinematics problem reduces to study the same system of kinematics equations but with unknowns playing different roles.

For the direct kinematics problem, the joint's unknowns play the role of parameters while the position unknowns play the role of variables: we study the structure (existence, number, multiplicity, etc.) of the positions w.r.t. the joint's values. For the inverse kinematics problem, the position unknowns play the role of parameters while the joint's unknowns play the role of variables: we study the structure (existence, number, multiplicity, etc.) of the joints w.r.t. the positions values.

In both cases, one has to consider an algebraic variety living in some ambient space of dimension $n = d + k$ with a variable subspace of dimension $d$ and a parameter subspace of dimension $k$. The goal is then to decompose the parameter space into regions above which the variables describe regular regions where, in short, the implicit function theorem could apply, and regions where something bad happens (solutions collaps, go to infinity, etc.). This means that, above the favorable regions, the direct (or the inverse) kinematics problem has a finite and constant number of solutions and each solution can easily be followed when the parameters vary.

## 3    The SIROPA Maple Library

SIROPA is a MAPLE library (see [8] for an eshaustive description) developed to analyze the singularities, workspace and joint space of serial and parallel manipulators as well as tensegrity structures [14]. There are two main parts of the library shown in Fig. 2, the first one provides the algebraic tools to solve the kinematic equations, including conversions of trigonometric equations into algebraic form. The other one, SIROPA, provides modeling, analyzing and plotting functions for different manipulators, shown in Fig. 3.

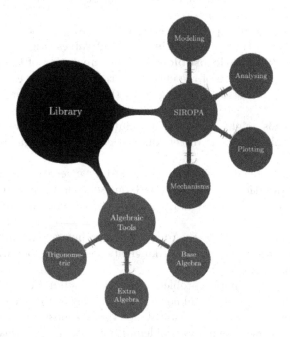

**Fig. 2.** Architecture of library

The main algebraic tools we need for our work are related to the study of parametric systems of polynomial equations with rational (or floating point) coefficients with generically finitely many complex solutions.

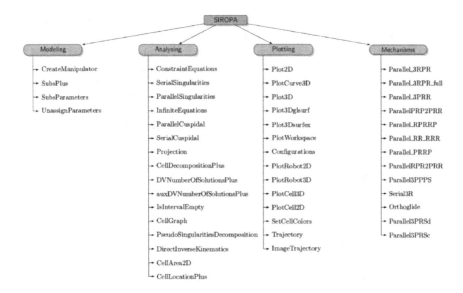

**Fig. 3.** List of all the defined functions in SIROPA library

One key point is to be able to provide a partition of the parameter space in regions/subsets over which a given parametric system has a constant number of real solutions and tools to characterize each region: computing the constant number of solutions over a given region, drawing the region when possible, deciding if two points are in the same region, etc.

Such a partition is naturally given by the so called *(minimal) Discriminant Variety* [10] associated with the projection onto the parameter space.

Let us denote by $S = \{f_1(T, X) = 0, \ldots, f_l(T, X) = 0\}$ the system we want to study, where $T = T_1, \ldots, T_d$ is the set of parameters, $X = X_1, \ldots, X_n$ is the set of unknowns, the $f_i, i = 1 \ldots l$ being polynomials in the indeterminates $T, X$, with rational coefficients.

We suppose that $S = \{f_1(T, X) = 0, \ldots, f_l(T, X) = 0\}$ has generically finitely many complex solutions, which is the case for our studies, that is to say, for almost all the d-uples $(t_1, \ldots, t_d) \in \mathbb{C}^d$, the system $S_{|T=t} = \{f_1(t, X) = 0, \ldots, f_l(t, X) = 0\}$ has finitely many complex solutions. The complex solutions of $S$ thus define an algebraic variety denoted by $V(I)$, with $I$ being the ideal of $\mathbb{Q}[T, X]$ generated by the polynomials $f_1, \ldots, f_l$, such that $\overline{\Pi_T(V(I))} = \mathbb{C}^d$, where $\Pi_T$ denotes the projection onto the parameter space and where $\overline{W}$ denotes the closure of any subset $W \subset \mathbb{C}^d$.

Let's consider the following definition (adapted from [10]). **The (minimal) Discriminant Variety** of $V(I)$ **wrt** $\Pi_T$, denoted $W_D$ in the sequel, is the smallest algebraic variety of $\mathbb{C}^d$ such that given any simply connected subset $\mathcal{U}$ of $\mathbb{R}^d \setminus W_D$, the number of real solutions of $\mathcal{S}$ is constant over $\mathcal{U}$.

In our context, the minimal Discriminant Variety of $V(I)$ wrt $\Pi_T$, will always be the union of

- $W_{\mathrm{sd}}$: the closure of the projection by $\Pi_T$ of the components of $V(I)$ of dimension $< d$
- $W_c$: the union of the closure of the critical values of $\Pi_T$ in restriction to $V(I)$ and of the projection of the singular values of $V(I)$
- $W_\infty$: the set of $u = (u_1, \ldots, u_d)$ such that $\Pi_T^{-1}(\mathcal{U}) \cap V(I)$ is not compact for any compact neighborhood $\mathcal{U}$ of $u$ in $\Pi_T(V(I))$.

An important remark is that if $l = n$ (say if the system has as many polynomials as unknowns), then $W_{\mathrm{sd}} = \emptyset$.

| DiscriminantVariety (sys, vars, pars) | | |
|---|---|---|
| DiscriminantVariety (eqs, ineqs, vars, pars) | | |
| sys | list of equations and strict inequalities between polynomials with rational coefficient | |
| vars | list of names; the indeterminates | |
| pars | (optional) list of names; the parameters | |
| eqs | list of polynomials $f$ with rational coefficients representing equations of the form $f = 0$ | |
| ineqs | list of polynomials $g$ with rational coefficients representing constraint inequalities of the form $0 < g$ | |

The function DiscriminantVariety(eqs, ineqs, vars, pars) available through the SIROPA Package computes a discriminant variety of the system

$$[f = 0, 0 < g]_{f \in eqs, \quad g \in ineqs} \tag{1}$$

of equations and inequalities with respect to the indeterminates vars and the parameters pars.

The input system must satisfy the following properties:

- There are at least as many equations as indeterminates.
- At least one and at most finitely many complex solutions exist for almost all complex parameter values (the system is generically solvable and generically zero-dimensional).

- For almost all complex parameter values, there are no solutions of multiplicity greater than one (the system is generically radical or, in other words, its Jacobian determinant with respect to the indeterminates does not vanish for almost all the parameter values). In particular, the input equations are square-free.

An error occurs if one of these three previous conditions is violated.

- The result is returned as a list of lists of polynomials in *pars* such that the discriminant variety is the union of the set of solutions of the polynomials in each inner list.
- If *pars* is not specified, it defaults to all the names in sys that are not indeterminates.
- This function attempts to find a minimal discriminant variety, but it may return a proper superset in the case that it does not succeed.
- The discriminant variety is computed using Gröbner basis techniques and thus, the choice of ordering of the variables might be critical for efficiency, some intermediate objects could be huge.

**A cylindrical algebraic decomposition (CAD)** of the n-dimensional real space is a partition of the whole space into connected semi-algebraic subsets such that the cells in the partition are cylindrically arranged, that is, the projection of any two cells onto any lower dimensional real space is either equal or disjoint. This decomposition is called F-invariant if, for any given cell, the sign of each polynomial in F does not change over the cell. CylindricalAlgebraicDecompose(F, R) returns an F-invariant CAD of the $n$-dimensional real space, where $n$ is the number of variables in R. This assumes that R has characteristic zero and no parameters, such that the base field of R is the field of rational numbers [1].

In our case, we make use of a partial CAD sometimes named **Open CAD** to decompose the parameter space of a parametric polynomial system into the union of the discriminant variety and a collection of cells in which the original system has a constant number of solutions (see [6] and [11] for similar contexts). The open CAD is thus the union of the discriminant variety and of the cells of maximum dimension of the CAD associated to the discriminant variety.

```
CellDecomposition (sys, vars, pars, options)
CellDecomposition (eqs, posineqs, vars, pars, options)
CellDecomposition (eqs, posineqs, nzineqs, vars,
                   pars, options)
```

| sys | list of equations and strict inequalities between polynomials with rational coefficients |
| vars | list of names; the indeterminates |
| pars | (optional) list of names; the parameters |
| eqs | list of polynomials $f$ with rational coefficients representing equations of the form $f = 0$ |
| posineqs | list of polynomials $g$ with rational coefficients representing constraint inequalities of the form $0 < g$ |
| nzineqs | list of polynomials $g$ with rational coefficients representing constraint inequations of the form $g \neq 0$ |
| options | sequence of optional equations of the form keyword=value where keyword is either output or method |

The function returns a data structure that can be used for (examples):

- Plotting the regions of the parameter space for which the system has a given number of solutions.
- Extracting sample points in the parameter space for which the system has a given number of solutions.
- Extracting boxes in the parameter space in which the system has a given number of solutions.

The record returned captures information about the solutions of the system depending on the parameter values, including:

- a discriminant variety;
- for each full-dimensional open cell, a sample point strictly in the interior of the cell; if possible, the coordinates of the sample point are chosen to be integers.

The input system must satisfy the same properties as for the discrimant variety.

## 4    Case Studies on Some 3-PPPS Manipulators

For illustrating this contribution, we will study some 3-PPPS manipulators: each leg is composed of three orthogonal prismatic joints (P) and one spherical joint (S), the first two prismatic joints being actuated. These 6-degree-of-freedom robots have the particularity to have independent sets of freedom motions i.e. 3 translations and 3 rotations and thus allows us to have 3 dimensional plots illustrating various kinds of singularities.

### 4.1    Joint Space and Workspace Analysis

For illustrating the way the SIROPA Library can be used for studying the workspace of a given manipulator, we chose a 3-PPPS parallel robot derived from [4] with an equilateral mobile platform and a U-shape base and use quaternion parameters to represent the aspects [3], i.e. the singularity free regions of the workspace. The three legs are identical and made with two actuated prismatic joints plus one passive prismatic joint and a spherical joint. The axes of first three joints form an orthogonal reference frame.

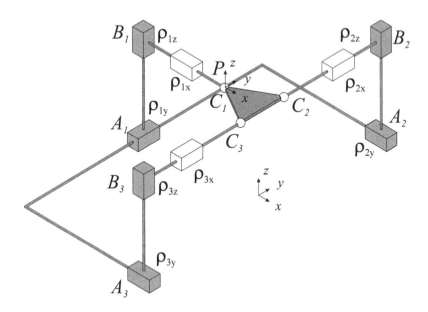

**Fig. 4.** The 3-PPPS parallel robot and its parameters in its "home" pose with the actuated prismatic joints in blue, the passive joints in white and the mobile platform in green (Color figure online)

The coordinates of the moving platform in the fixed reference frame can be expressed using a general rotation matrix as $\boldsymbol{W}_i = \boldsymbol{R}\boldsymbol{V}_i + \boldsymbol{P}$, where $(\boldsymbol{V}_i)_{i=1...3}$ are the 3 points describing the moving frame, $\boldsymbol{P} = [x, y, z]^T$ is a translation vector, and $\boldsymbol{R} = \begin{bmatrix} u_x & v_x & w_x \\ u_y & v_y & w_y \\ u_z & v_z & w_z \end{bmatrix}$ is a rotation which can be parametrized by unit quaternions as follows:

$$\boldsymbol{R} = \begin{bmatrix} 2q_1^2 + 2q_2^2 - 1 & -2q_1q_4 + 2q_2q_3 & 2q_1q_3 + 2q_2q_4 \\ 2q_1q_4 + 2q_2q_3 & 2q_1^2 + 2q_3^2 - 1 & -2q_1q_2 + 2q_3q_4 \\ -2q_1q_3 + 2q_2q_4 & 2q_1q_2 + 2q_3q_4 & 2q_1^2 + 2q_4^2 - 1 \end{bmatrix} \tag{2}$$

with $q_1 \geq 0$. We choose for $\boldsymbol{V}_1$ and $\boldsymbol{V}_2$ one corner and the median of the triangle modelling the moving platform, ($\boldsymbol{V}_3$ is then fixed by the two other points), say $\boldsymbol{V}_1 = [0, 0, 0]^T$, $\boldsymbol{V}_2 = [\sqrt{3}/2, 1/2, 0]^T$ and $\boldsymbol{V}_3 = [\sqrt{3}/2, -1/2, 0]^T$ appeared to be the best choice.

Obviously, one then has to consider additional constraints equations in order to describe the links and the geometry of the moving platform. The points $A_i$ and $C_i$ can be expressed w.r.t. to the joint variables

$$
\begin{aligned}
\mathbf{A}_1 &= [2, \rho_{1y}, \rho_{1z}]^T & \mathbf{C}_1 &= [\rho_{1x}, \rho_{1y}, \rho_{1z}]^T \\
\mathbf{A}_2 &= [-\rho_{2y}, 2, \rho_{2z}]^T & \mathbf{C}_2 &= [-\rho_{2y}, \rho_{2x}, \rho_{2z}]^T \\
\mathbf{A}_3 &= [\rho_{3y}, -2, \rho_{3z}]^T & \mathbf{C}_3 &= [\rho_{3y}, -\rho_{3x}, \rho_{3z}]^T
\end{aligned}
$$

and the fact that there are 3 passive joints $(\rho_{1,x}, \rho_{2,x}, \rho_{3,x})$ leads also to additional simple algebraic rules: $\rho_{1x} = x, \rho_{2x} = u_y\sqrt{3}/2 + v_y/2 + y$ and $\rho_{3x} = -u_y\sqrt{3}/2 + v_y/2 - y$. We finally obtain the following *kinematic equations* which are algebraic when adding the formal equation $\left(\sqrt{3}\right)^2 - 3 = 0$, considering $\sqrt{3}$ as a symbol:

$$
\begin{aligned}
\rho_{1y} - y &= 0 \\
\rho_{1z} - z &= 0 \\
(-2q_1^2 - 2q_2^2 + 1)\sqrt{3}/2 + q_1q_4 - q_2q_3 - x - \rho_{2y} &= 0 \\
\sqrt{3}(q_1q_3 - q_2q_4) - q_1q_2 - q_3q_4 + \rho_{2z} - z &= 0 \\
(-2q_1^2 - 2q_2^2 + 1)\sqrt{3}/2 - q_1q_4 + q_2q_3 - x + \rho_{3y} &= 0 \\
\sqrt{3}(q_1q_3 - q_2q_4) + q_1q_2 + q_3q_4 + \rho_{3z} - z &= 0
\end{aligned}
$$

**Joint Space Analysis**
Analyzing the joint space consists of computing a partition of this space in regions above which the DKP (Direct Kinematics Problem) has a fixed number of solutions. In fact, we are not interested in regions of dimension less than 3, so that we compute the discriminant variety of the kinematics variety w.r.t. the projection onto the joint space and then describe the complement of this variety using an open CAD (Fig. 5).

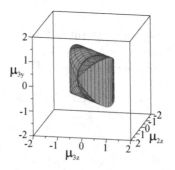

**Fig. 5.** Joint space of the 3-PPPS robot

In the present case we first make the following change of variables in order to reduce the joint space to a 3-dimensional subspace:

$$\mu_{1x} = \rho_{1x} - \rho_{2y} \quad \mu_{2x} = \rho_{2x} - \rho_{1y} \quad \mu_{3x} = \rho_{3x} + \rho_{2y}$$
$$\mu_{1y} = \rho_{1y} - \rho_{1y} = 0 \; \mu_{2y} = \rho_{2y} - \rho_{2y} = 0 \; \mu_{3y} = \rho_{3y} - \rho_{2y}$$
$$\mu_{1z} = \rho_{1z} - \rho_{1z} = 0 \; \mu_{2z} = \rho_{2z} - \rho_{1z} \quad \mu_{3z} = \rho_{3z} - \rho_{1z}$$

Then, the discriminant variety of the *kinematic system* w.r.t. the projection onto the joint space is given by the union of the following equations

$$\mu_{2z} - \mu_{3z} = 1 \quad 4(\mu_{2z}^2 - \mu_{2z}\mu_{3z} + \mu_{3z}^2) = 3$$
$$\mu_{2z} - \mu_{3z} = -1 \; (\mu_{2z} - \mu_{3z})^2 + \mu_{3y}^2 = 1$$

The cell description of the complement of the discriminant variety from the next table comes from [13]. For one variable, $[\mathcal{P}, n, \mu, \mathcal{Q}, m]$ means that the minimum value of $\mu$ is the $n^{th}$ root of $\mathcal{P}$ and the maximum value is $m^{th}$ root of $\mathcal{Q}$.

| $\mu_{2z}$ | $\mu_{3z}$ | $\mu_{3y}$ |
|---|---|---|
| $[\mathcal{P}_{1_{R_{2Z}}}, 1, \mu_{2z}, \mathcal{P}_{2_{R_{2Z}}}, 1]$ | $[\mathcal{P}_{3_{R_{3Z}}}, 1, \mu_{3z}, \mathcal{P}_{3_{R_{3Z}}}, 2]$ | $[\mathcal{P}_{1_{R_{3Y}}}, 1, \mu_{3y}, \mathcal{P}_{1_{R_{3Y}}}, 2]$ |
| $[\mathcal{P}_{2_{R_{2Z}}}, 1, \mu_{2z}, \mathcal{P}_{3_{R_{2Z}}}, 1]$ | $[\mathcal{P}_{3_{R_{3Z}}}, 1, \mu_{3z}, \mathcal{P}_{3_{R_{3Z}}}, 2]$ | $[\mathcal{P}_{1_{R_{3Y}}}, 1, \mu_{3y}, \mathcal{P}_{1_{R_{3Y}}}, 2]$ |
| $[\mathcal{P}_{3_{R_{2Z}}}, 1, \mu_{2z}, \mathcal{P}_{4_{R_{2Z}}}, 1]$ | $[\mathcal{P}_{3_{R_{3Z}}}, 1, \mu_{3z}, \mathcal{P}_{3_{R_{3Z}}}, 2]$ | $[\mathcal{P}_{1_{R_{3Y}}}, 1, \mu_{3y}, \mathcal{P}_{1_{R_{3Y}}}, 2]$ |

where

$$\mathcal{P}_{1_{R_{2Z}}} : \mu_{2z} + 1 = 0$$
$$\mathcal{P}_{2_{R_{2Z}}} : 2\mu_{2z} + 1 = 0$$
$$\mathcal{P}_{3_{R_{2Z}}} : 2\mu_{2z} - 1 = 0$$
$$\mathcal{P}_{4_{R_{2Z}}} : \mu_{2z} - 1 = 0$$
$$\mathcal{P}_{1_{R_{3Z}}} : \mu_{2z} - \mu_{3z} - 1 = 0$$
$$\mathcal{P}_{2_{R_{3Z}}} : \mu_{2z} - \mu_{3z} + 1 = 0$$
$$\mathcal{P}_{3_{R_{3Z}}} : 4(\mu_{2z}^2 - \mu_{2z}\mu_{3z} + \mu_{3z}^2) - 3 = 0$$
$$\mathcal{P}_{1_{R_{3Y}}} : (\mu_{2z} - \mu_{3z})^2 + \mu_{3y}^2 - 1 = 0$$

Taking one point in each of the cells and solving the related zero-dimensional system, it turns out that for the present example, the DKP (Direct Kinematics Problem) always admits 16 real roots which corresponds to eight assembly modes for the robot. This result is valid if there is no limit on the passive joints.

### Workspace Analysis

Analyzing the workspace consists of computing a partition of this space in regions where the IKP (Inverse Kinematics Problem) has a fixed number of solutions. As for the joint space analysis, we are only interested in regions of maximal dimension: we can compute the discriminant variety of the kinematics variety w.r.t. the projection onto the position space and then describe the complement of this variety using an open CAD.

The aim of the analysis is to determine the maximum regions without any singularities, i.e. the aspects of the robot. In these regions, the robot can perform any continuous trajectories. As the Jacobian w.r.t. the joints variables of the kinematic system can be factorized into two components, the discriminant variety splits the orientation space into four regions by using the sign of two components.

- Let PP be the regions where $q_2^2 + q_3^2 - 1/2 > 0$ and $q_2^2 + q_4^2 - 1/2 > 0$.
- Let NN be the regions where $q_2^2 + q_3^2 - 1/2 < 0$ and $q_2^2 + q_4^2 - 1/2 < 0$.
- Let PN be the regions where $q_2^2 + q_3^2 - 1/2 > 0$ and $q_2^2 + q_4^2 - 1/2 < 0$.
- Let NP be the regions where $q_2^2 + q_3^2 - 1/2 < 0$ and $q_2^2 + q_4^2 - 1/2 > 0$.

Each region can be defined by a set of cells (Fig. 6).

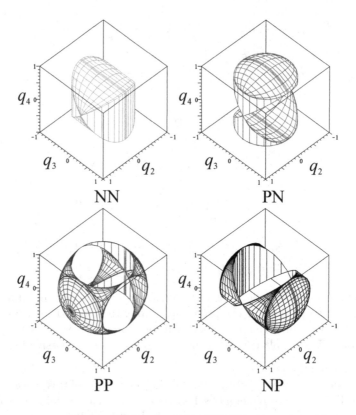

**Fig. 6.** Workspace of the 3-PPPS robot

## 4.2  Cuspidal Robots

The cuspidal character of a mechanism has a different meaning depending on its structure. For a serial mechanism, it is the capability to pass from one solution of the inverse kinematic problem to another one without crossing a singularity in the joint space (the Jacobian determinant of the kinematic system w.r.t. the joint variables does not vanish) while for a parallel mechanism it is the capability to pass from a solution of the direct kinematic problem to another one without crossing a singularity in the workspace (the Jacobian of the kinematic system w.r.t. the position variables does not vanish).

The term "cuspidal" is due to the fact that this capability is linked to the existence of a triple solution to the inverse kinematic problem in the case of a serial mechanism and to the existence of a triple solution to the direct kinematic problem in the case of a parallel mechanism.

For illustrating the way such a property is studied using the SIROPA library, we chose the following 3-PPPS Robot, which is a simplified kinematic version of the manipulator proposed in [5] (Fig. 7).

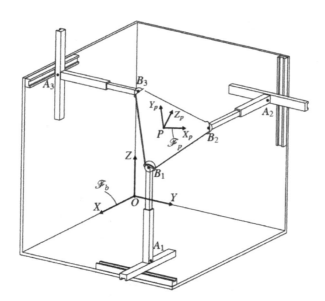

**Fig. 7.** A 3-PPPS robot with 3 orthogonal prismatic joints.

Let $B_1$, $B_2$ and $B_3$ be the corners of the moving platform (MP) of side length $r$. Let $\mathcal{F}_p$ $(P, X_p, Y_p, Z_p)$ be the frame attached to the moving platform, its origin $P$ being the centroid of the MP. $Y_p$ is parallel to line $(B_1 B_3)$ and $Z_p$ is normal to the MP. Accordingly,

$$\boldsymbol{b}_{1p} = \left[-r\sqrt{3}/6, -r/2, 0\right], \, \boldsymbol{b}_{2p} = \left[2r\sqrt{3}/6, 0, 0\right], \, \boldsymbol{b}_{3p} = \left[-r\sqrt{3}/6, r/2, 0\right] \quad (3)$$

are the Cartesian coordinate vectors of points $B_1$, $B_2$ and $B_3$ expressed in $\mathcal{F}_p$. Likewise, let $\mathcal{F}_b$ $(O, X, Y, Z)$ be the frame attached to the base and

$$a_{1b} = [x_1, y_1, 0], \quad a_{2b} = [0, y_2, z_2], \quad a_{3b} = [x_3, 0, z_3] \tag{4}$$

be the Cartesian coordinate vectors of points $A_1$, $A_2$ and $A_3$ expressed in $\mathcal{F}_b$.

Let $\boldsymbol{p} = \begin{bmatrix} p_x, p_y, p_z \end{bmatrix}^T$ be the Cartesian coordinate vector of point $P$, the centroid of the MP, expressed in $\mathcal{F}_b$ and let $r$ be equal to 1. The orientation space of the moving platform is fully represented with the variables $(\phi, \theta, \sigma)$, namely, the *azimuth*, *tilt* and *torsion* angles defined in [2]. The rotation matrix ${}^b\boldsymbol{Q}_p$ from $\mathcal{F}_b$ to $\mathcal{F}_p$ is expressed as follows:

$$
{}^b\boldsymbol{Q}_p = \begin{bmatrix}
C_\phi C_\psi - S_\phi C_\theta S_\psi & -C_\phi S_\psi - S_\phi C_\theta C_\psi & S_\phi S_\theta \\
S_\phi C_\psi + C_\phi C_\theta S_\psi & -S_\phi S_\psi + C_\phi C_\theta C_\psi & -C_\phi S_\theta \\
S_\theta S_\psi & S_\theta C_\psi & C_\theta
\end{bmatrix} \tag{5}
$$

$C$ and $S$ denoting the cosine and sine functions, respectively. Note that $\phi \in [-\pi, \pi]$, $\theta \in [0, \pi]$ and $\sigma \in [-\pi, \pi]$.

As a consequence, the following constraint equations characterize the geometric model of the 3-PPPS-manipulator and are obtained by considering the projection of the coordinates of points $B_i$ in the plane motion of the two actuated prismatic joints of the $i$th leg, $i = 1, \ldots, 3$:

$$p_x - x_1 = 0$$

$$p_y - y_1 = 0$$

$$3p_y - 3y_2 - 2\sqrt{3}C_\theta S_\sigma + 2\sqrt{3}C_\phi^2 C_\theta S_\sigma + 2\sqrt{3}C_\phi C_\sigma S_\phi$$
$$-\sqrt{6}S_\theta C_\sigma C_\phi - \sqrt{6}S_\theta S_\sigma S_\phi - 2\sqrt{3}S_\phi C_\theta C_\sigma C_\phi - 2\sqrt{3}C_\phi^2 S_\sigma = 0$$

$$3p_z - 3z_2 + \sqrt{3}C_\theta S_\sigma - \sqrt{3}C_\phi^2 C_\theta S_\sigma - \sqrt{3}C_\phi C_\sigma S_\phi$$
$$-\sqrt{6}S_\theta C_\sigma C_\phi - \sqrt{6}S_\theta S_\sigma S_\phi + \sqrt{3}S_\phi C_\theta C_\sigma C_\phi - 3C_\phi S_\sigma S_\phi$$
$$+3C_\phi^2 C_\theta C_\sigma + 3C_\sigma - 3C_\phi^2 C_\sigma + \sqrt{3}C_\phi^2 S_\sigma + 3S_\phi C_\theta S_\sigma C_\phi = 0$$

$$6p_x - 6x_3 + \sqrt{3}C_\theta S_\sigma + 2\sqrt{3}C_\phi^2 C_\theta S_\sigma + 2\sqrt{3}C_\phi C_\sigma S_\phi - \sqrt{6}S_\theta C_\sigma C_\phi$$
$$-\sqrt{6}S_\theta S_\sigma S_\phi + 3\sqrt{2}S_\theta S_\sigma C_\phi - 3\sqrt{2}S_\theta C_\sigma S_\phi - 2\sqrt{3}S_\phi C_\theta C_\sigma C_\phi$$
$$+3C_\theta C_\sigma - 6C_\phi^2 C_\theta C_\sigma + 6C_\phi S_\sigma S_\phi - 3C_\sigma + 6C_\phi^2 C_\sigma - 2\sqrt{3}C_\phi^2 S_\sigma$$
$$-6S_\phi C_\theta S_\sigma C_\phi + 3\sqrt{3}S_\sigma = 0$$

$$6p_z - 6z_3 + \sqrt{3}C_\theta S_\sigma - 4\sqrt{3}C_\phi^2 C_\theta S_\sigma - 4\sqrt{3}C_\phi C_\sigma S_\phi$$
$$-\sqrt{6}S_\theta C_\sigma C_\phi - \sqrt{6}S_\theta S_\sigma S_\phi + 3\sqrt{2}S_\theta S_\sigma C_\phi - 3\sqrt{2}S_\theta C_\sigma S_\phi$$
$$+4\sqrt{3}S_\phi C_\theta C_\sigma C_\phi + 3C_\theta C_\sigma + 3C_\sigma + 4\sqrt{3}C_\phi^2 S_\sigma - 3\sqrt{3}S_\sigma = 0$$

It is noteworthy that the translational and rotational motions of the moving platform of the 3-PPPS manipulator shown in Fig. 4 can be decoupled. In order to highlight this decoupling, the following change of variables is made:

$$X_1 = \frac{x_1+x_3}{2} \quad Y_1 = y_2 - y_1 \quad Z_2 = z_2 - z_3$$
$$X_3 = x_3 - x_1 \quad Y_2 = \frac{y_1+y_2}{2} \quad Z_3 = \frac{z_2+z_3}{2}$$

It is apparent that the translational motions of the MP depend only on variables $X_1$, $Y_2$ and $Z_3$, whereas its rotational motions depend only on variables $X_3$, $Y_1$ and $Z_2$. Due to the decoupling of the translational and rotational motions, one can, for example compute (and draw) the singularity surface in the joint space defined by the translational variables (Fig. 8).

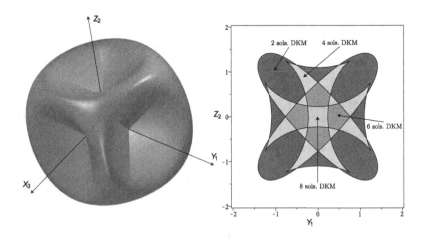

**Fig. 8.** Singularity surface and CAD-partition of the joint space

In the present case, the projection of this singularity surface onto the translational variables is equal to the discriminant variety of the kinematics equations w.r.t. the projection onto the translational variables and thus defines a partition of this space in region where the direct kinematic problem has a constant number of solutions.

# References

1. Arnon, D.S., Collins, G.E., McCallum, S.: Cylindrical algebraic decomposition I: the basic algorithm. SIAM J. Comput. **13**(4), 865–877 (1984)
2. Bonev, I.A., Ryu, J.: Orientation workspace analysis of 6-DOF parallel manipulators. In: Proceedings of the ASME (1999)
3. Chablat, D., Wenger, P.: Working modes and aspects in fully parallel manipulators. In: Proceedings of the 1998 IEEE International Conference on Robotics and Automation (Cat. No. 98CH36146), vol. 3, pp. 1964–1969. IEEE (1998)
4. Chen, C., Gayral, T., Caro, S., Chablat, D., Moroz, G., Abeywardena, S.: A six degree of freedom epicyclic-parallel manipulator. J. Mech. Robot. **4**(4), 41011 (2012)

5. Chen, C., Heyne, W.J., Jackson, D.: A new 6-DOF 3-legged parallel mechanism for force-feedback interface. In: Proceedings of 2010 IEEE/ASME International Conference on Mechatronic and Embedded Systems and Applications, pp. 539–544. IEEE (2010)
6. Corvez, S., Rouillier, F.: Using computer algebra tools to classify serial manipulators. In: Winkler, F. (ed.) ADG 2002. LNCS (LNAI), vol. 2930, pp. 31–43. Springer, Heidelberg (2004). https://doi.org/10.1007/978-3-540-24616-9_3
7. Husty, M.L.: An algorithm for solving the direct kinematics of general Stewart-Gough platforms. Mech. Mach. Theory **31**(4), 365–379 (1996)
8. Jha, R., Chablat, D., Baron, L., Rouillier, F., Moroz, G.: Workspace, joint space and singularities of a family of delta-like robot. Mech. Mach. Theory **127**, 73–95 (2018)
9. Khalil, W., Dombre, E.: Modeling, Identification and Control of Robots. Butterworth-Heinemann (2004)
10. Lazard, D., Rouillier, F.: Solving parametric polynomial systems. J. Symbolic Comput. **42**(6), 636–667 (2007)
11. Manubens, M., Moroz, G., Chablat, D., Wenger, P., Rouillier, F.: Cusp points in the parameter space of degenerate 3-RPR planar parallel manipulators. J. Mech. Robot. **4**(4), 41003 (2012)
12. Merlet, J.-P.: Parallel Robots. Springer, Cham (2006). https://doi.org/10.1007/1-4020-4133-0
13. Moroz, G., Rouiller, F., Chablat, D., Wenger, P.: On the determination of cusp points of 3-RPR parallel manipulators. Mech. Mach. Theory **45**(11), 1555–1567 (2010)
14. Wenger, P., Chablat, D.: Kinetostatic analysis and solution classification of a class of planar tensegrity mechanisms. Robotica **37**(7), 1214–1224 (2019)

# Studying Wythoff and Zometool Constructions Using Maple

Benoit Charbonneau$^{(\boxtimes)}$ ⓘ and Spencer Whitehead ⓘ

Department of Pure Mathematics, University of Waterloo,
200 University Avenue West, Waterloo, ON N2L 3G1, Canada
benoit@alum.mit.edu, snwhiteh@edu.uwaterloo.ca,
http://www.math.uwaterloo.ca/~bcharbon

**Abstract.** We describe a Maple package that serves at least four purposes. First, one can use it to compute whether or not a given polyhedral structure is Zometool constructible. Second, one can use it to manipulate Zometool objects, for example to determine how to best build a given structure. Third, the package allows for an easy computation of the polytopes obtained by the kaleiodoscopic construction called the Wythoff construction. This feature provides a source of multiple examples. Fourth, the package allows the projection on Coxeter planes.

**Keywords:** Geometry · Polytopes · Wythoff construction · 120-cell · Coxeter plane · Zome system · Zometool · Maple

## 1 Introduction

As geometry is a very visual field of mathematics, many have found it useful to construct geometric objects, simple and complex, either physically or with the help of a computer. The Maple package we present today allows for both. First, the package allows for the automatic construction of many of the convex uniform polytopes in any dimension by the kaleidoscopic construction known as the Wythoff construction. Combined with the plotting capabilities of Maple exploited by our package, we therefore provide an extension of Jeff Weeks's KaleidoTile software [19] to higher dimensions. This construction provides an impressive zoo of examples with which one can experiment on the second and most important aspect of our package, Zometool constructability.

The Zometool System as it currently exists was first released in 1992 by Marc Pelletier, Paul Hildebrant, and Bob Nickerson, based on the ideas of Steve Baer some twenty years earlier; see [1]. It is marketed by the company Zometool Inc. As a geometric building block with icosahedral symmetry and lengths based on the golden ratio, it is able to construct incredibly rich structures with a very small set of pieces (see a pictorial description of the system in Fig. 1, and examples in Fig. 2). The Zometool System finds use today in many of the sciences: modeling

BC supported by NSERC Discovery Grant. SW supported by NSERC USRA.

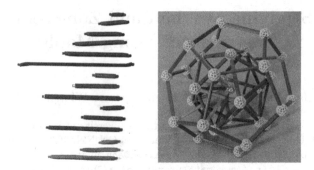

**Fig. 1.** Left are the pieces in the Zome system: the traditional red, blue, and yellow pieces, plus the newer green pieces that fit in the same holes as the reds. On the right is a construction showing the angles at which Zome pieces can meet: the endpoints of red struts form an icosahedron, while the yellow struts form the dual dodecahedron. Blue struts pass through the common midpoints of edges. (Color figure online)

of DNA [17], construction of Sierpiński superfullerenes [3], building models of quasicrystals (notably in the work of Nobel laureate Dan Shectman). Countless other mathematicians use the Zometool System for visualizing nuanced geometric structures easily (see for instance [11,15,16]). A notable example is Hall's companion [10] to the Lie algebra textbook [9]. Many have used Zometool to teach or raise interest in mathematics [12,13]. Zometool is also an invaluable tool for educators, who find great success in using it to teach geometry in a hands-on fashion.

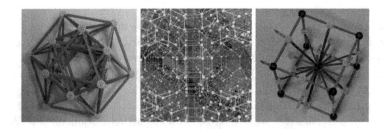

**Fig. 2.** A projection of the 24-cell, a close-up picture of the omnitruncated 120-cell along a 4-fold symmetry axis, and the root system of $B_3$.

Hart and Picciotto's book *Zome Geometry* [12] allows one to learn about geometric objects, polygons, polyhedra, and polytopes and their projections in a hands-on fashion. It contains instructions allowing one to build many projections. However, it can be difficult to verify that one is building the claimed structure from such instructions. Moreover, not all possible structures one might like to construct are included in [12], even when restricted to convex uniform polytopes. A list available from David Richter's website [14], shows real-world constructions of all the $H_4$ polytopes. At the time of writing, only the bitruncated 120/600-cell has not been constructed according to this page.

This issue highlights the need for a good computational framework. The only existing tool known to the authors for working with Zometool on a computer is Scott Vorthmann's *vZome* software [18], which can be used to construct and render models. vZome is very effective at constructing brand new solids from scratch, and at working with smaller projects, such as the regular polyhedra. For larger tasks, such as projections of 4-dimensional polytopes, it is more difficult to create a vZome construction. While vZome does support its own scripting language, Zomic, it does not possess all the analytic capabilities offered by Maple, and we found some operations we wanted to perform were not supported.

Another problem we concern ourselves with is the case when structures are *not* constructible in Zometool. When constructing things by hand or in the vZome software, it can be difficult to discern when a structure simply cannot be represented in Zometool, or if we are missing an idea in our construction.

A final issue with existing techniques is the need to be able to zoom in on parts of a model, and look at them in isolation. The approach used to construct a cube using 20 Zometool pieces is necessarily different from the one used to build the projection of an omnitruncated 120-cell, requiring 21,360 Zometool pieces. What is needed is a setting in which models can be broken into small workable components, and assembled easily from them, for example in layers.

When deciding constructability, Maple's symbolic nature is desirable, as it allows the output of our program to be taken as a formal proof in either case. Maple objects can be manipulated and broken apart easily, allowing one to construct a polytope by its individual cells. In this fashion, "recipes" for Zometool constructions can be designed in Maple, streamlining the real-world building process.

One additional note regarding the Wythoffian zoo of examples provided by our package is warranted before we end this introduction. The routines provided allows one to construct all the Wythoffian polytopes, excluding snub polytopes. The mathematical details of these polytopes are explained in Sect. 2. Thus 11 of the 13 Archimedean solids and 45 of the 47 non-prismatic convex uniform 4-dimensional polytopes described in [4] can be generated in this package and tested for Zometool constructability; missing are the snub cube and snub dodecahedron in the Archimedean case, and the snub 24-cell and grand antiprism in the 4-dimensional case. Finally, for this class of polytopes, we provide functions to perform projections onto Coxeter planes (see Fig. 3 for example).

In Sect. 2, the Wythoff construction is quickly described. In Sect. 3, the possibilities of our package related to Zometool constructions and constructibility are explored. In Sect. 5, the various projections provided by our package are explained and illustrated. The package is available at https://git.uwaterloo.ca/Zome-Maple/Zome-Maple.

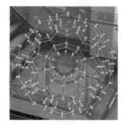 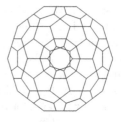

**Fig. 3.** The 120-cell projected on the $H_3$ Coxeter plane.

## 2   Wythoff Construction

To properly discuss this material, a few definitions must be given upfront. A *convex polytope* is a bounded region of $\mathbb{R}^n$ bounded by hyperplanes. The simplest polytopes are thus polygons (in $\mathbb{R}^2$) and polyhedra (in $\mathbb{R}^3$). The intersection of the polytope and its bounding hyperplanes are called *facets*. Hence facets of polygons are edges and facets of polyhedra are faces. A polygon is called *uniform* if it is regular, and a polytope in $n \geq 3$ dimensions is called *uniform* if all of its facets are uniform and the group of symmetries acts transitively on the vertices.

Wythoff had the brilliant idea in his 1918 paper [20] to construct uniform polytopes by first considering a finite group of reflections acting on $\mathbb{R}^n$, and then considering the polytopes obtained as the convex hull of the orbit of a point under the action of that group. If the initial point is equidistant to all reflecting hyperplanes in which it is not contained, the resulting polytope is uniform. This construction was polished by Coxeter's 1935 paper [7], following his observation that every finite group of reflections is (what we now call) a Coxeter group:

**Theorem 1 (Coxeter [5], Theorem 8).** *Every finite group of reflections has a presentation of the form $\langle r_1, \ldots, r_n \mid (r_i r_j)^{m_{ij}} = 1 \rangle$ for a symmetric matrix $m$ with $m_{ii} = 1$.*

Coxeter groups are conveniently expressed using *Coxeter diagrams*: labelled graphs whose vertices are the generating reflections $\{r_1, \ldots, r_n\}$ and where the edge $\{r_i, r_j\}$ is labelled $m_{ij}$ and exists only when $m_{ij} \geq 3$. The label 3 is the most common, and so it is usually omitted.

**Theorem 2 (Coxeter [6]).** *Every Coxeter diagram is a finite disjoint union of the Coxeter diagrams for the Coxeter groups $A_n, B_n, D_n, E_6, E_7, E_8, F_4, H_3, H_4,$ and $I_2(p)$.*

In this short paper, we won't describe all these graphs. They are classic and easily found. The upshot of this construction is that it is very easy to describe a polytope using a diagram. The recipe is easy, and one can compute easily the corresponding sub-objects recursively by following the simple algorithm explained in [2]. We defer the full explanation to this paper or to Coxeter's book [8] and simply illustrate it by few example.

In dimension three, imagine 3 mirrors passing through the origin and with dihedral angles $\frac{\pi}{2}, \frac{\pi}{3}, \frac{\pi}{5}$. Reflections in this mirror generate a finite group called $H_3$. This configuration is illustrated by the Coxeter diagram □——⁵——□————□. The point chosen is encoded by crossing boxes corresponding to mirror fixing the point. For instance, ⊠——⁵——⊠————□ is the polyhedra obtained by taking the convex hull of the orbit under the group $H_3$ of a point at distance 1 from the origin and on the intersection line of the first and second mirror: the icosahedron. This construction is implemented in Jeff Weeks's *KaleidoTile* software [19] with a visual interface allowing the choice of the point being reflected.

KaleidoTile does not allow one to compute the corresponding Wythoffian polytopes in higher dimension. Our package does, and allows one, for example, to compute vertices and edges of the 120-cell □——⁵——⊠————⊠————⊠, the 600-cell ⊠——⁵——⊠————⊠————□, and the more complicated omnitruncated 120-cell □——⁵——□————□————□.

An alteration to Wythoff's construction enables creation of the so-called *snub polytopes*. As before, their vertices are the orbit of a point under the action of a group. However, now the group is taken to be the rotary subgroup of a real reflection group. That is, the subgroup of any real reflection group consisting of the elements that are the product of an even number of reflections. Generally, the point chosen will not be the same as the points in Wythoff's construction.

## 3   Zometool Models

The package uses Maple's module system to work at three distinct levels of abstraction. The most basic data that is used is the vertices, provided as a list of $n$-tuples, and the edges, a list of unordered pairs of vertices. One level higher is the cell data: Maple's `ComputationalGeometry` package is used to convert the given skeleton into a list of 4-dimensional cells, which may be projected into

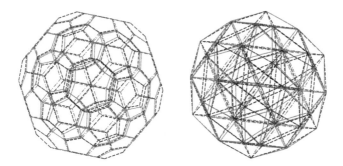

**Fig. 4.** The 120-cell (left) and 600-cell (right) projected cell-first and modeled in Maple. The view is from the $B_3$ and $H_3$ Coxeter planes respectively, and offset slightly to show 3D structure. A dashed line indicates a blue strut, a solid line a yellow strut, and alternating dashes and dots a red strut. (Color figure online)

3-dimensional space via a function in the package. Finally, cells are organized into *Zometool models*, which contain information describing how a set of cells can be physically realized in the Zometool system.

One feature is used to determine whether or not a model is Zometool constructible. For instance, taking the 120-cell and projecting vertex-first to $\mathbb{R}^3$, one finds a set of (normalized) edge lengths not compatible with the Zometool system. This set provides a certificate that this particular projection is not Zometool constructible. Regardless of constructibility, our package provides the ability to manipulate the object and display it if desired; see for instance Fig. 10.

Assuming the object is Zometool constructible, a list of projected cells can be assembled into a model. Figure 4 shows the 120-cell and 600-cell drawn as Zometool models, each projected into three dimensions cell-first. On its own, these image are too complicated to be of any use, although one point of interest is that they certify the fact that the 120-cell and 600-cell are constructible by Zometool. We can break apart the image by levels, for example to view the "core" of the model, or only the outermost cells, as in Fig. 5.

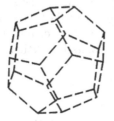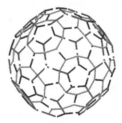

**Fig. 5.** Various components of the (cell-first projected) 120-cell. The core (left), and the upper half of the boundary (right).

After constructing the core, we can begin using some of the packaged utilities to determine what ought to be built next. For small models such as the 120-cell, building radially outwards is a standard strategy. Breaking by levels, we will show the model with its next layer of cells added. For convenience, the central

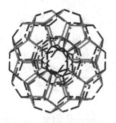

**Fig. 6.** The second layer of cells in a (cell-first projected) 120-cell. The type of cell added around the core is shown in the center. The right is the core with only one cell added.

part is drawn dotted, so the coloured edges are exactly the ones that must be added to the model. The cell can also be broken off entirely, so that it may be constructed on its own, or shown as the only cell in its layer adjoined to the previous layers. This makes the picture much less cluttered. An example is shown in Fig. 6.

We can continue this process for two more steps to get the full model. One useful feature is the ability to pass a filter function; for example, to cut away the cells in all but the positive orthant. When loaded in Maple, rotating the models is possible, making it somewhat easier to work with than static images. Using these two tools judiciously together allows one to effectively work with otherwise complicated models. The next step of the process looks like, with and without a filter function applied, is shown in Fig. 7.

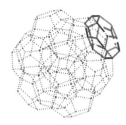

**Fig. 7.** The third layer of cells in a 120-cell. In the center is a piece cut from the left model using a filter. The right shows the previous step with only one cell added.

Since the 120-cell is uniform, if we understand how to build one part of the layer, we can repeat the construction elsewhere to finish it. Otherwise, we would have to be more careful with our filters, and handle each part of the layer individually. Finally, we can close up the last cells to get the full model, as seen in Fig. 8.

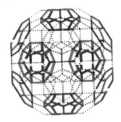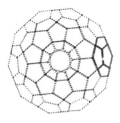

**Fig. 8.** Closing off the remaining cells with blue and red pieces completes the model of the 120-cell (Color figure online)

When considering large models such as the omnitruncated 120-cell, this sort of manipulation is quite helpful. Since it is impossible to build small-scale phys- ical copies of the model that can be disassembled and investigated (even the smallest incarnation possible in Zometool measures roughly 1.9 m in diameter and requires 21,360 pieces), the ability provided by this package to pick apart local features of the overall model is valuable for understanding how it should be constructed. For example, in order to understand how to suspend the model of the omnitruncated 120-cell from strings, we need to understand what the bottom half of the exterior looks like, to decide where strings should be placed. Some special paths formed by blue edges make good candidates for these string paths, and one could want to be able to isolate this feature. The results of both of these computations are shown in Fig. 9.

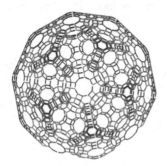 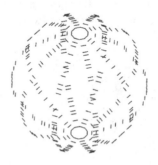

**Fig. 9.** Half of the boundary cells of the omnitruncated 120-cell (left), and the four "blue paths" in the omnitruncated 120-cell that occur on circles of constant longitude (right). (Color figure online)

These are not all the operations supported by the library, and generally it is easy to extend it to perform any other specific manipulations you might need. What we are trying to show is that by casting the question of Zometool modelling in the established framework of Maple, we get access to a powerful set of tools that can help in many aspects of a large-scale Zometool project.

One final use of this package that we shall point out is the ability to generate parts lists for a model. This is a rather long computation to run by hand, but simple to compute in Maple. Here is the list generated for the omnitruncated 120-cell, projected through a great rhombicosidodecahedral cell.

```
Balls = 7200
R2 = 2880
R1 = 2880
B2 = 3600
Y2 = 4800
```

Computations of this sort allow us to verify entries in Richter's list [15], for example.

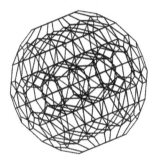

**Fig. 10.** Vertex-first and edge-first projections of the 120-cell

## 4   Computational Cost

Most operations on the package are performed fast enough that they will run on a desktop workstation with little issue. One of the most expensive operations is generating the reflection groups from which Wythoffian polytopes are constructed. All the 3 dimenisonal groups can be computed within a minute, and the four dimensional groups other than $H_4$ take no more than 10 min. On the other hand, $H_4$, which is of order 14400, takes approximately 12 h to generate on a desktop computer. After initial computation, groups can be saved and reused. Included with the project is the generated copy of $H_4$.

With the group $H_4$ available, to construct the 700 vertices of a 120-cell takes 12 s; the data has a size of 2 MB. The truncated 120-cell, having 2400 vertices, requires 2 min to compute, and has a size of 5 MB. The increase in time is largely due to the increased complexity of the symbolic expressions, as the generic point of the truncated 120-cell is more complicated than that of the 120-cell. If symbolic accuracy is not required, the corresponding floating point computation can be done in a much shorter time.

It is difficult to precisely analyse the complexity of constructing a Zome model, as this depends largely on the combinatorial structure of the vertices and edges. If a model is known to be constructible, it is possible to instantaneously create the model by providing the type of piece that the longest edge should be represented by in the construction. Otherwise, it will verify that the edge lengths are compatible with the edge length ratios of the Zometool system, and then verify that the neighbour configuration about each point is a possible one in the Zometool system. For smaller models, this process is very fast: the 4-hypercube (having 16 vertices) and 120-cell (having 700 vertices) both finished well within a minute. For the truncated 120-cell, this took 3 min.

## 5   Projections

In addition to constructing Zometool models, once cell data is constructed, it can be projected into three or two dimensional space and drawn, regardless of

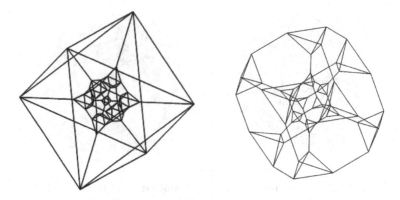

**Fig. 11.** A north pole stereographic projection of the truncated 16-cell and truncated hypercube.

Zometool constructability. The projections included in this package are orthogonal projections onto arbitrary bases, stereographic projections, and projections onto Coxeter planes. When creating Zometool constructions of 4-dimensional polytopes, the most useful of these is the orthogonal projection—stereographic projections distort distances too much, and Coxeter plane projections are two-dimensional. We include the other projections, as in many cases they are not available elsewhere, in some cases are very nice looking, and in general are useful to get a good grasp of the objects; see for instance in Fig. 11 the stereographic projections of the truncated 16-cell ($\boxtimes$——$^4$——$\boxtimes$————$\square$————$\square$) and truncated hypercube ($\square$——$^4$——$\square$————$\boxtimes$————$\boxtimes$).

Of particular interest are the vertex-first, edge-first, face-first, cell-first projections, see for instance Fig. 10. Shown in Fig. 12 are the $F_4$ and $H_4$ projections of the omnitruncated 120-cell computed by our package.

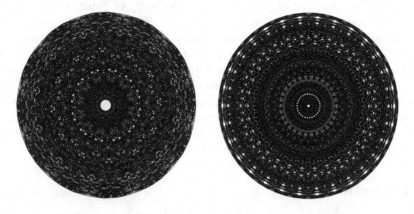

**Fig. 12.** The $F_4$ and $H_4$ Coxeter plane projection of the omnitruncated 120-cell

These projections do not occur as projections of the Zometool model of the omnitruncated 120-cell. So, in addition to allowing us to study Zometool models of Wythoffian polytopes, the Coxeter plane projections in this package can be used to enjoy some of the higher dimensional structure that is lost when projecting into three dimensions for the purposes of Zometool construction.

# References

1. A (not so brief) history of Zometool. https://www.zometool.com/about-us/
2. Champagne, B., Kjiri, M., Patera, J., Sharp, R.T.: Description of reflection-generated polytopes using decorated Coxeter diagrams. Canadian J. Phys. **73**, 566–584 (1995). https://doi.org/10.1139/p95-084
3. Chuang, C., Jin, B.Y.: Construction of Sierpiński superfullerenes with the aid of Zome geometry: application to beaded molecules. In: Hart, G.W., Sarhangi, R. (eds.) Proceedings of Bridges 2013: Mathematics, Music, Art, Architecture, Culture, pp. 495–498. Tessellations Publishing, Phoenix (2013). http://archive.bridgesmathart.org/2013/bridges2013-495.html
4. Conway, J.H., Guy, M.J.T.: Four-dimensional Archimedean polytopes. In: Proceedings of the Colloquium on Convexity, Copenhagen. Københavns Universitets Matematiske Institut, Copenhagen (1965)
5. Coxeter, H.S.M.: Discrete groups generated by reflections. Ann. Math. **35**(3), 588–621 (1934). http://www.jstor.org/stable/1968753
6. Coxeter, H.S.M.: The complete enumeration of finite groups of the form $r_i^2 = (r_i r_j)^{k_{ij}} = 1$. J. Lond. Math. Soc. **10**(1), 21–25 (1935). https://doi.org/10.1112/jlms/s1-10.37.21
7. Coxeter, H.S.M.: Wythoff's construction for uniform polytopes. Proc. Lond. Math. Soc. **2**(38), 327–339 (1935). https://doi.org/10.1112/plms/s2-38.1.327
8. Coxeter, H.S.M.: Regular Polytopes, 3rd edn. Dover Publications Inc., New York (1973)
9. Hall, B.C.: Lie Groups, Lie Algebras, and Representations. GTM, vol. 222. Springer, Cham (2015). https://doi.org/10.1007/978-3-319-13467-3
10. Hall, B.C.: The geometry of root systems: an exploration in the Zometool system. https://www3.nd.edu/~bhall/book/lie.htm
11. Hart, G.W.: Barn raisings of four-dimensional polytope projections. In: Proceedings of International Society of Art, Math, and Architecture (2007). http://www.georgehart.com/zome-polytopes-ISAMA07/hart-zome-polytopes.pdf
12. Hart, G.W., Picciotto, H.: Zome Geometry: Hands-On Learning with Zome Models. Key Curriculum Press (2001)
13. Hildebrandt, P.: Zome workshop. In: Sarhangi, R., Barrallo, J. (eds.) Bridges Donostia: Mathematics, Music, Art, Architecture, Culture, pp. 459–464. Tarquin Publications, London (2007). http://archive.bridgesmathart.org/2007/bridges2007-459.html
14. Richter, D.A.: H(4)-polychora with Zome. http://homepages.wmich.edu/~drichter/h4polychorazome.htm
15. Richter, D.A.: Two results concerning the Zome model of the 600-cell. In: Sarhangi, R., Moody, R.V. (eds.) Renaissance Banff: Mathematics, Music, Art, Culture, pp. 419–426. Bridges Conference, Southwestern College, Winfield (2005). http://archive.bridgesmathart.org/2005/bridges2005-419.html

16. Richter, D.A., Vorthmann, S.: Green quaternions, tenacious symmetry, and octa-hedreal Zome. In: Sarhangi, R., Sharp, J. (eds.) Bridges London: Mathematics, Music, Art, Architecture, Culture, pp. 429–436. Tarquin Publications, London (2006). http://archive.bridgesmathart.org/2006/bridges2006-429.html
17. Vörös, L.: A Zometool model of the B-DNA. In: Torrence, E., Torrence, B., Séquin, C., McKenna, D., Fenyvesi, K., Sarhangi, R. (eds.) Proceedings of Bridges 2016: Mathematics, Music, Art, Architecture, Education, Culture, pp. 435–438. Tessellations Publishing, Phoenix (2016). http://archive.bridgesmathart.org/2016/bridges2016-435.html
18. Vorthmann, S.: vZome, software. http://vzome.com/home/
19. Weeks, J.: KaleidoTile, software. http://geometrygames.org/KaleidoTile/index.html.en
20. Wijthoff, W.A.: A relation between the polytopes of the C600-family. Koninklijke Nederlandse Akademie van Wetenschappen Proc. Ser. B Phys. Sci. **20**, 966–970 (1918)

# Approximate GCD in a Bernstein Basis

Robert M. Corless and Leili Rafiee Sevyeri[(⊠)]

Ontario Research Centre for Computer Algebra, School of Mathematical
and Statistical Sciences, Western University, London, Canada
{rcorless,lrafiees}@uwo.ca

**Abstract.** We adapt Victor Y. Pan's root-based algorithm for finding approximate GCD to the case where the polynomials are expressed
in Bernstein bases. We use the numerically stable companion pencil of
Guðbjörn Jónsson to compute the roots, and the Hopcroft-Karp bipartite matching method to find the degree of the approximate GCD. We offer
some refinements to improve the process.

**Keywords:** Bernstein basis · Approximate GCD · Maximum matching ·
Bipartite graph · Root clustering · Companion pencil

## 1 Introduction

In general, finding the Greatest Common Divisor (GCD) of two exactly-known
univariate polynomials is a well understood problem. However, it is also known
that the GCD problem for *noisy* polynomials (polynomials with errors in their
coefficients) is an ill-posed problem. More precisely, a small error in coefficients of
polynomials $P$ and $Q$ with a non-trivial GCD generically leads to a trivial GCD. As
an example of such a situation, suppose $P$ and $Q$ are non constant polynomials
such that $P|Q$, then $\gcd(P,Q) = P$. Now for any $\epsilon > 0$, $\gcd(P,Q+\epsilon)$ is a
constant, since if $\gcd(P,Q+\epsilon) = g$, then $g|Q+\epsilon-Q = \epsilon$. This clearly shows
that the GCD problem is an ill-posed one. We note that the choice of polynomial
basis makes no difference to this difficulty.

At this point we have a good motivation to define something which can
play a similar role to the GCD of two given polynomials which is instead well-
conditioned. The idea is to define an *approximate GCD* [5]. There are various
definitions for approximate GCD which are used by different authors. All these
definitions respect "closeness" and "divisibility" in some sense.

In this paper an approximate GCD of a pair of polynomials $P$ and $Q$ is the
exact GCD of a corresponding pair of polynomials $\tilde{P}$ and $\tilde{Q}$ where $P$ and $\tilde{P}$
are "close" with respect to a specific metric, and similarly for $Q$ and $\tilde{Q}$ (see
Definition 1).

Finding the GCD of two given polynomials is an elementary operation needed
for many algebraic computations. Although in most applications the polynomials
are given in the power basis, there are cases where the input is given in other
bases such as the Bernstein basis. One important example of such a problem

© Springer Nature Switzerland AG 2020
J. Gerhard and I. Kotsireas (Eds.): MC 2019, CCIS 1125, pp. 77–91, 2020.
https://doi.org/10.1007/978-3-030-41258-6_6

is finding intersection points of Bézier curves, which are usually presented in a Bernstein basis. For computing the intersections of Bézier curves and surfaces the Bernstein resultant and GCD in the Bernstein basis comes in handy (see [3]).

One way to deal with polynomials in Bernstein bases is to convert them into the power basis. In practice poor stability of conversion from one basis to another and poor conditioning of the power basis essentially cancel the benefit one might get by using conversion to the simpler basis (see [10]).

The Bernstein basis is an interesting one for various algebraic computations, for instance, see [19,21]. There are many interesting results in approximate GCD including but not limited to [1,2,5,8,17,18,20,26] and [16]. In [27], the author has introduced a modification of the algorithm given by Corless, Gianni, Trager and Watt in [7], to compute the approximate GCD in the power basis.

Winkler and Yang in [25] give an estimate of the degree of an approximate GCD of two polynomials in a Bernstein basis. Their approach is based on computations using resultant matrices. More precisely, they use the singular value decomposition of Sylvester and Bézout resultant matrices. We do not follow the approach of Winkler and Yang here, because they essentially convert to a power basis. Owing to this difference we do not give a comparison of our algorithm with the results of [25].

Our approach is mainly to follow the ideas introduced by Pan in [21], working in the power basis. In distinction to the other known algorithms for approximate GCD, Pan's method does not algebraically compute a degree for an approximate GCD first. Instead it works in a reverse way. In [21] the author assumes the roots of polynomials $P$ and $Q$ are given as inputs. Having the roots in hand the algorithm generates a bipartite graph where one set of nodes contains the roots of $P$ and the other contains the roots of $Q$. The criterion for defining the set of edges is based on Euclidean distances of roots. When the graph is defined completely, a matching algorithm will be applied. Using the obtained matching, a polynomial $D$ with roots as averages of paired close roots will be produced which is considered to be an approximate GCD. The last step is to use the roots of $D$ to replace the corresponding roots in $P$ and $Q$ to get $\tilde{P}$ and $\tilde{Q}$ as close polynomials.

In this paper we introduce an algorithm for computing approximate GCD in the Bernstein basis which relies on the above idea. For us the inputs are the coefficient vectors of $P$ and $Q$. We use the correspondence between the roots of a polynomial $f$ in a Bernstein basis and generalized eigenvalues of a corresponding matrix pencil $(A_f, B_f)$. This idea for finding the roots of $f$ was first used in [14]. Then by finding the generalized eigenvalues we get the roots of $P$ and $Q$ (see [14, Section 2.3]). Using the roots and similar methods to [21], we form a bipartite graph and then we apply the maximum matching algorithm by Hopcroft and Karp [13] to get a maximum matching. Having the matching, the algorithm forms a polynomial which is considered as an approximate GCD of $P$ and $Q$. The last step is to construct $\tilde{P}$ and $\tilde{Q}$ for which we apply a generalization of the method used in [6, Example 6.10] (see Sect. 3).

Note that our algorithm, like that of Pan, does almost the reverse of the well-known algorithms for approximate GCD. Usually the algebraic methods do not try to find the roots. In [21] Pan assumes the polynomials are represented by their roots. In our case we do not start with this assumption. Instead, by computing the roots we can then apply Pan's method.

The second section of this paper is provided some background for concrete computations with polynomials in Bernstein bases which is needed for our purposes. The third section presents a method to construct a corresponding pair of polynomials to a given pair $(P, Q)$. More precisely, this section generalizes the method mentioned in [6, Example 6.10] (which is introduced for power basis) in the Bernstein basis. The fourth section introduces a new algorithm for finding an approximate GCD. In the final section we present numerical results based on our method.

## 2    Preliminaries

The Bernstein polynomials on the interval $0 \leq x \leq 1$ are defined as

$$B_k^n(x) = \binom{n}{k} x^k (1 - x)^{n-k} \tag{1}$$

for $k = 0, \ldots, n$, where the binomial coefficient is as usual

$$\binom{n}{k} = \frac{n!}{k!(n - k)!} . \tag{2}$$

More generally, in the interval $a \leq x \leq b$ (where $a < b$) we define

$$B_{a,b,k}^n(x) := \binom{n}{k} \frac{(x - a)^k (b - x)^{n-k}}{(b - a)^n} . \tag{3}$$

When there is no risk of confusion we may simply write $B_k^n$'s for the $0 \leq x \leq 1$ case. We suppose henceforth that $P(x)$ and $Q(x)$ are given in a Bernstein basis.

There are various definitions for approximate GCD. The main idea behind all of them is to find "interesting" polynomials $\tilde{P}$ and $\tilde{Q}$ close to $P$ and $Q$ and use $\gcd(\tilde{P}, \tilde{Q})$ as the approximate GCD of $P$ and $Q$. However, there are multiple ways of defining both "interest" and "closeness". To be more formal, consider the following weighted norm, for a vector $v$

$$\|v\|_{\alpha,r} = \left( \sum_{k=1}^n |\alpha_k v_k|^r \right)^{1/r} \tag{4}$$

for a given weight vector $\alpha \neq 0$ and a positive integer $r$ or $\infty$. The map $\rho(u, v) = \|u - v\|_{\alpha,r}$ is a metric and we use this metric to compare the coefficient vectors of $P$ and $Q$.

In this paper we define an approximate GCD using the above metric or indeed any fixed semimetric. More precisely, we define the *pseudogcd* set for the pair $P$ and $Q$ as

$$A_\rho = \left\{ g(x) \mid \exists \tilde{P}, \tilde{Q} \text{ with } \rho(P, \tilde{P}) \leq \sigma, \rho(Q, \tilde{Q}) \leq \sigma \text{ and } g(x) = \gcd(\tilde{P}, \tilde{Q}) \right\}.$$

Let

$$d = \max_{g \in A_\rho} \deg(g(x)). \tag{5}$$

**Definition 1.** *An approximate GCD for $P, Q$ which is denoted by $\mathrm{agcd}_\rho^\sigma(P, Q)$, is $G(x) \in A_\rho$ where $\deg(G) = d$ and $\rho(P, \tilde{P})$ and $\rho(Q, \tilde{Q})$ are simultaneously minimal in some sense. For definiteness, we suppose that the maximum of these two quantities is minimized.*

In Sect. 2.3 we will define another (semi) metric, that uses roots. In Sect. 4 we will see that the parameter $\sigma$ helps us to find approximate polynomials such that the common roots of $\tilde{P}$ and $\tilde{Q}$ have at most distance (for a specific metric) $\sigma$ to the associated roots of $P$ and $Q$ (see Sect. 4 for more details).

## 2.1    Finding Roots of a Polynomial in a Bernstein Basis

In this section we recount the numerically stable method introduced by Guðbjörn Jónsson for finding roots of a given polynomial in a Bernstein basis. We only state the method without discussing in detail its stability and we refer the reader to [14] and [15] for more details.

Consider a polynomial

$$P(x) = \sum_{i=0}^{n} a_i B_i^n(x) \tag{6}$$

in a Bernstein basis where the $a_i$'s are real scalars. We want to find the roots of $P(x)$ by constructing its companion pencil. In [14] Jónsson showed that this problem is equivalent to solving the following generalized eigenvalue problem. That is, the roots of $P(x)$ are the generalized eigenvalues of the corresponding companion pencil to the pair

$$\mathbf{A}_P = \begin{bmatrix} -a_{n-1} & -a_{n-2} & \cdots & -a_1 & -a_0 \\ 1 & 0 & & & \\ & 1 & 0 & & \\ & & 1 & 0 & \\ & & & \ddots & \ddots \\ & & & & 1 & 0 \end{bmatrix}, \quad \mathbf{B}_P = \begin{bmatrix} -a_{n-1} + \frac{a_n}{n} & -a_{n-2} & \cdots & -a_1 & -a_0 \\ 1 & \frac{2}{n-1} & & & \\ & 1 & \frac{3}{n-2} & & \\ & & & \ddots & \ddots \\ & & & & 1 & \frac{n}{1} \end{bmatrix}.$$

That is, $P(x) = \det(x\mathbf{B}_P - \mathbf{A}_P)$. In [14], the author showed that the above method is numerically stable.

**Theorem 1.** *[14, Section 2.3] Assume $P(x)$, $\mathbf{A}_P$ and $\mathbf{B}_P$ are defined as above. $z$ is a root of $P(x)$ if and only if it is a generalized eigenvalue for the pair $(\mathbf{A}_P, \mathbf{B}_P)$.*

*Proof.* We show

$$P(z) = 0 \quad \Leftrightarrow \quad (z\mathbf{B}_P - \mathbf{A}_P) \begin{bmatrix} B_{n-1}^n(z)(\frac{1}{1-z}) \\ \vdots \\ B_1^n(z)(\frac{1}{1-z}) \\ B_0^n(z)(\frac{1}{1-z}) \end{bmatrix} = 0. \tag{7}$$

We will show that all the entries of

$$(z\mathbf{B}_P - \mathbf{A}_P) \begin{bmatrix} B_{n-1}^n(z)(\frac{1}{1-z}) \\ \vdots \\ B_1^n(z)(\frac{1}{1-z}) \\ B_0^n(z)(\frac{1}{1-z}) \end{bmatrix} \tag{8}$$

are zero except for possibly the first entry:

$$(z - 1)B_1^n(z)(\frac{1}{z-1}) + nzB_0^n(z)(\frac{1}{z-1}) \tag{9}$$

since $B_1^n(z) = nz(1-z)^{n-1}$ and $B_0^n(z) = z^0(1-z)^n$ if $n \geq 1$, so Eq. (9) can be written as

$$-nz(1-z)^{n-1} + nz\frac{(1-z)^n}{(1-z)} = -nz(1-z)^{n-1} + nz(1-z)^{n-1} = 0. \tag{10}$$

Now for $k$-th entry:

$$\frac{(z-1)}{(1-z)}B_{n-k}^n(z) + \frac{k+1}{n-k}\frac{z}{(1-z)}B_{n-k-1}^n(z) \tag{11}$$

Again we can replace $B_{n-k}^n(z)$ and $B_{n-k-1}^n(z)$ by their definitions. We find that Eq. (11) can be written as

$$\frac{(z-1)}{(1-z)}\binom{n}{n-k}z^{n-k}(1-z)^{n-n+k} + \frac{k+1}{n-k}\frac{z}{(1-z)}\binom{n}{n-(k+1)}z^{n-k-1}(1-z)^{n-(n-k-1)}$$

$$= -\binom{n}{n-k}z^{n-k}(1-z)^k + \frac{k+1}{n-k}\frac{n!}{(n-(k+1))!(k+1)!}z^{n-k}(1-z)^k$$

$$= \frac{n!}{(n-k)!k!}z^{n-k}(1-z)^k + \frac{n!}{(n-k)(n-(k+1))!k!}z^{n-k}(1-z)^k = 0. \tag{12}$$

Finally, the first entry of Eq. (8) is

$$\frac{za_n}{n(1-z)}B_{n-1}^n(z) + \frac{a_{n-1}(z-1)}{(1-z)}B_{n-1}^n(z) + \sum_{i=0}^{n-2} a_i B_{n-1}^n(z) \tag{13}$$

In order to simplify the Eq. (13), we use the definition of $B_{n-1}^n(z)$ as follows:

$$\frac{za_n}{n(1-z)}B_{n-1}^n(z) = \frac{z}{n}\binom{n}{n-1}z^{n-1}\frac{1-z}{1-z} = \frac{z^n a_n}{n}\binom{n}{n-1} = a_n B_n^n(z) \quad (14)$$

So the Eq. (13) can be written as:

$$a_n B_n^n(z) + (a_{n-1})B_{n-1}^n(z) + \sum_{i=0}^{n-2} a_i B_{n-1}^n(z) \quad (15)$$

This is just $P(z)$ and so

$$(z\mathbf{B}_P - \mathbf{A}_P)\begin{bmatrix} B_{n-1}^n(z)(\frac{1}{1-z}) \\ \vdots \\ B_1^n(z)(\frac{1}{1-z}) \\ B_0^n(z)(\frac{1}{1-z}) \end{bmatrix} = \begin{bmatrix} 0 \\ \vdots \\ 0 \\ 0 \end{bmatrix} \quad (16)$$

if and only if $P(z) = 0$.

□

This pencil (or rather its transpose) has been implemented in MAPLE since 2004.

*Example 1.* Suppose $P(x)$ is given by its list of coefficients

$$[42.336, 23.058, 11.730, 5.377, 2.024] \quad (17)$$

Then by using Theorem 1, we can find the roots of $P(x)$ by finding the eigenvalues of its corresponding companion pencil namely:

$$\mathbf{A}_P := \begin{bmatrix} -5.377 & -11.73 & -23.058 & -42.336 \\ 1 & 0 & 0 & 0 \\ 0 & 1 & 0 & 0 \\ 0 & 0 & 1 & 0 \end{bmatrix} \quad (18)$$

and

$$\mathbf{B}_P := \begin{bmatrix} -4.871 & -11.73 & -23.058 & -42.336 \\ 1 & .6666666666 & 0 & 0 \\ 0 & 1 & 1.5 & 0 \\ 0 & 0 & 1 & 4 \end{bmatrix} \quad (19)$$

Now if we solve the generalized eigenvalue problem using MAPLE for pair of $(\mathbf{A}_P, \mathbf{B}_P)$ we get:

$$[5.59999999999989, 3.00000000000002, 2.1, 1.2] \quad (20)$$

Computing residuals, we have exactly[1] $P(1.2) = 0$, $P(2.1) = 0$, $P(3) = 0$, and $P(5.6) = 0$ using de Casteljau's algorithm (see Sect. 2.4).

## 2.2 Clustering the Roots

In this brief section we discuss the problem of having multiplicities greater than 1 for roots of our polynomials. Since we are dealing with approximate roots, for an specific root $r$ of multiplicity $m$, we get $r_1, \ldots, r_m$ where $|r - r_i| \leq \sigma$ for $\sigma \geq 0$. Our goal in this section is to recognize the *cluster*, $\{r_1, \ldots, r_m\}$, for a root $r$ as $\tilde{r}^m$ where $|\tilde{r} - r| \leq \sigma$ in a constructive way.

Assume a polynomial $f$ is given by its roots as $f(x) = \prod_{i=1}^{n}(x - r_i)$. Our goal is to write $f(x) = \prod_{i=1}^{s}(x - t_i)^{d_i}$ such that $(x - t_i) \nmid f(x)/(x - t_i)^{d_i}$. In other words, $d_i$'s are multiplicities of $t_i$'s. In order to do so we need a parameter $\sigma$ to compare the roots. If $|r_i - r_j| \leq \sigma$ then we replace both $r_i$ and $r_j$ with their average.

For our purposes, even the naive method, *i.e.* computing distances of all roots, works. This idea is presented as Algorithm 1. It is worth mentioning that for practical purposes a slightly better way might be a modification of the well known divide and conquer algorithm for solving the closest pair problem in plane [28, Section 33.4].

---

**Algorithm 1.** ClusterRoots$(P, \sigma)$

---

**Input:** $P$ is a list of roots
**Output:** $[(\alpha_1, d_1), \ldots, (\alpha_m, d_m)]$ where $\alpha_i$ is considered as a root with multiplicity $d_i$
    $temp \leftarrow EmptyList$
    $C \leftarrow EmptyList$
    $p \leftarrow \mathbf{size}(P)$
    $i \leftarrow 1$
    while $i \leq p$ do
        $\mathbf{append}(temp, P[i])$
        $j \leftarrow i + 1$
        while $j \leq p$ do
            if $|P[i] - P[j]| \leq s$ then
                $\mathbf{append}(temp, P[j])$
                $\mathbf{remove}(P, j)$
                $p \leftarrow p - 1$
            else
                $j \leftarrow j + 1$
        $i \leftarrow i + 1$
        $\mathbf{append}(C, [\mathbf{Mean}(temp), \mathbf{size}(temp)])$
    return $C$

---

---

[1] In some sense the exactness is accidental; the computed residual is itself subject to rounding errors. See [6] for a backward error explanation of how this can happen.

## 2.3   The Root Marriage Problem

The goal of this section is to provide an algorithmic solution for solving the following problem:

**The Root Marriage Problem (RMP):** Assume $P$ and $Q$ are polynomials given by their roots. For a given $\sigma > 0$, for each root $r$ of $P$, (if it is possible) find a unique root of $Q$, say $s$, such that $|r - s| \leq \sigma$.

A solution to the RMP can be achieved by means of graph theory algorithms. We recall that a maximum matching for a bipartite graph $(V, E)$, is $M \subseteq E$ with two properties:

- every node $v \in V$ appears as an end point of an edge in $E'$ at most once.
- $E'$ has the maximum size among the subsets of $E$ satisfying the previous condition.

We invite the reader to consult [4] and [24] for more details on maximum matching.

There are various algorithms for solving the maximum matching problem in a graph. Micali and Vazirani's matching algorithm is probably the most well-known. However there are more specific algorithms for different classes of graphs. In this paper, as in [21], we use the Hopcroft-Karp algorithm for solving the maximum matching problem in a bipartite graph which has a complexity of $O((m + n)\sqrt{n})$ operations.

Now we have enough tools for solving the RMP. The idea is to reduce the RMP to a maximum matching problem. In order to do so we have to associate a bipartite graph to a pair of polynomials $P$ and $Q$. For a positive real number $\sigma$, let $G_{P,Q}^{\sigma} = (G_P^{\sigma} \cup G_Q^{\sigma}, E_{P,Q}^{\sigma})$ where

- $G_P = \mathtt{ClusterRoots}(\text{the set of roots of } P, \sigma)$,
- $G_Q = \mathtt{ClusterRoots}(\text{the set of roots of } Q, \sigma)$,
- $E_{P,Q}^{\sigma} = \left\{ (\{r, s\}, \min(d_t, d_s)) : r \in G_P, \text{ with multiplicity } d_r, s \in G_Q \text{ with multiplicity } d_s, \left| r[1] - s[1] \right| \leq \sigma \right\}$

Assuming we have access to the roots of polynomials, it is not hard to see that there is a naive algorithm to construct $G_{P,Q}^{\sigma}$ for a given $\sigma > 0$. Indeed it can be done by performing $O(n^2)$ operations to check the distances of roots where $n$ is the larger degree of the given pair of polynomials.

The last step to solve the RMP is to apply the Hopcroft-Karp algorithm on $G_{P,Q}^{\sigma}$ to get a maximum matching. The complexity of this algorithm is $O(n^{\frac{5}{2}})$ which is the dominant term in the total cost. Hence we can solve RMP in time $O(n^{\frac{5}{2}})$.

As was stated in Sect. 2, we present a semi-metric which works with polynomial roots in this section. For two polynomials $R$ and $T$, assume $m \leq n$ and $\{r_1, \ldots, r_m\}$ and $\{t_1, \ldots, t_n\}$ are respectively the sets of roots of $R$ and $T$. Moreover assume $S_n$ is the set of all permutations of $\{1, \ldots, n\}$. We define

$$\rho(R, T) = \min_{\tau \in S_n} \left\| [r_1 - t_{\tau(1)}, \ldots, r_m - t_{\tau(m)}] \right\|_{\alpha, r},$$

where $\alpha$ and $r$ are as before.

*Remark 1.* The cost of computing this semi-metric by this definition is $O(n!)$, and therefore prohibitive. However, once a matching has been found then

$$\rho(R,T) = \|[r_1 - s_{\text{match}(1)}, r_2 - s_{\text{match}(2)}, \ldots, r_m - s_{\text{match}(m)}]\|_{\alpha,r}$$

where the notation $s_{\text{match}(k)}$ indicates the root found by the matching algorithm that matches $r_k$.

## 2.4   de Casteljau's Algorithm

Another component of our algorithm is a method which enables us to evaluate a given polynomial in a Bernstein basis at a given point. There are various methods for doing that. One of the most popular algorithms, for its convenience in Computer Aided Geometric Design (CAGD) applications and its numerical stability [9], is de Casteljau's algorithm which for convenience here is presented as Algorithm 2.

---

**Algorithm 2.** de Casteljau's Algorithm

---

**Input:** C: a list of coefficients of a polynomial $P(x)$ of degree $n$ in a Bernstein basis
   of size $n + 1$
   $\alpha$: a point
**Output:** $P(\alpha)$
 1: $c_{0,j} \leftarrow C_j$ for $j = 0 \ldots n$.
 2: recursively define
    $c_{i,j} \leftarrow (1 - \alpha) \cdot c_{i-1,j} + \alpha \cdot c_{i-1,j+1}$.
    for $i = 1 \ldots n$ and $j = 1 \ldots n - i$.
 3: return $c_{n,0}$.

---

We note that the above algorithm uses $O(n^2)$ operations for computing $P(\alpha)$. In contrast, Horner's algorithm for the power basis, Taylor polynomials, or the Newton basis, and the Clenshaw algorithm for orthogonal polynomials, and the barycentric forms[2] for Lagrange and Hermite interpolational basis cost $O(n)$ operations.

## 3   Computing Approximate Polynomials

This section is a generalization of [6, Example 6.10] in Bernstein bases. The idea behind the algorithm is to create a linear system from coefficients of a given polynomial and the values of the polynomial at the approximate roots.

---

[2] Assuming that the barycentric weights are precomputed.

Now assume

$$P(x) = \sum_{i=0}^{n} p_i B_i^n(x) \tag{21}$$

is given with $\alpha_1, \ldots, \alpha_t$ as its approximate roots with multiplicities $d_i$. Our aim is to find

$$\tilde{P}(x) = (P + \Delta P)(x) \tag{22}$$

where

$$\Delta P(x) = \sum_{i=0}^{n} (\Delta p_i) B_i^n(x) \tag{23}$$

so that the set $\{\alpha_1, \ldots, \alpha_t\}$ appears as exact roots of $\tilde{P}$ with multiplicities $d_i$ respectively. On the other hand, we do want to have some control on the coefficients in the sense that the new coefficients are related to the previous ones. Defining $\Delta p_i = p_i \delta p_i$ (which assumes $p_i$'s are non-zero) yields

$$\tilde{P}(x) = \sum_{i=0}^{n} (p_i + p_i \delta p_i) B_i^n(x) \tag{24}$$

Representing $P$ as above, we want to find $\{\delta p_i\}_{i=0}^{n}$. It is worth mentioning that with our assumptions, since perturbations of each coefficient, $p_i$ of $P$ are proportional to itself, if $p_i = 0$ then $\Delta p_i = 0$. In other words we have assumed zero coefficients in $P$ will not be perturbed.

In order to satisfy the conditions of our problem we have

$$\tilde{P}(\alpha_j) = \sum_{i=0}^{n} (p_i + p_i \delta p_i) B_i^n(\alpha_j) = 0, \tag{25}$$

for $j = 1, \ldots, t$. Hence

$$\tilde{P}(\alpha_j) = \sum_{i=0}^{n} p_i B_i^n(\alpha_j) + \sum_{i=0}^{n-1} p_i \delta p_i B_i^n(\alpha_j) = 0, \tag{26}$$

or equivalently

$$\sum_{i=0}^{n-1} p_i \delta p_i B_i^n(\alpha_j) = -P(\alpha_j). \tag{27}$$

Having the multiplicities, we also want the approximate polynomial $\tilde{P}$ to respect multiplicities. More precisely, for $\alpha_j$, a root of $P$ of multiplicity $d_j$, we expect that $\alpha_j$ has multiplicity $d_j$ as a root of $\tilde{P}$. As usual we can state this fact by means of derivatives of $\tilde{P}$. We want

$$\tilde{P}^{(k)}(\alpha_j) = 0 \text{ for } 0 \leq k \leq d \tag{28}$$

More precisely, we can use the derivatives of Eq. (27) to write

$$\left( \sum_{i=0}^{n-1} p_i \delta p_i B_i^n \right)^{(k)} (\alpha_j) = -P^{(k)}(\alpha_j). \tag{29}$$

In order to find the derivatives in (29), we can use the differentiation matrix $\mathbf{D_B}$ in the Bernstein basis which is introduced in [29]. We note that it is a sparse matrix with only 3 nonzero elements in each column [29, Section 1.4.3]. So for each root $\alpha_i$, we get $d_i$ equations of the type (29). This gives us a linear system in the $\delta p_i$'s. Solving the above linear system using the Singular Value Decomposition (SVD) one gets the desired solution.

Algorithm 3 gives a numerical solution to the problem. For an analytic solution for one single root see [12, 22, 23] and [11].

---

**Algorithm 3.** Approximate-Polynomial$(P, L)$

---

**Input:** $P$ : list of coefficients of a polynomial of degree $n$ in a Bernstein basis
$\qquad$ $L$ : list of pairs of roots with their multiplicities.
**Output:** $\tilde{P}$ such that for any $(\alpha, d) \in L$, $(x - \alpha)^d | \tilde{P}$.
$\quad$ 1: $Sys \leftarrow EmptyList$
$\quad$ 2: $D_B \leftarrow$ Differentiation matrix in the Bernstein basis of size $n + 1$
$\quad$ 3: $X \leftarrow \begin{bmatrix} x_1 & \cdots & x_{n+1} \end{bmatrix}^t$
$\quad$ 4: $T \leftarrow$ EntrywiseProduct(Vector$(P), X$)
$\quad$ 5: for $(\alpha, d) \in L$ do
$\qquad$ $A \leftarrow I_{n+1}$
$\qquad$ for $i$ from 0 to $d - 1$ do
$\qquad\qquad$ $A \leftarrow D_B \cdot A$
$\qquad\qquad$ $eq \leftarrow$ DeCasteljau$(A \cdot T, \alpha) = -$DeCasteljau$(A \cdot$ Vector$(P), \alpha)$
$\qquad\qquad$ append$(Sys, eq)$
$\quad$ 6: Solve $Sys$ using SVD to get a solution with minimal norm (such as 4), and return the result.

---

Although Algorithm 3 is written for one polynomial, in practice we apply it to both $P$ and $Q$ separately with the appropriate lists of roots with their multiplicities to get $\tilde{P}$ and $\tilde{Q}$.

## 4 Computing Approximate GCD

Assume the polynomials $P(x) = \sum_{i=0}^{n} a_i B_i^n(x)$ and $Q(x) = \sum_{i=0}^{m} b_i B_i^m(x)$ are given by their lists of coefficients and suppose $\alpha \geq 0$ and $\sigma > 0$ are given. Our goal here is to compute an approximate GCD of $P$ and $Q$ with respect to the given $\sigma$. Following Pan [21] as mentioned earlier, the idea behind our algorithm is to match the close roots of $P$ and $Q$ and then based on this matching find approximate polynomials $\tilde{P}$ and $\tilde{Q}$ such that their GCD is easy to compute. The parameter $\sigma$ is our main tool for constructing the approximate polynomials. More precisely, $\tilde{P}$ and $\tilde{Q}$ will be constructed such that their roots are respectively approximations of roots of $P$ and $Q$ with $\sigma$ as their error bound. In other words, for any root $x_0$ of $P$, $\tilde{P}$(similarly for $Q$) has a root $\tilde{x}_0$ such that $|x_0 - \tilde{x}_0| \leq \sigma$.

For computing approximate GCD we apply graph theory techniques. In fact the parameter $\sigma$ helps us to define a bipartite graph as well, which is used to construct the approximate GCD before finding $\tilde{P}$ and $\tilde{Q}$.

We can compute an approximate GCD of the pair $P$ and $Q$, which we denote by $\mathrm{agcd}_\rho^\sigma(P(x), Q(x))$, in the following 5 steps.

**Step 1. Finding the roots:** Apply the method of Sect. 2.1 to get $X = [x_1, x_2, \ldots, x_n]$, the set of all roots of $P$ and $Y = [y_1, y_2, \ldots, y_m]$, the set of all roots of $Q$.

**Step 2. Forming the graph of roots $G_{P,Q}$:** With the sets $X$ and $Y$ we form a bipartite graph, $G$, similar to [21] which depends on parameter $\sigma$ in the following way:

If $|x_i - y_j| \leq 2\sigma$ for $i = 1, \ldots, n$ and $j = 1, \ldots, m$, then we can store that pair of $x_i$ and $y_j$.

**Step 3. Find a maximum matching in $G_{P,Q}$:** Apply the Hopcroft-Karp algorithm [13] to get a maximum matching $\{(x_{i_1}, y_{j_1}), \ldots, (x_{i_r}, y_{j_r})\}$ where $1 \leq k \leq r$, $i_k \in \{1, \ldots, n\}$ and $j_k \in \{1, \ldots, m\}$.

**Step 4. Forming the approximate GCD:**

$$\mathrm{agcd}_\rho^\sigma(P(x), Q(x)) = \prod_{s=1}^{r} (x - z_s)^{t_s} \tag{30}$$

where $z_s = \dfrac{1}{2}(x_{i_s} + y_{j_s})$ and $t_s$ is the minimum of multiplicities of $x_s$ and $y_s$ for $1 \leq s \leq r$.

**Step 5. Finding approximate polynomials $\tilde{P}(x)$ and $\tilde{Q}(x)$:** Apply Algorithm 2 with $\{z_1, \ldots, z_r, x_{r+1}, \ldots, x_n\}$ for $P(x)$ and $\{z_1, \ldots, z_r, y_{r+1}, \ldots, y_m\}$ for $Q(x)$.

For steps 2 and 3 one can use the tools provided in Sect. 2.3. We also note that the output of the above algorithm is directly related to the parameter $\sigma$ and an inappropriate $\sigma$ may result in an unexpected result.

## 5    Numerical Results

In this section we show small examples of the effectiveness of our algorithm (using an implementation in MAPLE ) with two low degree polynomials in a Bernstein basis, given by their list of coefficients:

$$P := [5.887134, 1.341879, 0.080590, 0.000769, -0.000086]$$

and

$$Q := [-17.88416, -9.503893, -4.226960, -1.05336]$$

defined in MAPLE using Digits := 30 (we have presented the coefficients with fewer than 30 digits for readability). So $P(x)$ and $Q(x)$ are seen to be

$$P(x) := 5.887134 \left(1 - x\right)^4 + 5.367516\, x \left(1 - x\right)^3$$
$$+ 0.483544\, x^2 \left(1 - x\right)^2 + 0.003076\, x^3 \left(1 - x\right)$$
$$- 0.000086\, x^4$$

and

$$Q(x) := -17.88416\,(1-x)^3 - 28.51168\,x\,(1-x)^2$$
$$-12.68088\,x^2\,(1-x) - 1.05336\,x^3$$

Moreover, the following computations is done using parameter $\sigma = 0.7$, and unweighted norm-2 as a simple example of Eq. (4), with $r = 2$ and $\alpha = (1, \ldots, 1)$. Using Theorem 1, the roots of $P$ are, printed to two decimals for brevity,

$$\left[5.3 + 0.0\,i,\, 1.09 + 0.0\,i,\, 0.99 + 0.0\,i,\, 1.02 + 0.0\,i\right]$$

This in turn is passed to `ClusterRoots` (Algorithm 1) to get

$$P_{\texttt{ClusterRoots}} := [[1.036 + 0.0\,i, 3], [5.3 + 0.0\,i, 1]]$$

where 3 and 1 are the multiplicities of the corresponding roots. Similarly for $Q$ we have:

$$\left[1.12 + 0.0\,i,\, 4.99 + 0.0\,i,\, 3.19 + 0.0\,i\right]$$

which leads to

$$Q_{\texttt{ClusterRoots}} := [[3.19 + 0.0\,i, 1], [4.99 + 0.0\,i, 1], [1.12 + 0.0\,i, 1]]$$

Again the 1's are the multiplicities of the corresponding roots.

Applying the implemented maximum matching algorithm in MAPLE (see Sect. 2.3), a maximum matching for the clustered sets of roots is

$$T_{\texttt{MaximumMatching}} := [[\{4.99, 5.30\}, 1], [\{1.03, 1.12\}, 1]]$$

This clearly implies we can define (see Step 4 of our algorithm in Sect. 4)

$$\mathrm{agcd}_\rho^{0.7}(P, Q) := (x - 5.145)(x - 1.078)$$

Now the last step of our algorithm is to compute the approximate polynomials having these roots, namely $\tilde{P}$ and $\tilde{Q}$. This is done using Algorithm 3 which gives

$$\tilde{P} := [6.204827, 1.381210, 0.071293, 0.000777, -0.000086]$$

and

$$\tilde{Q} := [-17.202067, -10.003156, -4.698063, -0.872077]$$

Note that

$$\|P - \tilde{P}\|_{\alpha,2} \approx 0.32 \le 0.7 \quad \text{and} \quad \|Q - \tilde{Q}\|_{\alpha,2} \approx 0.68 \le 0.7$$

We remark that in the above computations we used the built-in function `LeastSquares` in MAPLE to solve the linear system to get $\tilde{P}$ and $\tilde{Q}$, instead of using the SVD ourselves. This equivalent method returns a solution to the system which is minimal according to norm-2. This can be replaced with any other solver which uses SVD to get a minimal solution with the desired norm.

As the last part of experiments we have tested our algorithm on several random inputs of two polynomials of various degrees. The resulting polynomials $\tilde{P}$ and $\tilde{Q}$ are compared to $P$ and $Q$ with respect to 2-norm (as a simple example of our weighted norm) and the root semi-metric which is defined in Sect. 2.3. Some of the results are displayed in Table 1.

**Table 1.** Distance comparison of outputs and inputs of our approximate GCD algorithm on randomly chosen inputs.

| $\max_{\deg}\{P, Q\}$ | $\deg(agcd^\sigma_\rho(P, Q))$ | $\|P - \tilde{P}\|_2$ | $\rho(P, \tilde{P})$ | $\|Q - \tilde{Q}\|_2$ | $\rho(Q, \tilde{Q})$ |
|---|---|---|---|---|---|
| 2 | 1 | 0.00473 | 0.11619 | 0.01199 | 0.05820 |
| 4 | 3 | 1.08900 | 1.04012 | 0.15880 | 0.15761 |
| 6 | 2 | 0.80923 | 0.75634 | 0.21062 | 0.31073 |
| 7 | 2 | 0.02573 | 0.04832 | 0.12336 | 0.02672 |
| 10 | 5 | 0.165979 | 0.22737 | 0.71190 | 0.64593 |

## 6   Concluding Remarks

In this paper we have explored the computation of approximate GCD of polynomials given in a Bernstein basis, by using a method similar to that of Pan [21]. We first use the companion pencil of Jónsson to find the roots; we cluster the roots as Zeng does to find the so-called pejorative manifold. We then algorithmically match the clustered roots in an attempt to find agcd$^\sigma_\rho$ where $\rho$ is the *root distance semi-metric*. We believe that this will give a reasonable solution in the Bernstein coefficient metric; in future work we hope to present analytical results connecting the two.

## References

1. Beckermann, B., Labahn, G.: When are two numerical polynomials relatively prime? J. Symb. Comput. **26**(6), 677–689 (1998)
2. Beckermann, B., Labahn, G., Matos, A.C.: On rational functions without Froissart doublets. Numerische Mathematik **138**(3), 615–633 (2018)
3. Bini, D.A., Marco, A.: Computing curve intersection by means of simultaneous iterations. Numer. Algorithms **43**(2), 151–175 (2006)
4. Bondy, J.A., Murty, U.S.R.: Graph Theory (Graduate Texts in Mathematics 244). Springer, New York (2008)
5. Botting, B., Giesbrecht, M., May, J.: Using Riemannian SVD for problems in approximate algebra. In: Proceedings of the International Workshop of Symbolic-Numeric Computation, pp. 209–219 (2005)
6. Corless, R.M., Fillion, N.: A Graduate Introduction to Numerical Methods: From the Viewpoint of Backward Error Analysis. Springer, New York (2013). https://doi.org/10.1007/978-1-4614-8453-0
7. Corless, R.M., Gianni, P.M., Trager, B.M., Watt, S.M.: The singular value decomposition for polynomial systems. In: Proceedings of the 1995 International Symposium on Symbolic and Algebraic Computation, ISSAC 1995, pp. 195–207. ACM, New York (1995)
8. Farouki, R.T., Goodman, T.N.T.: On the optimal stability of the Bernstein basis. Math. Comput. **65**, 1553–1566 (1996)
9. Farouki, R.T., Rajan, V.T.: On the numerical condition of polynomials in Bernstein form. Comput. Aided Geom. Des. **4**(3), 191–216 (1987)
10. Farouki, R.T., Rajan, V.T.: Algorithms for polynomials in Bernstein form. Comput. Aided Geom. Des. **5**(1), 1–26 (1988)

11. Hitz, M.A., Kaltofen, E., Flaherty, J.E.: Efficient algorithms for computing the nearest polynomial with constrained roots. In: Proceedings of the 1998 International Symposium on Symbolic and Algebraic Computation, ISSAC 1998, pp. 236–243. ACM (1998)

12. Hitz, M.A., Kaltofen, E., Lakshman, Y.N.: Efficient algorithms for computing the nearest polynomial with a real root and related problems. In: Proceedings of the 1999 International Symposium on Symbolic and Algebraic Computation, ISSAC 1999, pp. 205–212. ACM (1999)

13. Hopcroft, J.E., Karp, R.M.: An $n^{5/2}$ algorithm for maximum matching in bipartite graphs. SIAM J. Comput. **2**, 225–231 (1973)

14. Jónsson, G.F.: Eigenvalue methods for accurate solution of polynomial equations. Ph.D. dissertation, Center for Applied Mathematics, Cornell University, Ithaca, NY (2001)

15. Jónsson, G.F., Vavasis, S.: Solving polynomials with small leading coefficients. SIAM J. Matrix Anal. Appl. **26**(2), 400–414 (2004)

16. Kaltofen, E., Yang, Z., Zhi, L.: Structured low rank approximation of a Sylvester matrix. In: Wang, D., Zhi, L. (eds.) Symbolic-Numeric Computation. TM, pp. 69–83. Springer, Heidelberg (2007). https://doi.org/10.1007/978-3-7643-7984-1_5

17. Karmarkar, N.K., Lakshman, Y.N.: Approximate polynomial greatest common divisors and nearest singular polynomials. In: Proceedings of the 1996 International Symposium on Symbolic and Algebraic Computation, ISSAC 1996, pp. 35–39. ACM, New York (1996)

18. Karmarkar, N.K., Lakshman, Y.N.: On approximate gcds of univariate polynomials. J. Symb. Comput. **26**(6), 653–666 (1998)

19. Mackey, D.S., Perović, V.: Linearizations of matrix polynomials in Bernstein bases. Linear Algebra Appl. **501**, 162–197 (2016)

20. Nakatsukasa, Y., Sàte, O., Trefethen, L.: The AAA algorithm for rational approximation. SIAM J. Sci. Comput. **40**(3), A1494–A1522 (2018)

21. Pan, V.Y.: Numerical computation of a polynomial gcd and extensions. Inf. Comput. **167**, 71–85 (2001)

22. Rezvani, N., Corless, R.M.: The nearest polynomial with a given zero, revisited. ACM SIGSAM Bull. **39**(3), 73–79 (2005)

23. Stetter, H.J.: The nearest polynomial with a given zero, and similar problems. ACM SIGSAM Bull. **33**(4), 2–4 (1999)

24. West, D.B.: Introduction to Graph Theory. Prentice Hall Inc., Upper Saddle River (1996)

25. Winkler, J.R., Yang, N.: Resultant matrices and the computation of the degree of an approximate greatest common divisor of two inexact Bernstein basis polynomials. Comput. Aided Geom. Des. **30**(4), 410–429 (2013)

26. Zeng, Z.: The numerical greatest common divisor of univariate polynomials. In: Randomization, Relaxation, and Complexity in Polynomial Equation Solving, vol. 556, pp. 187–217 (2011)

27. Zeng, Z.: The approximate GCD of inexact polynomials. Part I: a univariate algorithm. In: Proceedings of the 2004 International Symposium on Symbolic and Algebraic Computation, ISSAC 2004, pp. 320–327. ACM (2004)

28. Cormen, T.H., Leiserson, C.E., Rivest, R.L., Stein, C.: The Knuth-Morris-Pratt Algorithm. In: Introduction to Algorithms, 2nd edn., pp. 923–932 (2001). Chap. 32.4

29. Amiraslani, A., Corless, R.M., Gunasingam, M.: Differentiation matrices for univariate polynomials. Numer. Algorithms **83**, 1–31 (2018)

# Using Maple to Compute the Intersection Curve of Two Quadrics: Improving the `Intersectplot` Command

Laureano Gonzalez-Vega[1]([⊠]) [ID] and Alexandre Trocado[2] [ID]

[1] Universidad de Cantabria, Santander, Spain
`laureano.gonzalez@unican.es`
[2] Universidade Aberta, Lisbon, Portugal
`mail@alexandretrocado.com`

**Abstract.** The Maple `intersectplot` command plots the intersection curve in three-dimensional space between a pair of two-dimensional surfaces. We will present the implementation in Maple of a new algorithm computing the intersection curve between two quadrics in 3D that improves the results produced by the `intersectplot` command.

**Keywords:** Quadrics · Intersection curve · Resultant · Maple

## 1 Introduction

We present here the implementation in Maple of a new method to determine the intersection curve of two quadrics through projection onto a plane and lifting. In some cases, it will be possible to determine the exact parameterisation of the intersection curve (involving radicals if needed) and, in others, the output (topologically correct) will be presented as the lifting of the discretisation of the branches of the projection curve once its singular points have been fully determined. The way the lifting will be made is the main criteria followed to analyse the cutcurve, the projection of the intersection curve to be computed.

The introduction of algorithms for computing the intersection of two quadrics dates back to the late 1970s. The representation and the definition of quadrics' intersection has been a relevant problem to solve over the last decades. Levin in 1976 and 1979 [6,7] introduced a method failing when the intersection curve is singular and even generates results that are not topologically correct. Levin's method was improved by Wang et al. [12] making it capable of computing geometric and structural information; besides, Dupont et al. (see [2]) succeeded in finding parameterizations that overcame the fact that Levin's method generated formulas that were not suited for further symbolic processing. On the other hand,

First author is partially supported by the Spanish Ministerio de Economía y Competitividad and by the European Regional Development Fund (ERDF), under the project MTM2017-88796-P.

ⓒ Springer Nature Switzerland AG 2020
J. Gerhard and I. Kotsireas (Eds.): MC 2019, CCIS 1125, pp. 92–100, 2020.
https://doi.org/10.1007/978-3-030-41258-6_7

Mourrain et al. [8] studied a sweeping algorithm for computing the arrangement of a set of quadrics in $\mathbb{R}^3$ that reduces the intersection of two quadrics to a dynamic two-dimensional problem. Dupont et al. [3–5] proposed algorithms that enable to compute in practice an exact form of the parameterization of two arbitrary quadrics with rational coefficients. These algorithms represented a substantial improvement over Levin's pencil method and its subsequent refinements. A different approach is based on the analysis of the projection of the intersection curve by using resultants [9, 10]. It is the algorithm in Trocado and Gonzalez-Vega [10] the one we have used to analyse the behaviour of the Maple `intersectplot` command plotting the intersection curve in three-dimensional space of a pair of two-dimensional surfaces when these two surfaces are quadrics.

This paper is organised as follows: Sect. 2 describes briefly the algorithm implemented in Maple, Sect. 3 compares this implementation with the Maple `intersectplot` command and the last section draws several conclusions.

## 2   The Algorithm

We introduce here a brief presentation of the algorithm in [10] that we have implemented in Maple in order to analyse the efficiency and accuracy of the Maple `intersectplot` command plotting the intersections in three-dimensional space of a pair of two-dimensional surfaces when these two surfaces are quadrics. The algorithm in [10] performs a very efficient Cylindrical Algebraic Decomposition [1] for the particular case of two quadrics since, in this case, the projection of the intersection curve has several properties making its analysis easier than expected.

Let $f$ and $g$ be the two polynomials in $\mathbb{R}[x, y, z]$

$$f(x, y, z) = z^2 + p_1(x, y)z + p_0(x, y) \qquad g(x, y, z) = z^2 + q_1(x, y)z + q_0(x, y)$$

with $\deg(p_1) \leq 1$, $\deg(p_0) \leq 2$, $\deg(q_1) \leq 1$ and $\deg(q_0) \leq 2$.

Let $\Delta_{\mathcal{E}_1}(x, y) = p_1(x, y)^2 - 4p_0(x, y)$ and $\Delta_{\mathcal{E}_2}(x, y) = q_1(x, y)^2 - 4q_0(x, y)$ be the discriminants of $f(x, y, z)$ and $g(x, y, z)$ (respectively) with respect to $z$. Let $\mathbf{S}_0(x, y)$ the resultant of $f$ and $g$, with respect to $z$.

Computing the intersection of the two quadrics defined by $f$ and $g$ is equivalent to solving in $\mathbb{R}^3$ the polynomial system of equations

$$f(x, y, z) = 0, \qquad g(x, y, z) = 0.$$

The solution set to be computed, when non empty, may include curves and isolated points. Analyzing $\mathbf{S}_0(x, y) = 0$ in $\mathbb{R}^2$ will be called the projection step and moving the information obtained in $\mathbb{R}^2$ to $\mathbb{R}^3$ will be called the lifting step.

The curve in $\mathbb{R}^2$ defined by $\mathbf{S}_0(x, y) = 0$ is called the cutcurve of $\mathcal{E}_1$ and $\mathcal{E}_2$ and the curve in $\mathbb{R}^2$ defined by $\Delta_{\mathcal{E}_i}(x, y) = 0$ the silhouette of $\mathcal{E}_i$.

Let $\mathcal{E}_1$ and $\mathcal{E}_2$ be two quadrics in $\mathbb{R}^3$ defined by $f(x, y, z) = 0$ and $g(x, y, z) = 0$ respectively. The cutcurve of $\mathcal{E}_1$ and $\mathcal{E}_2$ is the set

$$\left\{ (x, y) \in \mathbb{R}^2 : \mathbf{S}_0(x, y) = 0, \Delta_{\mathcal{E}_1}(x, y) \geq 0, \Delta_{\mathcal{E}_2}(x, y) \geq 0 \right\}.$$

The main ingredients of this approach are a detailed analysis of the cutcurve, its singular points and of its relation with the silhouette curves together with the using of an uniform way to perform the lifting of the cutcurve to the intersection curve of the two considered quadrics.

Concerning the analysis of the cutcurve we classify its singular points in two different types depending on how they will be lifted. Those belonging to the line $p_1(x, y) = q_1(x, y)$ are easy to compute and difficult to lift (but just solving a degree two equation) and those not in that line are more complicated to be determined but easier to lift.

This approach is not intended to classify the intersection curve between the two considered quadrics. Its main goal is to produce in a very direct way a description of the intersection curve which is topologically correct. This is the reason why we allow in the lifting of the cutcurve, when possible, the use of radicals or we rely on the discretisation of the branches of the cutcurve (easily to determine by knowing the points computed in that curve). Next example shows one case where the intersection curve is presented by a parameterisation involving radicals.

*Example 1.* Let $f$ and $g$ be the polynomials

$$f(x, y, z) = z^2 + xz + y \qquad g(x, y, z) = z^2 + yz + x$$

defining two hyperbolic paraboloids, $\mathcal{E}_1$ and $\mathcal{E}_2$, whose intersection curve is to be computed. To characterise the intersection curve of $\mathcal{E}_1$ and $\mathcal{E}_2$ we must analyse the cutcurve

$$\mathbf{S}_0(x, y) = (x - y)^2(x + y + 1) = 0$$

and its lifting. Lifting the line $x + y + 1 = 0$ is easy since these points are regular points of the cutcurve and it is enough to solve the equation $f - g = 0$ with respect to $z$ (which, in this case, is the subresultant of index 1 of $f$ and $g$ with respect to $z$). Lifting the line $x - y = 0$, requires to solve $f = 0$ or $g = 0$ with respect to $z$.

By using the following functions:

- For $x \in \mathbb{R}$, we define:

$$h_1(x) = x \qquad h_2(x) = -x - 1$$

- Let $e_1$ and $e_2$ be the functions defined by:

$$e_1(x, y) = -\frac{y}{2} + \frac{\sqrt{y^2 - 4x}}{2} \qquad e_2(x, y) = -\frac{y}{2} - \frac{\sqrt{y^2 - 4x}}{2}$$

the parameterisation of the intersection curve is given by the following three components:

- For $x \in \ ]-\infty, 0] \cup [4, +\infty[$: $(x, h_1(x), e_1(x, h_1(x)))$.
- For $x \in \ ]-\infty, 0] \cup [4, +\infty[$: $(x, h_1(x), e_2(x, h_1(x)))$.
- For $x \in \mathbb{R}$: $(x, h_2(x), 1)$ (the lifting the line $x + y + 1 = 0$).

In Fig. 1 it is shown, both, the cutcurve and the intersection curve of the two considered quadrics. It is worth to remark here that the cutcurve contains two half-lines and that the intersection curve shows two connected components and one self-intersection point.

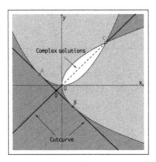 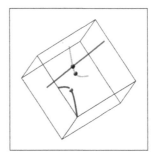

**Fig. 1.** (Left) The cutcurve contains two half-lines. (Right) Intersection curve of $f = 0$ and $g = 0$: in red the lifting of the singular points of the cutcurve (including one complete line) and in gray those points where the parameterisation equations change. (Color figure online)

## 3  The Implementation and the Comparison with the Intersectplot Command

The algorithm introduced in [10] and presented in the previous section has been implemented in Maple and its performance has been compared with the `intersectplot` command. We show next the advantages of our algorithm when compared with the `intersectplot` command.

First issue to mention is that `intersectplot` command works locally requiring to decide in advance the region where the intersection curve is to be computed. This means that some connected components of the intersection curve may be missing. Our algorithm does not miss any connected component of the intersection curve. Figure 2 shows how the `intersectplot` command needs to be applied several times in order to recover all the connected components of the intersection curve of the quadrics in Example 1 but with no guarantee that more connected components are missing (Fig. 1 shows the intersection curve as produced by the implementation of our algorithm).

Second issue to mention is that the `intersectplot` command may miss some points when the intersection curve is a finite set of isolated points. For example the quadrics defined by

$$4z^2 + 4x^2 - 4xy - 8xz + 3y^2 - 8y + 8z + 8 = 0$$

$$6z^2 + 4x^2 + 4xy - 8xz - 3y^2 - 16x + 8y + 16z + 8 = 0$$

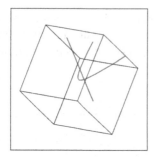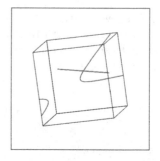

**Fig. 2.** Computing the intersection curve of $z^2 + xz + y = 0$ and $z^2 + xz + y = 0$ by using the `intersectplot` command. Left: in the box $[-3, 3]^3$. Right: in the box $[-5, 5]^3$.

only intersect at the points $(-1, -2/3, -2)$ and $(-1, 2, -2)$. Using the `intersectplot` command in the boxes $[-3, 3]^3$ and $[-1.5, -0.5] \times [-1, 2.5] \times [-2.5, -1.5]$ produces the empty plot in both cases while our implementation produces the only two points in the intersection curve.

Third issue to mention concerns the quality of the output produced by the `intersectplot` command when the box does not fit adequately with the intersection curve. Figure 3 shows the output produced by the `intersectplot` command when computing the intersection curve of the quadrics

$$p(x, y, z) = -16xy + 24xz + 4y^2 + 8yz - 16z^2 + 16x + 16y - 40z - 32 = 0$$

$$q(x, y, z) = -8xy + 4xz + 2y^2 + 8yz - 8z^2 - 8x + 16y - 20z - 16 = 0$$

considering different boxes. We can conclude here that, depending on the size of the box, the quality and accuracy of the output change drastically.

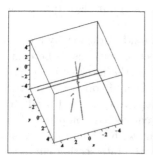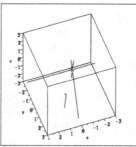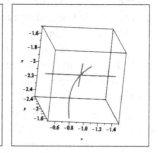

**Fig. 3.** Computing the intersection curve of $p(x, y, z) = 0$ and $q(x, y, z) = 0$ by using the `intersectplot` command in three different boxes.

The output of our Maple implementation when computing the intersection curve of $p(x, y, z) = 0$ and $q(x, y, z) = 0$ in shown in Fig. 4. Procedure `par` tries to compute a closed form for some of the components of the intersection curve and, when involving radicals, determines the those intervals where such a description can be evaluated. Procedure sing computes three lists of points: the first one contains, if any, the tangential intersection points, the second one the lifting to the intersection curve of the singular points of the cutcurve, if any, and the third one a discretisation for the components of the intersection curve without a parameterisation in closed form available.

```
par(-16*x*y+24*x*z+4*y^2+8*y*z-16*z^2+16*x+16*y-40*z-32,-8*x*y+4*x*z+2*y^2+8*y*z-8*z^2-8*
x+16*y-20*z-16)
                        [[[x,4x+2,-2],[x,-2,-2]]]
sing(-16*x*y+24*x*z+4*y^2+8*y*z-16*z^2+16*x+16*y-40*z-32,-8*x*y+4*x*z+2*y^2+8*y*z-8*z^2-8*
x+16*y-20*z-16)
```

$$\left[ [\,], \left[ \left[ x, 2x, \frac{5x}{4} - \frac{5}{4} + \frac{\sqrt{9x^2 - 2x - 7}}{4} \right], \left[ x, 2x, \frac{5x}{4} - \frac{5}{4} - \frac{\sqrt{9x^2 - 2x - 7}}{4} \right], \left[ \left[ -\frac{16}{9}, -\frac{7}{9} \right], [1,2] \right] \right], [\,] \right]$$

**Fig. 4.** Maple output when computing the intersection curve of $p(x, y, z) = 0$ and $q(x, y, z) = 0$ by using our Maple implementation.

Figure 5 plots the intersection curve of $p(x, y, z) = 0$ and $q(x, y, z) = 0$ by using as input the information presented in Fig. 4 together with two different views placing the intersection curve onto the two considered quadrics.

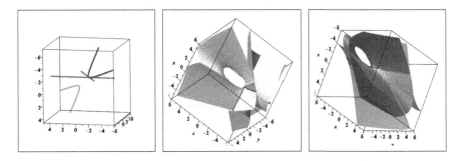

**Fig. 5.** The intersection curve of $p(x, y, z) = 0$ and $q(x, y, z) = 0$ by using our Maple implementation together with the two quadrics (two views).

Finally, Fig. 6 presents the output of our Maple implementation when computing the intersection curve for three pairs of quadrics. Last example shows a concrete case where the intersection curve is discretised.

Figure 7 shows, for two concrete examples, the shape of the intersection curve when some (or all) of the components of the cutcurve have been discretised.

```
par((-16*x^2+8*x*z+8*y^2-8*z^2+48*x-32*y-24*z)*(-1/8), (-12*x*z+4*y^2+2*z^2+8*x-16*y-4*z)*(1/2));
```

$$\left[\left[\left[x, 2+2\sqrt{x^2+1}, 2x\right], \left[x, 2-2\sqrt{x^2+1}, 2x\right], \left[x, 2+\frac{2\sqrt{4x^2-10x+4}}{3}, \frac{2x}{3}-\frac{4}{3}\right], \left[x, 2-\frac{2\sqrt{4x^2-10x+4}}{3}, \frac{2x}{3}-\frac{4}{3}\right], \left[\left[-\frac{1}{2}, \frac{1}{2}\right], [2,3]\right]\right]\right]$$

```
sing((-16*x^2+8*x*z+8*y^2-8*z^2+48*x-32*y-24*z)*(-1/8), (-12*x*z+4*y^2+2*z^2+8*x-16*y-4*z)*(1/2));
```

$$\left[\left[\;\right], \left[\left[-1, 2-2\sqrt{2}, -2\right], \left[-1, 2+2\sqrt{2}, -2\right], [\;]\right]\right]$$

```
> sing(32*x^2-48*x*y+64*x*z+14*y^2-34*z^2+32*x-8*y-72*z-16, 48*x^2-48*x*y+24*x*z+9*y^2-9*z^2-16*x+12*y-12*z)
```

$$\left[\left[[0.3819660113, 0.3519093630, -0.1573786518], [0.6666666667, 2.000000000, -0.6666666672], [1.500000000, 2.000000000, -6.629999999\,10^{-10}],\right.\right.$$

$$\left.[2.618033989, 6.314757311, 2.824045318]\right], \left[\left[-1, -\frac{14}{3}+\frac{4\sqrt{5}}{3}, -\frac{10}{3}+\frac{4\sqrt{5}}{3}\right], \left[-1, -\frac{14}{3}+\frac{4\sqrt{5}}{3}, -\frac{2}{3}-\frac{4\sqrt{5}}{3}\right], \left[-1, -\frac{14}{3}-\frac{4\sqrt{5}}{3}, -\frac{2}{3}\right.\right.$$

$$\left.\left.+\frac{4\sqrt{5}}{3}\right], \left[-1, -\frac{14}{3}-\frac{4\sqrt{5}}{3}, -\frac{10}{3}-\frac{4\sqrt{5}}{3}\right]\right], [\;]\right]$$

```
> par((32*x^2-48*x*y+64*x*z+14*y^2-34*z^2+32*x-8*y-72*z-16, 48*x^2-48*x*y+24*x*z+9*y^2-9*z^2-16*x+12*y-12*z)
```

$$\left[\left[\left[x, 4x-\frac{2}{3}+\frac{4\sqrt{5}x}{5}-\frac{8\sqrt{5}}{15}-\frac{2}{3}+\frac{(-12x+8)\sqrt{5}}{15}\right], \left[x, 4x-\frac{2}{3}-\frac{4\sqrt{5}x}{5}+\frac{8\sqrt{5}}{15}, -\frac{2}{3}+\frac{(12x-8)\sqrt{5}}{15}\right], \left[x, \frac{8x}{3}-2+\frac{8\sqrt{5}x}{15}-\frac{4\sqrt{5}}{5},\right.\right.\right.$$

$$\left.\left.\left.\frac{(8x-12)\sqrt{5}}{15}+\frac{4x}{3}-2\right], \left[x, \frac{8x}{3}-2-\frac{8\sqrt{5}x}{15}+\frac{4\sqrt{5}}{5}, \frac{(-8x+12)\sqrt{5}}{15}+\frac{4x}{3}-2\right]\right]\right]$$

```
> par(48*x^2-96*x*y+8*x*z+44*y^2-4*y*z+4*z^2+80*x-72*y+16*z+48, -24*x^2+52*x*z+2*y^2-10*y*z-18*z^2+40*x-4*
y-40*z-24)
```

$$[[\;]]$$

```
> sing(48*x^2-96*x*y+8*x*z+44*y^2-4*y*z+4*z^2+80*x-72*y+16*z+48, -24*x^2+52*x*z+2*y^2-10*y*z-18*z^2+40*
x-4*y-40*z-24)
```

$$[[\;], [\;], [[-4.344526811, -4.048555471, -2.388763429], [-4.344526811, -3.982907043, -2.388004965], [-4.297136261, -4.008781568,$$
$$-2.368323547], [-4.297136261, -3.918472459, -2.367251752], [-4.249745710, -3.965228355, -2.347843176], [-4.249745710, -3.857809940,$$
$$-2.346534136], [-4.202355159, -3.919538652, -2.327339500], [-4.202355159, -3.799277195, -2.325835208], [-4.154964609, -3.872346607,$$
$$-2.306818594], [-4.154964609, -3.742239614, -2.305148742], [-4.107574058, -3.823974265, -2.286283225], [-4.107574058, -3.686374787,$$
$$-2.284471828], [-4.060183507, -3.774615326, -2.265735067], [-4.060183507, -3.631489666, -2.263803136], [-4.012792957, -3.724381755,$$
$$-2.245174690], [-4.012792957, -3.577471891, -2.243142028], [-3.965402406, -3.673342909, -2.224602254], [-3.965402406, -3.524252141,$$
$$-2.222488401], [-3.918011856, -3.621533607, -2.204017390], [-3.918011856, -3.471795200, -2.201842483], [-3.870621305, -3.568968925,$$
$$-2.183419630], [-3.870621305, -3.420086052, -2.181204924], [-3.823230754, -3.515641398, -2.162808175], [-3.823230754, -3.369132566,$$
$$-2.160576723], [-3.775840204, -3.461512538, -2.142181529], [-3.775840204, -3.318972671, -2.139959226], [-3.728449653, -3.406517395,$$
$$-2.121537836], [-3.728449653, -3.269671349, -2.119354412], [-3.681059102, -3.350549380, -2.100874519], [-3.681059102, -3.221335421,$$
$$-2.098765127], [-3.633668552, -3.293422865, -2.080187297], [-3.633668552, -3.174150194, -2.078195539], [-3.586278001, -3.234820415,$$
$$-2.059469443], [-3.586278001, -3.128433026, -2.057652492], [-3.538887450, -3.174121845, -2.038708540], [-3.538887450, -3.084804478,$$
$$-2.037148758], [-3.491496900, -3.109712611, -2.017873131], [-3.491496900, -3.044878505, -2.016715622], [-3.444106349, -3.025333350,$$
$$-1.996662147], [-3.444106349, -3.024914767, -1.996654509]]]$$

**Fig. 6.** Three examples of using Maple to compute the intersection curve of two quadrics.

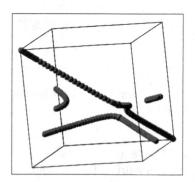 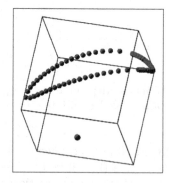

**Fig. 7.** Plotting the intersection curve with our Maple implementation when the components are discretised. Red points: lifting of the singular points of the cutcurve. (Color figure online)

# 4  Conclusions

We have presented our Maple implementation of the algorithm in [10] for computing the intersection curve of two quadrics. Compared with the Maple `intersectplot` command it brings, for this particular case, several advantages concerning the accuracy of the result. Its output brings more information since the cutcurve is completely characterised allowing for example to identify the tangential intersection points among many other things. Regarding efficiency our algorithm is slower than the Maple `intersectplot` command but in most cases our implementation requires less than one second when computing the intersection curve of the two considered quadrics. It is worth to remark here that there are lot of potential improvements in the implementation we have reported here.

We think that the Maple `intersectplot` command should improve its accuracy by adding several particular cases where the intersection curve is determined by an ad-hoc algorithm and the case of two quadrics is one of these possibilities.

In [11] we have reported our implementation in GeoGebra of the algorithm in [10]. Compared with the implementation in Maple reported here it is worth to mention that Maple is clearly more adequate that GeoGebra thanks to how easy is to perform the symbolic and numerical analysis of the cutcurve which is the cornerstone of the algorithm we are dealing with. The only point where GeoGebra outperforms Maple concerns the lifting of the branches of the cutcurve around a tangential intersection point: the determination of these points is quite complicated but they are very easy to lift in GeoGebra without computing them explicitly (in Maple this is needed): by continuity the lifting of the points produced by the `Locus` GeoGebra function around these singular points produces automatically their lifting to the intersection curve (and it is not necessary to compute them explicitly).

# References

1. Basu, S., Pollack, R., Roy, M.-F.: Algorithms in Real Algebraic Geometry. Springer, Heidelberg (2006). https://doi.org/10.1007/3-540-33099-2
2. Dupont, L., Lazard, S., Lazard, D., Petitjean, S.: Near-optimal parameterization of the intersection of quadrics. In: Proceedings of the Annual Symposium on Computational Geometry (2003). https://doi.org/10.1145/777829.777830
3. Dupont, L., Lazard, D., Lazard, S., Petitjean, S.: Near-optimal parameterization of the intersection of quadrics: I. The generic algorithm. J. Symb. Comput. (2008). https://doi.org/10.1016/j.jsc.2007.10.006
4. Dupont, L., Lazard, D., Lazard, S., Petitjean, S.: Near-optimal parameterization of the intersection of quadrics: II. A classification of pencils. J. Symb. Comput. (2008). https://doi.org/10.1016/j.jsc.2007.10.012
5. Dupont, L., Lazard, D., Lazard, S., Petitjean, S.: Near-optimal parameterization of the intersection of quadrics: III. Parameterizing singular intersections. J. Symb. Comput. (2008). https://doi.org/10.1016/j.jsc.2007.10.007
6. Levin, J.: A parametric algorithm for drawing pictures of solid objects composed of quadric surfaces. Commun. ACM (1976). https://doi.org/10.1145/360349.360355

7. Levin, J.Z.: Mathematical models for determining the intersections of quadric surfaces. Comput. Graph. Image Process. (1979). https://doi.org/10.1016/0146-664X(79)90077-7

8. Mourrain, B., Técourt, J.P., Teillaud, M.: On the computation of an arrangement of quadrics in 3D. Comput. Geom. Theory Appl. (2005). https://doi.org/10.1016/j.comgeo.2004.05.003

9. Schömer, E., Wolpert, N.: An exact and efficient approach for computing a cell in an arrangement of quadrics. Comput. Geom. Theory Appl. (2006). https://doi.org/10.1016/j.comgeo.2004.02.007

10. Trocado, A., Gonzalez-Vega, L.: On the intersection of two quadrics (2018, submitted). arXiv:1903.06983v2

11. Trocado, A., Gonzalez-Vega, L., Dos Santos, J.M.: Intersecting two quadrics with GeoGebra. In: Ćirić, M., Droste, M., Pin, J.É. (eds.) CAI 2019. LNCS, vol. 11545, pp. 237–248. Springer, Cham (2019). https://doi.org/10.1007/978-3-030-21363-3_20

12. Wang, W., Goldman, R., Tu, C.: Enhancing Levin's method for computing quadric-surface intersections. Comput. Aided Geom. Des. (2003). https://doi.org/10.1016/S0167-8396(03)00081-5

# Exact Parametric Solutions for the Intersections of Quadric Surfaces Using MAPLE

Samir Hamdi[1,2(✉)], David I. W. Levin[1], and Brian Morse[3]

[1] Department of Computer Science, University of Toronto,
Toronto, ON M5S 3G4, Canada
`samir.hamdi@utoronto.ca`
[2] Department of Mathematical and Computational Sciences,
University of Toronto Mississauga, Mississauga, ON L5L 1C6, Canada
[3] Department of Civil and Water Engineering, Laval University,
Quebec G1V 0A6, Canada

**Abstract.** Quadric surfaces play a very important role in solid geometric modeling and in the design and fabrication of mechanical and industrial parts. Solving the intersection curve between two quadrics is a fundamental problem in computer graphics and solid modeling. We present a new analytical method for parameterizing the intersection curve of two quadrics, which are represented by implicit quadratic equations in 3D. The method is based on the observation that the intersection curve of two quadrics comprises all the points that satisfy a parametric second order polynomial system. We show that the computation of the intersection problem of two general quadrics can be reduced to the solution of quartic polynomials. In particular, we show that the intersection problem of two quadric surfaces that are expressed in canonical forms can be reduced to the solution of quadratic polynomials. All the exact parametric solutions for the intersections of quadric surfaces are implemented in the Computer Algebra System MAPLE. Several previously published test problems of the intersection of quadric surfaces are presented and discussed.

**Keywords:** Quadric surface · Intersection · Parametric polynomial system · Quartic polynomial · Exact solution · Computer Algebra System

## 1 Introduction

Computing the intersection curve of two surfaces is a fundamental problem in solid modeling and computer graphics. Solving the intersection problem is essential for computing convex hulls of patches, the representation of complex objects, computer animation and numerical control machining for trimming off the region bounded by the self-intersection curves of offset surfaces [15]. It is also useful for Boolean operations in constructive solid modeling [15].

© Springer Nature Switzerland AG 2020
J. Gerhard and I. Kotsireas (Eds.): MC 2019, CCIS 1125, pp. 101–113, 2020.
https://doi.org/10.1007/978-3-030-41258-6_8

Quadric surfaces play a very important role in solid geometric modeling, the design and fabrication of mechanical and industrial parts. Patches of quadrics (planes, cones, spheres, and cylinders) and tori represent 95% of all mechanical and industrial parts [8,15].

Parametric solutions for the intersections of quadric surfaces were pioneered by Levin [9,10] and further developed by Sarraga [16]. An enhanced Levin's method for computing quadric-surface intersections was presented in [18,19]. Near-optimal parametrization of the intersection of quadrics was studied by Dupont et al. [2–4]. Efficient and exact implementations for determining the intersection of quadrics were developed by [8,17]. A sweeping algorithm for computing the arrangement of a set of quadrics in 3D was studied by Mourrain et al. [13]. Fioravanti et al. [5] introduced a new algebraic approach for computing the intersection of two ruled surfaces in implicit/parametric form by finding the zero set of a bivariate equation which represents the parameter values of the intersection curve, as a subset of the other surface.

In this study, we present a new and simple analytical method for parameterizing the intersection curve of two quadrics, which are represented by implicit quadratic equations in 3D.

The method is based on the observation that the intersection curve of two quadrics comprises all the points that satisfy a parametric second order polynomial system. In particular, we show that the intersection problem of two quadric surfaces that are expressed in canonical forms can be reduced to the solution of quadratic polynomials.

Several previously published test problems of the intersection of quadric surfaces such as spheres, ellipsoids, cylinders, cones and hyperboloids are presented and discussed. All the exact parametric solutions for the intersections of quadric surfaces are implemented in the Computer Algebra System MAPLE.

## 2   Problem Statement of the Intersection of Quadric Surfaces

We consider the intersection of two quadratic surfaces specified implicitly in general form as

$$\begin{aligned}
\mathcal{S}_1 : F(\mathbf{X}) &\equiv a_1 x^2 + b_1 y^2 + c_1 z^2 + 2f_1 yz + 2g_1 xz + 2h_1 xy \\
&\quad + 2p_1 x + 2q_1 y + 2r_1 z + d_1 = 0
\end{aligned} \tag{1}$$

and

$$\begin{aligned}
\mathcal{S}_2 : G(\mathbf{X}) &\equiv a_2 x^2 + b_2 y^2 + c_2 z^2 + 2f_2 yz + 2g_2 xz + 2h_2 xy \\
&\quad + 2p_2 x + 2q_2 y + 2r_2 z + d_2 = 0
\end{aligned} \tag{2}$$

The intersection curve of the two quadrics $\mathcal{S}_1$ and $\mathcal{S}_2$ comprises all the points that satisfy both Eqs. (1) and (2). If we consider one of the independent variables $x$, $y$ or $z$ as a parameter, the intersection curve can be determined by solving the parametric polynomial system (1) and (2). It is a quadratic system, which can be solved analytically to obtain a parametric representation of the intersection curve.

## 3   Exact Solutions for the Intersection of Quadric Surfaces in General Forms

In this section, we will present a simple elimination method for finding exact real roots of the quadratic system (1) and (2) in the case of general quadrics. If we consider one of the independent variables $x$, $y$ or $z$ as a parameter, the intersection curve can be determined by solving the resulting parametric polynomial system [6]. The choice of the parameter is arbitrary, but in practice, we should consider a parametrisation that leads to a simple solution. For simplicity and without loss of generality, we set the coefficients of $y^2$ equal to unity ($b_1 = b_2 = 1$), such that

$$S_1 : F(\mathbf{X}) \equiv a_1x^2 + y^2 + c_1z^2 + 2f_1yz + 2g_1xz + 2h_1xy \\ + 2p_1x + 2q_1y + 2r_1z + d_1 = 0 \tag{3}$$

and

$$S_2 : G(\mathbf{X}) \equiv a_2x^2 + y^2 + c_2z^2 + 2f_2yz + 2g_2xz + 2h_2xy \\ + 2p_2x + 2q_2y + 2r_2z + d_2 = 0 \tag{4}$$

which are quadratic polynomials for the unknowns $x$ and $y$ and parameter $z$.

First, we solve for $y^2$ using Eq. (3)

$$y^2 = - a_1x^2 - c_1z^2 - 2f_1yz - 2g_1xz - 2h_1xy \\ - 2p_1x - 2q_1y - 2r_1z - d_1 \tag{5}$$

substituting the expression of $y^2$ in (4), we obtain an equation that is first order in $y$.

$$-2(f_{12}z + h_{12}x + q_{12})\,y = (a_{12}x^2 + c_{12}z^2 + 2g_{12}xz + 2p_{12}x + 2r_{12}z + d_{12}) \tag{6}$$

if $(f_{12}z + h_{12}x + q_{12}) = 0$, we obtain

$$a_{12}x^2 + (2g_{12}z + 2p_{12})x + c_{12}z^2 + 2r_{12}z + d_{12} = 0 \tag{7}$$

which is a quadratic equation that can be easily solved for the unknown $x(z)$ as a function of the parameter $z$.

If $(f_{12}z + h_{12}x + q_{12}) \neq 0$, it follows that

$$y = \frac{-(a_{12}x^2 + c_{12}z^2 + 2g_{12}xz + 2p_{12}x + 2r_{12}z + d_{12})}{2(f_{12}z + h_{12}x + q_{12})} \tag{8}$$

in which

$$a_{12} = a_1 - a_2, \ c_{12} = c_1 - c_2, \ g_{12} = g_1 - g_2, \\ p_{12} = p_1 - p_2, \ r_{12} = r_1 - r_2, \ d_{12} = d_1 - d_2, \\ f_{12} = f_1 - f_2, \ h_{12} = h_1 - h_2, \ q_{12} = q_1 - q_2 \tag{9}$$

A back substitution of the expression (8) of $y$ in (4), and after clearing the denominators, leads to a quartic polynomial $Q(x)$ for the unknown $x$ that is parameterized by $z$.

$$Q(x) = ax^4 + bx^3 + cx^2 + dx + e = 0 \tag{10}$$

The coefficient $a$ is a constant. The coefficients $b$, $c$, $d$ and $e$ are polynomials that depend continuously on $z$, with increasing degrees ranging from one to four. The expression of the coefficient $a$ is explicitly given by

$$a = -a_{12}h_{12}h_2 + a_2 h_{12}^2 + \frac{1}{4}a_{12}^2 \tag{11}$$

The coefficient $b(z)$ is a first order polynomial in $z$. It is expressed as

$$b(z) = \beta_1 z + \beta_0 \tag{12}$$

where

$$\begin{aligned}\beta_1 = {}& 2g_2 h_{12}^2 + (-a_{12}f_2 + 2a_2 f_{12} - 2g_{12}h_2)h_{12} \\ & + a_{12}(-f_{12}h_2 + g_{12})\end{aligned} \tag{13}$$

and

$$\begin{aligned}\beta_0 = {}& 2p_2 h_{12}^2 + (-a_{12}q_2 + 2a_2 q_{12} - 2h_2 p_{12})h_{12} \\ & + a_{12}(-h_2 q_{12} + p_{12})\end{aligned} \tag{14}$$

The coefficient $c(z)$ is a second order polynomial in $z$. It is given by

$$c(z) = \gamma_2 z^2 + \gamma_1 z + \gamma_0 \tag{15}$$

where

$$\begin{aligned}\gamma_2 = {}& c_2 h_{12}^2 + (\frac{1}{2}(-2c_{12}h_2 + 8f_{12}g_2 - 4f_2 g_{12}))h_{12} \\ & + a_2 f_{12}^2 (\frac{1}{2}(-2a_{12}f_2 - 4g_{12}h_2))f_{12} + \frac{1}{2}a_{12}c_{12} + g_{12}^2\end{aligned} \tag{16}$$

$$\begin{aligned}\gamma_1 = {}& (\frac{1}{2}(8f_{12}p_2 - 4f_2 p_{12} - 4g_{12}q_2 + 8g_2 q_{12} - 4h_2 r_{12}))h_{12} \\ & + 2r_2 h_{12}^2 + (\frac{1}{2}(-2a_{12}q_2 + 4a_2 q_{12} - 4h_2 p_{12}))f_{12} \\ & + (\frac{1}{2}(-2a_{12}f_2 - 4g_{12}h_2))q_{12} + a_{12}r_{12} + 2g_{12}p_{12}\end{aligned} \tag{17}$$

$$\begin{aligned}\gamma_0 = {}& h_{12}^2 d_2 + (\frac{1}{2}(-2d_{12}h_2 - 4p_{12}q_2 + 8p_2 q_{12}))h_{12} + q_{12}^2 a_2 \\ & - (\frac{1}{2}(2a_{12}q_2 + 4h_2 p_{12}))q_{12} + \frac{1}{2}a_{12}d_{12} + p_{12}^2\end{aligned} \tag{18}$$

The coefficient $d(z)$ is a third order polynomial in $z$. It is given by

$$d(z) = \delta_3 z^3 + \delta_2 z^2 + \delta_1 z + \delta_0 \tag{19}$$

where

$$\begin{aligned}\delta_3 = {}& 2g_2 f_{12}^2 + (-c_{12}h_2 + 2c_2 h_{12} \\ & - 2f_2 g_{12})f_{12} + c_{12}(-f_2 h_{12} + g_{12})\end{aligned} \tag{20}$$

$$\begin{aligned}\delta_2 = {}& (-2f_2 p_{12} - 2g_{12}q_2 + 4g_2 q_{12} + 4h_{12}r_2 - 2h_2 r_{12})f_{12} \\ & + 2p_2 f_{12}^2 + (-c_{12}h_2 + 2c_2 h_{12} - 2f_2 g_{12})q_{12} \\ & - (c_{12}q_2 + 2f_2 r_{12})h_{12} + c_{12}p_{12} + 2g_{12}r_{12}\end{aligned} \tag{21}$$

$$\begin{aligned}
\delta_1 = {}& (-d_{12}h_2 + 2d_2h_{12} - 2p_{12}q_2 + 4p_2q_{12})f_{12} \\
& + 2q_{12}^2g_2 + (-2f_2p_{12} - 2g_{12}q_2 + 4h_{12}r_2 - 2h_2r_{12})q_{12} \\
& + (-d_{12}f_2 - 2q_2r_{12})h_{12} + d_{12}g_{12} + 2p_{12}r_{12}
\end{aligned} \tag{22}$$

$$\begin{aligned}
\delta_0 = {}& 2q_{12}^2p_2 + (-d_{12}h_2 + 2d_2h_{12} - 2p_{12}q_2)q_{12} \\
& - d_{12}(h_{12}q_2 - p_{12})
\end{aligned} \tag{23}$$

The coefficient $e(z)$ is a fourth order polynomial in $z$. It is given by

$$e(z) = \epsilon_4 z^4 + \epsilon_3 z^3 + \epsilon_2 z^2 + \epsilon_1 z + \epsilon_0 \tag{24}$$

where

$$\epsilon_4 = \frac{1}{4}(-4c_{12}f_{12}f_2 + 4c_2f_{12}^2 + c_{12}^2) \tag{25}$$

$$\begin{aligned}
\epsilon_3 = {}& \big[(2r_2f_{12}^2 + (2c_2q_{12} - c_{12}q_2 - 2f_2r_{12})f_{12} \\
& + c_{12}(r_{12} - f_2q_{12}))\big]
\end{aligned} \tag{26}$$

$$\begin{aligned}
\epsilon_2 = {}& \big[f_{12}^2d_2 + (4q_{12}r_2 - d_{12}f_2 - 2q_2r_{12})f_{12} \\
& + q_{12}^2c_2 + (r_{12}^2 - c_{12}q_2 - 2f_2r_{12})q_{12} + \frac{1}{2}c_{12}d_{12}\big]
\end{aligned} \tag{27}$$

$$\begin{aligned}
\epsilon_1 = {}& \big[(2d_2q_{12} - d_{12}q_2)f_{12} + 2r_2q_{12}^2 \\
& + d_{12}r_{12} - (d_{12}f_2 + 2q_2r_{12})q_{12}\big]
\end{aligned} \tag{28}$$

$$\epsilon_0 = \frac{1}{4}d_{12}^2 - d_{12}q_{12}q_2 + d_2q_{12}^2 \tag{29}$$

Once all the five coefficients $a$, $b$, $c$, $d$ and $e$ of the quartic polynomial $Q(x)$ are computed using the previous relations it is possible to compute, in closed–form, its real roots, which depend continuously on the parameter $z$. The classical formulae for the four exact roots of the quartic are expressed analytically in radicals as

$$x_{1,2} = -\frac{b}{4a} - S \pm \frac{1}{2}\sqrt{-4S^2 - 2p + \frac{q}{S}} \tag{30}$$

and

$$x_{2,4} = -\frac{b}{4a} + S \pm \frac{1}{2}\sqrt{-4S^2 - 2p - \frac{q}{S}} \tag{31}$$

where $p$ and $q$ are given by the rational expressions

$$p = \frac{8ac - 3b^2}{8a^2} \quad \text{and} \quad q = \frac{b^3 - 4abc + 8a^2d}{8a^3} \tag{32}$$

and where $S$ is given by

$$S = \frac{1}{2}\sqrt{-\frac{2}{3}p + \frac{1}{3a}\left(\frac{\Delta_2^2 + \Delta_0}{\Delta_2}\right)} \tag{33}$$

in which

$$\Delta_2 = \sqrt[3]{\frac{\Delta_1 + \sqrt{\Delta_1^2 - 4\Delta_0^3}}{2}} \tag{34}$$

with

$$\Delta_0 = c^2 - 3bd + 12ae \tag{35}$$

and

$$\Delta_1 = 2c^3 - 9bcd + 27b^2e + 27ad^2 - 72ace \tag{36}$$

Once the exact solution $x(z)$ of the quartic equation $Q(x)$ is computed using the previous analytical formulae (30) and (31), the exact solution for $y(z)$ is determined using a back substitution of $x(z)$ in the expression (8). Explicit classical quartic expressions such as (30) and (31) are already built in MAPLE and available in the solver solve.

The discriminant $\Delta$ of the quartic polynomial $Q(x)$ is given by

$$\Delta = \frac{4\Delta_0^3 - \Delta_1^2}{27} \tag{37}$$

The nature of the roots (real or complex) is mainly determined by the sign of the discriminant $\Delta$ as outlined by Rees and Lazard [7,14]. Only real values of $x(z)$ and $y(z)$ correspond to solution points of the intersection curve. Using MAPLE with regular chains library for polynomial rings [1,11,20–23], it is possible to determine the conditions for obtaining real solutions for the quartic as follows:

```
R := PolynomialRing([x, a, b, c, d, e]);
F := [a*x^4+b*x^3+c*x^2 + d*x + e];N := [];P := [];H := [];
rrc := RealRootClassification(F, N, P, H, 5, 1 .. k, R);
```

As a result of real root classification, it follows that the quartic equation (10) has given number of real solution(s) if and only if

$$\Delta > 0 \,\wedge\, D \geq 0 \,\wedge\, P \leq 0 \tag{38}$$

or

$$\Delta < 0 \,\wedge\, D \leq 0 \,\wedge\, P \geq 0 \tag{39}$$

or

$$\Delta < 0 \,\wedge\, P \leq 0 \tag{40}$$

where the polynomials $P$ and $D$ are defined by

$$P = 8ac - 3b^2 \tag{41}$$

and

$$D = 16a^2ce - 18a^2d^2 - 6ab^2e + 14abcd - 4ac^3 - 3b^3d + b^2c^2 \tag{42}$$

The discriminants $\Delta$, $D$ and $P$ are polynomials in $z$ of constant coefficients. The polynomial $\Delta$ is of order 12. The polynomial $D$ is a sextic polynomial in $z$ and the polynomial $P$ is a quadratic polynomial in $z$. Their signs can be easily determined from the locations of their real roots. Alternately, using the inequality solver `solve` in MAPLE, it is possible to find the range of values for $z$ for which the solutions $x(z)$, and $y(z)$ have real values.

The range of variation for the parameter $z$ is defined by the interval $I_z$, which is determined by identifying the intervals $I_\Delta$, $I_P$, and $I_D$ over which the sign of the discriminant $\Delta$ and the signs of $P$ and $D$ give rise to real points. The parametric curve is constructed piecewise by combining all the segments of the solutions.

$$I_z = I_\Delta \cap I_P \cap I_D \tag{43}$$

An easier and direct approach for determining the range of values for $z$ consists of solving real root classification problem for the system of polynomial equations $F$ and $G$ for the quadrics (1) and (2):

```
R := PolynomialRing([x, y, z]);
N := [];P := [];H := [];
rrc := RealRootClassification([F,G], N, P, H, 5, 1 .. k, R);
```

where the coefficients for the quadrics (1) and (2) have given numerical values. The resulting polynomial inequality in $z$ can be easily solved in MAPLE using the solver `solve` to determine the interval $I_z$, for which the solutions $x(z)$, and $y(z)$ have real values. More implementation details for real root classification for parametric polynomials with the regularchains library in MAPLE can be found in [1, 11, 20–23].

## 4    Exact Solutions for the Intersection of Quadric Surfaces in Canonical Forms

When the quadratic system (1) and (2) of the quadric surfaces is in canonical form, the elimination method for finding exact real roots of the system is greatly simplified. As in the previous section, we will consider one of the independent variables $x$, $y$ or $z$ as a parameter, so that the intersection curve can be determined by solving the resulting parametric polynomial system. We will consider $z$ as parameter, but we could also choose $x$ or $y$ for parameterizing the system. In practice, we should consider a parametrisation that leads to a simple solution of the system. As previously, in order to simplify the elimination procedure and without loss of generality, we will also set the coefficients of $y^2$ equal to unity $(b_1 = b_2 = 1)$, such that

$$S_1 : F(\mathbf{X}) \equiv a_1 x^2 + y^2 + c_1 z^2 + d_1 = 0 \tag{44}$$

and

$$S_2 : G(\mathbf{X}) \equiv a_2 x^2 + y^2 + c_2 z^2 + d_2 = 0 \tag{45}$$

which is a quadratic polynomial system that is parameterised by $z$ for the unknowns $x$ and $y$.

Similarly, as in the case of general quadrics, we first solve for $y^2$ using Eq. (44)

$$y^2 = -a_1 x^2 - c_1 z^2 - d_1 \tag{46}$$

substituting the expression of $y^2$ in (45), we obtain a parametric equation in $z$ that is second order for the unknown $x$.

$$(a_1 - a_2)x^2 + (c_1 - c_2)z^2 + d_1 - d_2 = 0 \tag{47}$$

The exact solution of this quadratic equation is given by

$$x_{1,2} = \pm \sqrt{\frac{(c_1 - c_2)z^2 + (d_1 - d_2)}{a_2 - a_1}} \tag{48}$$

The exact solution $y(z)$ is obtained using a back substitution of $x(z)$ in (46),

$$y_{1,2} = \pm \sqrt{\frac{a_2(c_1 z^2 + d_1) - a_1(c_2 z^2 + d_2)}{a_1 - a_2}} \tag{49}$$

Only real values of $x(z)$ and $y(z)$ will be considered as solution points of the intersection curve. The range of variation for the parameter $z$ is defined by the interval $I_z$, which is determined by identifying the intervals $I_x$ and $I_y$ for which the expressions (48) and (49) give rise to real solutions.

$$I_z = I_x \cap I_y \tag{50}$$

The parametric curve is constructed piecewise by combining all the segments of the solutions.

The interval $I_z$ is also determined by solving a real root classification problem for the system of polynomial equations $F$ and $G$ for the quadrics (44) and (45):

```
R := PolynomialRing([x, y, z, a1, c1, d1, a2, c2, d2]);
N := [];P := [];H := [];
rrc := RealRootClassification([F,G], N, P, H, 7, 1 .. k, R);
```

To obtain real solutions for $x(z)$ and $y(z)$, the parameter $z$ should satisfy the following conditions

$$R_1 < 0 \land R_2 > 0 \land R_3 > 0 \tag{51}$$

or

$$R_1 > 0 \land R_2 < 0 \land R_3 < 0 \tag{52}$$

where

$$R_1 = a_{12}, \; R_2 = c_{12}z^2 + d_{12} \text{ and } R_3 = (a_1 c_2 - a_2 c_1)z^2 + a_1 d_2 - a_2 d_1 \tag{53}$$

These systems of polynomial inequalities (51) and (52) can be easily solved for $z$ using solve in MAPLE.

## 5    Implementation and Examples

In this section, several illustrative examples will be presented to demonstrate the application of the analytical solutions for solving the intersection problem of two quadrics. These examples correspond to test problems that are taken from [8,12,18,19,24], for the intersection of quadric surfaces such as spheres, ellipsoids, cylinders, cones and hyperboloids. A plethora of additional test problems can be found at https://gamble.loria.fr/qi/server/, which allows the online computation of the intersection curve of two quadrics and also the comparison with our proposed MAPLE implementation.

We consider first the intersection of two ellipsoids defined by

$$\mathcal{S}_1 : F(\mathbf{X}) \equiv x^2 + y^2 + 3z^2 + 2yz - 2xz - 1 = 0 \qquad (54)$$

and

$$\mathcal{S}_2 : G(\mathbf{X}) \equiv x^2 + 12y^2 + 17z^2 + 24yz - 2xz + 2x - 2z - 3 = 0 \qquad (55)$$

First, we solve for $y^2$ using Eq. (3), and we substitute $y^2$ in (4),

```
Y2 := solve(F, y^2);
F2 := subs(y^2 = Y2, G);
```

Since $(f_{12}z + h_{12}x + q_{12}) = 0$, we use the quadratic equation (7)

```
Qx := F2;
solution_x := solve(Qx, x);    solution_y := solve(Y2-y^2, y);
```

Only real values of $x(z)$ and $y(z)$ will be considered as solution points of the intersection curve. The interval $I_z$ is computed by solving a real root classification problem for the system of polynomial equations $F$ and $G$ for the two ellipsoids

```
R := PolynomialRing([x, y, z]);
rrc := RealRootClassification([F,G], [], [], [], 1, 1 .. k, R);
R[1] := 9*z^2-8;
Iz := solve({R[1]<=0}, {z});
Intersection_Curve_Range := lhs(Iz[2])..rhs(Iz[1]);
```

The intersection curve for these two ellipsoids is represented in Fig. 1. The complete MAPLE implementation for solutions is provided in the next page.

The intersection of two very similar ellipsoids are represented in Fig. 2. The intersection of a sphere and an ellipsoid and the intersection of a sphere and a cone with a cusp are illustrated in Figs. 3 and 4 respectively.

The graphical representations of singular and nonsingular intersections of a sphere and a cylinder are given in Figs. 5 and 6 respectively.

Similar examples for the intersection of an elliptical cylinder and an ellipsoid and the intersection of an elliptical cylinder and a hyperboloid of one sheet are depicted in Figs. 7 and 8 respectively.

```
# Intersection of two ellipsoids (Figure 1)
with(plots): with(plots,intersectplot): with(RegularChains):
with(ParametricSystemTools): with(SemiAlgebraicSetTools):
with(SolveTools):
F := x^2-2*x*z+y^2+2*y*z+3*z^2-1;
G := x^2-2*x*z+12*y^2+24*y*z+17*z^2+2*x-2*z-3;
# Solving for y^2 using (5) and substituting y^2 in (4)
Y2 := solve(F,y^2);
F2 := subs(y^2=Y2,G);
# Since (f12#z + h12*x + q12)=0, we use quadratic equation (7)
Qx := F2;
solution_x := solve(Qx, x);
x1 := solution_x[1];    x2 := solution_x[2];
solution_y := solve(Y2-y^2, y):
y1_1 := simplify(subs(x=x1,solution_y[1]));
y1_2 := simplify(subs(x=x1,solution_y[2]));
y2_1 := simplify(subs(x=x2,solution_y[1]));
y2_2 := simplify(subs(x=x2,solution_y[2]));
R := PolynomialRing([x, y, z]);
infolevel[RegularChains] := 1;
rrc := RealRootClassification([F,G], [], [], [], 1, 1 .. k, R);
R[1] := 9*z^2-8;
Iz := solve({R[1]<=0}, {z});
Intersection_Curve_Range := lhs(Iz[2])..rhs(Iz[1]);
# Only solutions x2, y2_1 and y2_2 are real for the range Iz
Intersection_Curve := spacecurve({ [x2, y2_1,z], [x2, y2_2,z] },
z=Intersection_Curve_Range,color=red,thickness=3,transparency=0,
labels=["x","y","z"], labelfont = ["TimesNewRoman", 26] ,
axesfont = ["TimesNewRoman", 22]):

plot_ellipsoid1 := implicitplot3d( F =0, x=-1.5..1.5,y=-1.5..1.5,
z=-1.5..1.5 , grid=[50,50,50] ,axes=boxed, labels=["x","y","z"],
scaling=constrained, style=surface,color=blue,transparency=0.7,
lightmodel=light1,labels=["x","y","z"],
labelfont=["TimesNewRoman",26], axesfont=["TimesNewRoman", 22]):
plot_ellipsoid2 := implicitplot3d( G  =0, x=-3.5..2.0,y=-1.5..1.5,
z=-1.5..1.5, grid=[50,50,50], axes=boxed, labels=["x","y","z"],
scaling=constrained, style=surface,color=green,transparency=0.7,
lightmodel=light1,labels=["x","y","z"],
labelfont=["TimesNewRoman",26], axesfont=["TimesNewRoman", 22]):
plots[display](plot_ellipsoid1,plot_ellipsoid2,Intersection_Curve):
```

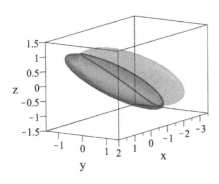

**Fig. 1.** Intersection of two ellipsoids

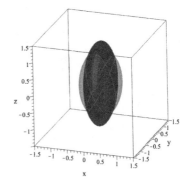

**Fig. 2.** Intersection of two very similar ellipsoids

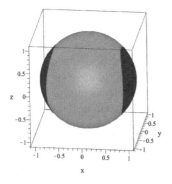

**Fig. 3.** Intersection of a sphere and an ellipsoid

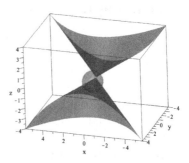

**Fig. 4.** Intersection of a sphere and a cone with a cusp

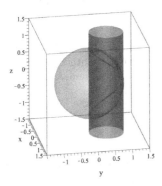

**Fig. 5.** Singular intersection of a sphere and a cylinder

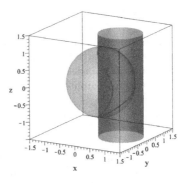

**Fig. 6.** Nonsingular intersection of a sphere and a cylinder

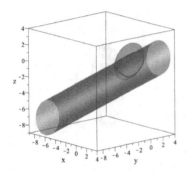

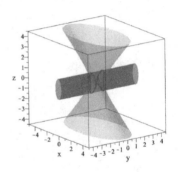

**Fig. 7.** Intersection of an elliptical cylinder and an ellipsoid

**Fig. 8.** Intersection of an elliptical cylinder and a hyperboloid of one sheet

## 6   Concluding Remarks

The primary contribution of this research work is the new derivation of simple, analytical and parametric solutions for the intersection of quadric surfaces. The solution procedure that we devised for finding the intersections curves was based on classical exact roots of quadratic polynomials and quartic polynomials. The parametric representation of the intersection curves is compact, resolution independent, efficient and exact up to the precision of the machine floating point arithmetic.

The implementation of the analytical representation of the intersection curves is based on the particular capabilities of the computer algebra system MAPLE. The illustrative examples of the solution method, that were presented in the previous section, were implemented in MAPLE and are available from the first author upon request.

**Acknowledgements.** This research was supported in part by the Natural Sciences and Engineering Research Council of Canada (NSERC). The first author would like to thank Professor Ken Jackson, Dr. Joanne Dunstall, Julia Martyn and Andrew Wang of the University of Toronto for their assistance.

## References

1. Chen, C., et al.: Computing the real solutions of polynomial systems with the regularchains library in maple. ACM Commun. Comput. Algebra **45**(3/4), 166–168 (2012)
2. Dupont, L., Lazard, D., Lazard, S., Petitjean, S.: Near-optimal parameterization of the intersection of quadrics: I. The generic algorithm. J. Symb. Comput. **43**(3), 168–191 (2008)
3. Dupont, L., Lazard, D., Lazard, S., Petitjean, S.: Near-optimal parameterization of the intersection of quadrics: II. A classification of pencils. J. Symb. Comput. **43**(3), 192–215 (2008)

4. Dupont, L., Lazard, D., Lazard, S., Petitjean, S.: Near-optimal parameterization of the intersection of quadrics: III. Parameterizing singular intersections. J. Symb. Comput. **43**(3), 216–232 (2008)
5. Fioravanti, M., Gonzalez-Vega, L., Necula, I.: Computing the intersection of two ruled surfaces by using a new algebraic approach. J. Symb. Comput. **41**(11), 1187–1205 (2006)
6. Gerhard, J., Jeffrey, D.J., Moroz, G.: A package for solving parametric polynomial systems. ACM Commun. Comput. Algebra **43**(3/4), 61–72 (2010)
7. Lazard, D.: Quantifier elimination: optimal solution for two classical examples. J. Symb. Comput. **5**(1), 261–266 (1988)
8. Lazard, S., Peñanda, L.M., Petitjean, S.: Intersecting quadrics: an efficient and exact implementation. Comput. Geom. **35**(1), 74–99 (2006)
9. Levin, J.: A parametric algorithm for drawing pictures of solid objects composed of quadric surfaces. Commun. ACM **19**(10), 555–563 (1976)
10. Levin, J.Z.: Mathematical models for determining the intersections of quadric surfaces. Comput. Graph. Image Process. **11**(1), 73–87 (1979)
11. Liang, S., Jeffrey, D.J.: An algorithm for computing the complete root classification of a parametric polynomial. In: Calmet, J., Ida, T., Wang, D. (eds.) AISC 2006. LNCS (LNAI), vol. 4120, pp. 116–130. Springer, Heidelberg (2006). https://doi.org/10.1007/11856290_12
12. Maekawa, T.: Self-intersections of offsets of quadratic surfaces: Part II, implicit surfaces. Eng. Comput. **14**(1), 14–22 (1998)
13. Mourrain, B., Técourt, J.-P., Teillaud, M.: On the computation of an arrangement of quadrics in 3D. Comput. Geom. **30**(2), 145–164 (2005). Special Issue on the 19th European Workshop on Computational Geometry
14. Rees, E.L.: Graphical discussion of the roots of a quartic equation. Am. Math. Mon. **29**(2), 51–55 (1922)
15. Requicha, A.A.G., Voelcker, H.B.: Solid modeling: a historical summary and contemporary assessment. IEEE Comput. Graph. Appl. **2**(2), 9–24 (1982)
16. Sarraga, R.F.: Algebraic methods for intersections of quadric surfaces in GMSOLID. Comput. Vis. Graph. Image Process. **22**(2), 222–238 (1983)
17. Schömer, E., Wolpert, N.: An exact and efficient approach for computing a cell in an arrangement of quadrics. Comput. Geom. **33**(1), 65–97 (2006). Robust Geometric Applications and their Implementations
18. Wang, W., Goldman, R., Changhe, T.: Enhancing Levin's method for computing quadric-surface intersections. Comput. Aided Geom. Des. **20**(7), 401–422 (2003)
19. Wang, W., Joe, B., Goldman, R.: Computing quadric surface intersections based on analysis of plane cubic curves. Graph. Models **64**(6), 335–367 (2002)
20. Xia, B., Yang, L.: Automated Inequality Proving and Discovering. World Scientific, Singapore (2016)
21. Yang, L., Hou, X., Xia, B.: A complete algorithm for automated discovering of a class of inequality-type theorems. Sci. China Ser. F Inf. Sci. **44**(1), 33–49 (2001)
22. Yang, L., Xia, B.: Quantifier elimination for quartics. In: Calmet, J., Ida, T., Wang, D. (eds.) AISC 2006. LNCS (LNAI), vol. 4120, pp. 131–145. Springer, Heidelberg (2006). https://doi.org/10.1007/11856290_13
23. Yang, L., Xia, B.: Deciding nonnegativity of polynomials by MAPLE (2013). CoRR, abs/1306.4059
24. Ye, X., Maekawa, T.: Differential geometry of intersection curves of two surfaces. Comput. Aided Geom. Des. **16**(8), 767–788 (1999)

# Decomposing the Parameter Space of Biological Networks via a Numerical Discriminant Approach

Heather A. Harrington[1], Dhagash Mehta[2], Helen M. Byrne[1], and Jonathan D. Hauenstein[2(✉)]

[1] Mathematical Institute, The University of Oxford, Oxford OX2 6GG, UK
{harrington,helen.byrne}@maths.ox.ac.uk
https://www.maths.ox.ac.uk/people/{heather.harrington,helen.byrne}
[2] Department of Applied and Computational Mathematics and Statistics,
University of Notre Dame, Notre Dame, IN 46556, USA
{dmehta,hauenstein}@nd.edu
https://www.nd.edu/~{dmehta,jhauenst}

**Abstract.** Many systems in biology (as well as other physical and engineering systems) can be described by systems of ordinary differential equation containing large numbers of parameters. When studying the dynamic behavior of these large, nonlinear systems, it is useful to identify and characterize the steady-state solutions as the model parameters vary, a technically challenging problem in a high-dimensional parameter landscape. Rather than simply determining the number and stability of steady-states at distinct points in parameter space, we decompose the parameter space into finitely many regions, the number and structure of the steady-state solutions being consistent within each distinct region. From a computational algebraic viewpoint, the boundary of these regions is contained in the discriminant locus. We develop global and local numerical algorithms for constructing the discriminant locus and classifying the parameter landscape. We showcase our numerical approaches by applying them to molecular and cell-network models.

**Keywords:** Parameter landscape · Numerical algebraic geometry · Discriminant locus · Cellular networks

## 1 Introduction

The dynamic behavior of many biophysical systems can be mathematically modeled with systems of differential equations that describe how the state variables interact and evolve over time. The differential equations typically include parameters that represent physical processes such as kinetic rate constants, the strength of cell-cell interactions, and external stimuli. The qualitative behavior of the state variables may change as the parameters vary. Typically, determining and classifying all steady-state solutions of such nonlinear systems, as a function of the

© Springer Nature Switzerland AG 2020
J. Gerhard and I. Kotsireas (Eds.): MC 2019, CCIS 1125, pp. 114–131, 2020.
https://doi.org/10.1007/978-3-030-41258-6_9

parameters, is a difficult problem. However, when the equations are polynomial, or can be translated into polynomials (e.g, rational functions), which is the case for many biological systems (as well as other physical and engineering systems), computing the steady-state solutions becomes a problem in computational algebraic geometry. Thus, it is possible to compute the regions of the parameter space that give rise to different numbers of steady-state solutions.

## 1.1  Previous Work

Due to the ubiquity of such problems, many methods have been proposed for identifying and characterizing steady-state solutions over a parameter space. A standard approach to understand changes in qualitative behavior of differential equations as a parameter is varied is to study bifurcations (singularities). Many standard bifurcation techniques focus on local behavior in the phase space near a structurally unstable object (e.g., fixed point) and the analysis is algebraic by focusing on the normal form [20,22]. Numerical bifurcation techniques as implemented in, for example, AUTO [15] and MATCONT [14], require an initial starting point, use a root finding solver to find a fixed point, and then continue along a branch (e.g., via arc-length continuation). However, these methods are nearly all local in the phase space in that one "continues" (or "sweeps" [37]) from a given initial point. Thus, studying a larger phase space requires sampling of different initial conditions and parameter values. In recent years, computation of bifurcation diagrams of disconnected branches, so-called deflation continuation methods, have been developed [16], however, these do not guarantee finding all solutions at a particular parameter value.

We take a geometric approach and do not restrict ourselves to a local area of the phase space (e.g., no initial condition or guess) nor do we start our analysis by solving (e.g., using Newton's method) for a single fixed point. We focus on where the discriminant vanishes – called the *discriminant locus* – in which roots merge along the discriminant as a parameter varied. Recall that when solving the equation $ax^2 + bx + c = 0$, where $a$, $b$, and $c$ are parameters, and $x$ is the variable, the discriminant locus defined by $\Delta := b^2 - 4ac = 0$ is the boundary separating regions in which the two distinct solutions for $x$ are real ($\Delta > 0$) and nonreal ($\Delta < 0$). The discriminant locus, when arising from a system of ODEs, is often called the bifurcation variety [1]. A parametrization of the discriminant set (variety) can sometimes be computed explicitly, e.g., [8], but this is generally a difficult problem for systems with more than a handful of variables and parameters. Moreover, most of these methods, even those that can systematically 'globally' divide the parameter plane are local in the sense of the phase space [38]. Other symbolic methods are global in terms of phase space include using a cylindrical algebraic decomposition [12] with related variants [9,32,46] and computing the ideal of the discriminant locus using resultants or Gröbner basis methods, e.g., see [13,18,34,42]. Unfortunately, each of these methods has potential drawbacks due to their algorithmic complexity, symbolic expression swell, and inherent sequential nature.

By using homotopy continuation and, more generally, numerical algebraic geometry (see [5, 39, 45]), all solutions over the complex numbers $\mathbb{C}$ to a system of polynomial equations can be computed. In this sense, numerical algebraic geometry permits the computation, with probability one, of *all* real steady-state solutions over a chosen region of parameter space effectively capturing the global behavior of the dynamical system and even detecting disconnected branches of solutions. Such methods have been implemented in software packages including Bertini [6], HOM4PS-3 [10], and PHCpack [43] with Paramtopy [3] extending Bertini to study the solutions at many points in parameter space. Typically, these methods work over $\mathbb{C}$ while the solutions of interest in biological models are in a subset of the real numbers $\mathbb{R}$, e.g., one is interested in steady-states in the positive orthant where the variables are biologically meaningful.

## 1.2    Problem Setup

The general framework of problems under consideration are autonomous systems of differential equations of the form

$$\frac{d}{dt}\boldsymbol{x} = \boldsymbol{f}(\boldsymbol{x}, \boldsymbol{p}) \tag{1}$$

where $\boldsymbol{x} = (x_1, \ldots, x_N)$ denotes the state variables, $\boldsymbol{p} = (p_1, \ldots, p_s)$ denotes the system parameters, and $\boldsymbol{f}(\boldsymbol{x}, \boldsymbol{p})$ is a system of $N$ functions. For $\boldsymbol{p} \in \mathbb{R}^s$, since we aim to compute the *steady-state solutions* to Eq. 1, which are $\boldsymbol{x} \in \mathbb{R}^N$ such that $\boldsymbol{f}(\boldsymbol{x}, \boldsymbol{p}) = 0$. By using numerical algebraic geometry, we additionally require that $\boldsymbol{f}(\boldsymbol{x}, \boldsymbol{p}) = 0$ can be translated into solving polynomial equations, e.g., $\boldsymbol{f}(\boldsymbol{x}, \boldsymbol{p})$ consists of polynomial or rational functions. Moreover, we are particularly interested in the typical situation for biological networks where, for almost all $\boldsymbol{p}$, the system $\boldsymbol{f}(\boldsymbol{x}, \boldsymbol{p}) = 0$ has finitely many distinct (isolated) solutions, all of which are nonsingular, i.e., every eigenvalue of the Jacobian matrix $J_x \boldsymbol{f}(\boldsymbol{x}, \boldsymbol{p})$ of $\boldsymbol{f}$ with respect to the state variables is nonzero. Therefore, certified techniques are used to distinguish between real and nonreal solutions [29].

We consider the parameter space $\mathcal{P} \subset \mathbb{R}^s$ for Eq. 1 to consist of those parameter values $\boldsymbol{p}$ that are biologically meaningful, e.g., $\mathbb{R}^s$ or positive orthant in $\mathbb{R}^s$. The quantitative behavior of the steady-state solutions, that is, the number of them, not necessarily the value of the steady-state, is constant on subregions in $\mathcal{P}$, e.g., the number of physically realistic steady-state solutions is the same for all parameter values in a region. One can also refine the quantitative behavior, by restricting, for example, to only positive steady-state solutions that are locally stable. The points forming the boundaries of these regions are called *critical points* and collectively form the *discriminant locus*, which is called the *minimal discriminant variety* in [32]. The discriminant locus is contained in a hypersurface in $\mathcal{P}$.

Suppose that $\boldsymbol{p} \in \mathcal{P}$ is such that $\boldsymbol{f}(\boldsymbol{x}, \boldsymbol{p}) = 0$ is in the interior of a subregion in the complement of the discriminant locus. The implicit function theorem yields that the solutions can be extended to an open neighborhood containing $\boldsymbol{p}$. One

---

**Algorithm 1.** Perturbed sweeping

---

**Input:** Parameterized equations $f(x, p) = 0$ which can be translated into solving polynomial
 equations with parameter space $\mathcal{P} \subset \mathbb{R}$, perturbation $\epsilon \in \mathbb{R} \setminus \{0\}$, and description of
 the discriminant $\Delta$ associated with quantitative behavior of interest.
**Output:** Description of the intervals in the parameter space $\mathcal{P}$ with the same quantitative
 behavior.
Randomly select $p^* \in \mathcal{P}$ and compute the solution set $S \subset \mathbb{C}^N$ of $f(x, p^* + \epsilon\sqrt{-1}) = 0$.
Track each smooth solution path parameterized by $p \in \mathcal{P}$ defined by $f(x, p + \epsilon\sqrt{-1}) = 0$
 with start points $S$ at $p = p^*$.
Use the solution paths to approximate all values of $p \in \mathcal{P}$ where a solution path becomes
 ill-conditioned and refine, e.g., using [21, 23], to identify the critical points $C \subset \mathcal{P}$ where
 the quantitative behavior of interest changes.
Return the set of intervals of $\mathcal{P}$ whose endpoints are consecutive points in $C$.

---

can keep increasing the size of this neighborhood in the parameter space until
it touches the discriminant locus.

## 1.3   Contribution and Organization of Paper

In Sect. 2, we present a *numerical discriminant locus* method for decomposing the
parameter space into distinct solution regions effectively stratifying the parameter space. We propose three methods for decomposing the parameter space that
build upon advances in real numerical algebraic geometry. A schematic is given
in Fig. 1. The first (Algorithm 1) is for one-dimensional parameter spaces in
which case the discriminant locus consists of finitely many points. We enhance
sweeping approaches such as [27, 37] with a perturbation and use all solutions
simultaneously to locate the finitely many regions, which are open intervals in
this case, where the number of steady-state solutions is consistent. The second
(Algorithm 2) is for low-dimensional parameter spaces and provides a complete
decomposition of the parameter space into finitely many regions after decomposing the discriminant locus. Since computing and decomposing the discriminant locus may be impractical for high-dimensional parameter spaces, our third
method (Algorithm 3) uses the sweeping approach to compute a local decomposition of the parameter space near a given point in the parameter space. When
decomposing a high-dimensional parameter space is desirable, one could boot-
strap together local analyses to generate a more complete, or global, view of the
parameter space.

   In Sect. 3, we apply these algorithms to two biological models. The first is a
detailed ODE model involving rational functions of gene and protein signaling
network that induces long-term memory proposed in [36]. We demonstrate our
method goes beyond singularity theory results in [40]. The second is a new
network model of cell fate specification in a population of interacting stem cells
where the algorithms provide insight into the qualitative behaviors that the
model can exhibit.

   The paper concludes in Sect. 4.

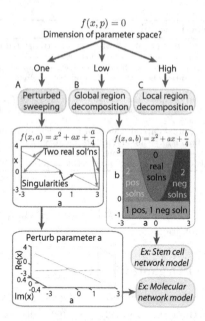

**Fig. 1.** Flow chart of methods. (A) Perturbed sweeping. (B) Global method for decomposing $(a, b)$ parameter space into regions based on number of real steady-states. (C) Local method for high dimensional parameter space analysis.

---

**Algorithm 2.** Global decomposition of 2-dimensional parameter space

---

**Input:** Parameterized equations $\boldsymbol{f}(\boldsymbol{x}, \boldsymbol{p}) = 0$ which can be translated into solving polynomial equations, parameter space $\mathcal{P} \subset \mathbb{R}^2$, and description of the discriminant $\Delta$ associated with quantitative behavior of interest.

**Output:** Description of regions in the parameter space $\mathcal{P}$ with the same quantitative behavior.

Randomly select $\boldsymbol{\alpha} \in \mathbb{R}^2$ and $\mu^* \in \mathbb{R}$, and compute the simultaneous solution set $S \subset \mathbb{C}^N \times \mathbb{C}^2$ of $\boldsymbol{f}(\boldsymbol{x}, \boldsymbol{p}) = 0$, $\boldsymbol{\alpha} \cdot \boldsymbol{p} = \mu^*$, and the discriminant locus $\Delta = 0$.

Use isosingular deflation [30] as needed to permit tracking on the discriminant locus intersected with the line $\boldsymbol{\alpha} \cdot \boldsymbol{p} = \mu$ parameterized by $\mu$.

Apply perturbed sweeping where $\mu \in \mathbb{R}$ is the parameter to compute the critical points $C \subset \mathbb{R}$ of the discriminant locus intersected with the line $\boldsymbol{\alpha} \cdot \boldsymbol{p} = \mu$.

If $\mathcal{P}$ has a boundary, append to $C$ the values of $\boldsymbol{\alpha} \cdot \boldsymbol{p}$ such that $\boldsymbol{p}$ lies at the intersection of the discriminant locus and the boundary of $\mathcal{P}$.

Between two consecutive values in $C$, say $\mu_1 < \mu_2$, pick $\mu = (\mu_1 + \mu_2)/2$ and use perturbed sweeping to compute a decomposition into intervals along the line $\boldsymbol{\alpha} \cdot \boldsymbol{p} = \mu$ inside of $\mathcal{P}$. Connect the boundaries of these intervals to the endpoints $\mu_1$ and $\mu_2$ to create a region decomposition between critical points. Optionally, merge regions across slices $\boldsymbol{\alpha} \cdot \boldsymbol{p} = \mu_j$ for $j = 1, 2$ which have the same quantitative behavior.

Return the regions of $\mathcal{P}$.

---

## 2 Decomposition Using Numerical Algebraic Geometry

This section reviews the required ingredients from numerical algebraic geometry with expanded details provided in, e.g., [5, 39, 45]. Traditionally, symbolic approaches such as [12, 32] describe the regions using both equations that vanish on the discriminant locus and inequalities, e.g., see [24, 26, 31, 35] for applications

---

**Algorithm 3.** Local decomposition of parameter space

---

**Input:** Parameterized equations $f(x, p) = 0$ which can be translated into solving
polynomial equations, parameter space $\mathcal{P} \subset \mathbb{R}^s$, description of the discriminant $\Delta$
associated with quantitative behavior of interest, and point $p^* \in \mathcal{P}$ that is not
contained in the discriminant locus.

**Output:** Description of some regions of the parameter space $\mathcal{P}$ with the same quantitative
behavior.

Randomly select a direction $\alpha^* \in \mathbb{R}^s$ and apply perturbed sweeping to the system
$f(x, p^* + \mu \alpha^*) = 0$ parameterized by $\mu$.

Use isosingular deflation [30] as needed to permit tracking on the discriminant locus
intersected with the linear space parameterized by $p^* + \mu \alpha$ as $\alpha$ varies.

Apply path tracking to vary $\alpha$ to trace out boundaries of regions inside of $\mathcal{P}$ with the same
quantitative behavior.

Return the regions of $\mathcal{P}$ whose boundaries were traced out.

---

to biology. The key observation from numerical algebraic geometry is to replace computing equations and inequalities with geometric descriptions as described below. We can use numerical algebraic geometry techniques presented in Sect. 2.4 to compute every point in the intersection of a line with the discriminant locus and then compute the boundary of the region. Finding sample points in the interior of the regions (which is not on the discriminant locus) has been suggested, e.g., [2,7,33,34]. *Isosingular deflation* developed in [30] allows one to construct a system for tracking along the discriminant locus thereby tracing out the corresponding boundary. Adaptive multiprecision path tracking [4] is used to ensure reliable numerical computations, especially near the discriminant locus. The computations described in Sect. 3 adaptively changed between double, 64-bit, and 96-bit precision.

### 2.1 Computing All Solutions

From algebraic geometry, a parameterized system of equations $f(x, p) = 0$ which are polynomial or can be translated into polynomials has a *generic* behavior for parameters $p$ over the complex numbers. For example, the number of distinct solutions of $f(x, p) = 0$ for almost all $p \in \mathbb{C}^s$ are equal. Therefore, a random choice of $p \in \mathbb{C}^s$, say $p^*$, will have the generic behavior *with probability one*. Classical homotopy continuation, e.g., see [5,39], permits one to compute all distinct solutions of $f(x, p^*) = 0$. One can then continue the solutions of $f(x, p^*) = 0$ via a *parameter homotopy* to solve $f(x, p) = 0$ for any other parameter value $p$. This improves computational efficiency since solving at other parameter values is typically much faster than the *ab initio* solving of $f(x, p^*) = 0$.

### 2.2 Perturbed Sweeping

For one-dimensional parameter spaces, i.e., $s = 1$, the discriminant locus consists of at most finitely many points. Thus, the parameter $p$ parameterizes solutions paths $x(p)$ defined by $f(x(p), p) = 0$ which can be tracked. In particular, one can *sweep* [27,37] as $p$ varies and locate all values of $p$ where a solution path is not smooth, i.e., where $J_x f(x(p), p)$ has a zero eigenvalue. Since each solution path $x(p)$ satisfies

$$\frac{dx}{dp} = -J_x f(x, p)^{-1} \cdot J_p f(x, p), \tag{2}$$

numerical ill-conditioning will occur near the discriminant locus. In fact, Eq. 2 will become stiff since $J_x f(x, p)$ is not invertible on the discriminant locus.

Rather than attempting to track through the discriminant locus, we propose a *perturbed sweeping* approach that guarantees smoothness of the path for easier tracking while still observing some ill-conditioning for identifying the discriminant locus. An example of this is shown in Fig. 2 in Sect. 2.5.

**Theorem 1.** *For $i = \sqrt{-1}$ and $\epsilon \in \mathbb{R}$, we consider the perturbed solution paths $x_\epsilon(p)$ defined by $f(x_\epsilon(p), p + \epsilon i) = 0$. With the setup described above, for all but finitely many $\epsilon \in \mathbb{R}$, all perturbed solution curves are smooth.*

*Proof.* Since there are only finitely many points in the discriminant locus over the complex numbers, there can be only finitely many values of $\epsilon \in \mathbb{R}$ such that there exists $\delta \in \mathbb{R}$ with $\delta + \epsilon i$ in the discriminant locus.

Since $x_\epsilon(p) \to x(p)$ as $\epsilon \to 0$, we can recover information about the actual solution curves by monitoring the condition number as in [27] with the distinct numerical advantage of tracking *smooth* solution curves. If further refinement is needed, additional efficient local computations can be employed, e.g., [21,23].

## 2.3  Global Region Decomposition

We now build upon the perturbed sweeping approach to decompose parameter spaces which are not one-dimensional. This approach, called a *global region decomposition*, is applicable for low-dimensional parameter spaces which mixes projections, critical sets, and perturbed sweeping. For illustration, we start in the two-dimensional case, i.e., $s = 2$, for which the discriminant locus is contained in a curve. Given $\alpha \in \mathbb{R}^2$ and $\mu \in \mathbb{R}$, we consider intersecting the discriminant locus with the line $\pi_\alpha(p) := \alpha \cdot p = \mu$. For almost all choices of $(\alpha, \mu) \in \mathbb{R}^2 \times \mathbb{R}$, i.e., with probability one for randomly selected $(\alpha, \mu) \in \mathbb{R}^2 \times \mathbb{R}$, there are at most finitely many values of $p$ such that the line $\pi_\alpha(p) = \mu$ intersects the discriminant locus. Hence, we can use the perturbed sweeping approach along this line to compute these values. For example, an initial plot of the global region decomposition can be made by simply selecting various values of $\alpha$ and $\mu$ and plotting the various regions along the various lines $\pi_\alpha(p) = \mu$.

To create a complete global region decomposition, we follow a modification of [33]. First, we compute all $p \in \mathbb{C}^2$ such that the line defined by $\pi_\alpha(p) = \mu$ intersects the discriminant locus via homotopy continuation. We solve in $\mathbb{C}^2$ here since the number of real intersection points need not be constant whereas the complex numbers ensures that we will locate every real component. One then uses isosingular deflation [30] as needed to construct a system which permits the tracking along the discriminant locus intersected with the line $\pi_\alpha(p) = \mu$.

Perturbed sweeping viewing $\mu$ as a parameter moves the line in parallel sweeping out the entire plane yielding critical points of the discriminant locus with respect to $\mu$. Hence, if $\mu_1 < \mu_2$ are two consecutive critical points, we know that the topology of the global region decomposition along the line $\pi_\alpha(p) = \mu$ for $\mu \in (\mu_1, \mu_2)$ is equivalent, e.g., same number of intervals which connect up to form regions. Hence, one simply connects the regions across the critical points to form the global regions. An example of this is presented below in Sect. 2.5.

For higher-dimensional spaces, a global region decomposition is computed by applying global region decomposition to smaller-dimensional spaces. For example, one can compute a global region decomposition on a plane inside of a high-dimensional parameter space using the two-dimensional method described above. By selecting various planes, one obtains an initial plot of the global region decomposition. To have a complete picture, one can utilize projections into lower-dimensional spaces computing critical sets of the discriminant locus to locate all areas where the quantitative behavior can change. For example, in the three-dimensional case, $s = 3$, with linear map $\pi_{\alpha,\beta}(p) = (\alpha \cdot p, \beta \cdot p)$ where $\alpha, \beta \in \mathbb{R}^3$, one first considers the parameter space of $\mu \in \mathbb{R}^2$ where $\pi_{\alpha,\beta}(p) = \mu$. By computing a global region decomposition for $\mu \in \mathbb{R}^2$ with respect to the critical curve of the original discriminant locus, one can then stitch together a global region decomposition in the original three-dimensional space as follows. Upon fixing $\mu \in \mathbb{R}^2$ inside of a region, one obtains a curve in the original three-dimensional parameter space where the finitely many points on the discriminant locus can be found using perturbed sweeping. Then, applying isosingular deflation [30] as needed permits the tracking of the original discriminant locus as one moves $\mu \in \mathbb{R}^2$ inside of its corresponding region to connect neighboring regions at the critical points.

## 2.4   Local Region Decomposition

Since a global region decomposition is not practical for high-dimensional parameter spaces, we propose a local region decomposition method by combining perturbed sweeping with the classical approach of ray tracing, e.g., see [19]. Given a point $p \in \mathcal{P}$ not contained in the discriminant locus, which happens for a random point with probability one, the codimension-one components of the discriminant locus can be obtained by using the perturbed sweeping approach along lines emanating from $p$, say in the direction $\alpha$, yielding the real values of $\mu$ for which the corresponding line parameterized by $p + \mu\alpha$ intersects the discriminant locus. As above, once points on the discriminant locus are found, applying isosingular deflation [30] as needed permits the tracking along the discriminant locus tracing the region boundaries as one changes $\alpha$.

This method is local in the sense that one is only tracing along the real points obtained by the intersection of the codimension one components of the discriminant locus with the line parameterized by $p + \mu\alpha$. As mentioned in Sect. 2.3, the number of such real points can change as one changes $\alpha$. To overcome this, one could first compute all such complex intersection points and track all of the corresponding paths as one changes $\alpha$. Thus, one can be sure to obtain all real points of intersection along any other direction emanating from $p$. Even though lower-dimensional boundaries of the regions could be missed with such an approach, it avoids the expense of computing critical points of projections. Nonetheless, local decompositions starting from various $p$ with various $\alpha$ can provide a reasonable plot of the main features of a global decomposition of the parameter space.

## 2.5   Quadratic Example

To illustrate the perturbed sweeping and global region decomposition approach, we consider two examples of a parameterized quadratic equation. The first has one parameter, namely $f(x,a) = x^2 + ax + a/4$. The classical discriminant for quadratic polynomials yields $\Delta = a^2 - a$ with discriminant locus $\{0,1\}$. In particular, $f = 0$ has two distinct real solutions when $a < 0$ or $a > 1$, two distinct complex (i.e., nonreal) solutions when $0 < a < 1$, and a multiplicity 2 real solution when $a = 0$ or $a = 1$. The perturbed sweeping method avoids tracking through the the singularities to have two smooth paths $x_\epsilon(a)$ defined by $f(x_\epsilon(a), a+\epsilon i) = 0$ for any $\epsilon \in \mathbb{R} \setminus \{0\}$ and $a \in \mathbb{R}$. For example, with $\epsilon = 10^{-6}$, we sweep along the smooth $x_\epsilon(a)$ and observe the expected solution behavior as shown in Fig. 1A where the number of real solutions changes at the singularities $a = 0$ and $a = 1$, which are clearly observed in Fig. 2.

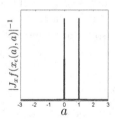

**Fig. 2.** Illustration of using perturbed sweeping to locate singularities at $a = 0, 1$.

The second example is $f(x,a,b) = x^2 + ax + b/4$ which has two parameters and is shown in Fig. 1B. We aim to decompose the parameter space where the quantitative behavior of interest is the number of real and positive solutions, which is typical in biological problems. The discriminant locus for this situation corresponds with the closure of $(a,b)$ such that there exists $x$ such that $f(x,a,b) = 0$ and $x(2x+a) = 0$. In particular, the corresponding discriminant locus consists of two irreducible curves, defined by $b = 0$ and $a^2 = b$, which cut the parameter space $(a,b) \in \mathbb{R}^2$ into four regions where the number of real, positive, and negative solutions are constant on these regions as shown in Fig. 1B. The following describes the essence of computing a global region decomposition.

Consider taking $\alpha = (1,0)$ so that $\pi_\alpha(a,b) = a$. Fixing, say, $\mu^* = 0.5$, we use perturbed sweeping along the line defined by $\pi_\alpha(a,b) = a = \mu^* = 0.5$. This locates the two real points on the discriminant locus, namely $b = 0$ and $b = 0.25$. We do not need to apply isosingular deflation since both are nonsingular solutions with respect to the discriminant system $f(x, a, b) = 0$ and $x(2x + a) = 0$.

Next, we use perturbed sweeping with these two solutions parameterized by $\mu$ starting at $\mu^* = 0.5$ to locate critical points of the discriminant locus. This locates the critical point of the discriminant locus when $\pi_\alpha(a, b) = a = \mu = 0$.

Finally, we simply need to put everything together. At the critical point of the discriminant locus at $a = 0$, there are two regions in terms of $b$, namely $b < 0$ and $b > 0$. For any $a < 0$ or $a > 0$, there are three regions in terms of $b$, namely $b < 0$, $0 < b < a^2$, and $b > a^2$. Hence, can merge together the regions $b < 0$ and $b > a^2$ for $a < 0$ and $a > 0$ at the critical point $a = 0$. Therefore, this yields a global region decomposition consisting of 4 distinct regions shown in Fig. 1B.

# 3    Results from Biological Models

We showcase our methods by applying them to two biological models. First, we analyze a detailed ODE model of the gene and protein signaling network that induces long-term memory proposed by Pettigrew et al. [36]. We demonstrate that our method can be applied to rational functions and reproduce known bifurcation results. Moreover, we find an additional, disconnected branch solution using our method. The second model is a network of cell fate specification in a population of interacting stem cells with complicated dynamics. Since cellular decision making often depends on the number of accessible (stable) steady-states that a system exhibits, we seek to identify distinct regions of parameter space that can elicit different system behavior.

## 3.1    Molecular Network Model

A gene and protein network for long-term memory was proposed by Pettigrew et al. [36] and investigated using bifurcation and singularity analysis by Song et al. [40]. The model is of the form of Eq. 1 where $f$ consists of 15 rational functions, $x \in \mathbb{R}^{15}$ is the vector of model variables, and $p \in \mathbb{R}^{40}$ is the vector of parameters. The following summarizes the structure of the 15 rational functions:

| numerator degree | denominator degree | number of functions in $f$ |
|:---:|:---:|:---:|
| 1 | 0 | 2 |
| 2 | 0 | 2 |
| 2 | 1 | 1 |
| 3 | 1 | 4 |
| 3 | 2 | 2 |
| 4 | 2 | 3 |
| 5 | 4 | 1 |

For a random choice of parameters $p$, the system of equations $f(x, p) = 0$ has 432 isolated nonsingular solutions.

We demonstrate that our discriminant method can (1) reproduce their results as a proof-of-principle, (2) handle rational functions, and (3) we find an additional solution branch not previously located. For this system, the denominators do not vanish near the regions of interest so they do not have any impact on the behavior of the solutions. If the denominator also vanished when finding a solution to the system of equations from the numerators, then the parameter values for which this occurs would be added into the discriminant locus.

We fix the model parameters using the values from [40]. This leaves two parameters to investigate, namely $\lambda$ which represents the extracellular stimulus [5-HT] and $k_{\text{ApSyn}}$. The variable of biological interest required for long-term facilitation is the steady-state of protein kinase A (PKA) in response to the extracellular stimulus parameter $\lambda$. Our aim is to demonstrate the perturbed sweeping method on a large model to reproduce results from Fig. 5 of [40]. In particular, we verify all of their solution branches but also find another solution not reported, which is the top branch shown in Fig. 3. This demonstrates the power of this method to ensure all real solutions are computed. On inspection, this additional steady-state is not on the same branch, but is not biologically feasible so we can reject it as nonphysical. However, by using such an exhaustive first step, we can identify all steady-states, and then systematically characterize and check each solution.

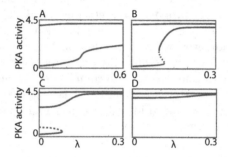

**Fig. 3.** Results of perturbed sweeping method applied to molecular network model of long-term memory that plots the behavior of protein kinase A (PKA) in response to changes in $\lambda$, which is the extracellular stimulus [5-HT], for elected values of $k_{\text{ApSyn}}$ as in [40, Fig. 5]. Solid lines denote a stable steady-state while dashed lines denote an unstable steady-state. (A) Parameter $k_{\text{ApSyn}} = 0.0022$. (B) Parameter $k_{\text{ApSyn}} = 0.015$. (C) Parameter $k_{\text{ApSyn}} = 0.03$. (D) Parameter $k_{\text{ApSyn}} = 0.1$.

## 3.2   Cellular Network Model

Most multicellular organisms emerge from a small number of stem-like cells which become increasingly specialized as they proliferate until they transition to one of a finite number of differentiated states [11]. We propose a caricature model of cell fate specification for a ring of cells, and investigate how cell-cell interactions, mediated by diffusive exchange of a key growth factor, may affect the number of (stable) configurations or patterns that the differentiated cells may adopt. The model serves as a good test case for these discriminant locus methods since, by construction, there is an upper bound on the number of feasible steady-states ($2^N$ stable solutions for a ring of $N$ cells) and some of these patterns are equivalent due to symmetries inherent in the governing equations.

We consider a ring of $N$ interacting cells and denote by $x_i(t) \geq 0$ the concentration within cell $i$ of a growth factor or protein (e.g., notch), whose value determines that cell's differentiation status [17,25,41,44,47]. For $0 < \varepsilon < a < 1$, the subcellular dynamics of $x_i$ are represented by a phenomenological function $q(x_i) = (x - \varepsilon)(x - a)(1 - x)$. This function guarantees bistability of each cell in the absence of cell-cell communication. The bistability represents two distinct cell fates, e.g., high and low levels of notch, which may be associated with differentiation of intestinal epithelial cells into secretory and absorptive phenotypes [11,17,41]. We assume further that cell $i$ communicates with its nearest neighbors (cells $i \pm 1$) via diffusive exchange of $x_i$ and denote by a parameter $g \geq 0$ that describes the coupling strength. Thus, our cell-network model is

$$\frac{dx_i}{dt} = q(x_i) + g \cdot \sum_{j=i-1}^{i+1} (x_j - x_i), \text{ for all } i = 1, \ldots, N. \tag{3}$$

This model assumes uniform coupling $g$ for all nearest neighbors as well as periodic boundary conditions ($x_{N+1} \equiv x_1$ and $x_0 \equiv x_N$), as shown in Fig. 4A.

We analyze the model by using the global region decomposition method for $N = 3, 4, 5$ cells and construct classification diagrams in $(a, g)$ parameter space (Fig. 4B). In addition to decomposing the parameter space into regions based on the number of steady-state solutions, the method also provides valuable information about how solution structure and stability changes as the system parameters vary. For example, in Fig. 4C for the $N = 3$ cell-network, we show how the values and stability of the steady-states for $(x_1, x_2, x_3)$ change as $a$ varies with $g = 0.025$ and as $g$ varies with $a = 0.4$ where $\varepsilon = 0.01$. In Fig. 4D, we plot bifurcation diagrams as $a$ and $g$ vary as before for the $N = 4$ cell-network. In this plot, instead of presenting particular components $x_i$ ($i = 1, 2, 3, 4$), we plot the 2-norm ($\|x\|_2 = (x_1^2 + x_2^2 + x_3^2 + x_4^2)^{1/2}$) to capture the multiplicity of solutions. We note that for $N = 3$ and $N = 4$ there are always two stable and one unstable steady-states, independent of $(a, g)$ parameter values (as shown by the black and red points in Fig. 4C, and by the solid blue lines in Fig. 4D).

For $N = 4$, the classical discriminant locus for this model can be shown to have degree 72 using homotopy continuation. That is, there is a degree 72 polynomial $\Delta(a, g)$ such that the classical discriminant locus is defined by $\Delta = 0$. Even though we were unable to compute this polynomial explicitly, the advantage of using numerical algebraic geometry, as first described in [28], is that computations can be performed on this discriminant locus without having explicit defining equations. In particular, with $a = 0.4$ as in Fig. 4D, the univariate polynomial equation $\Delta(0.4, g) = 0$ has 45 complex solutions, 25 of which are real. Of these 25 real solutions, 15 are positive with only 4 of them corresponding to where change in the number of stable steady-state solutions occur. The regions (intervals) for $g \geq 0$ when $a = 0.4$ are approximately:

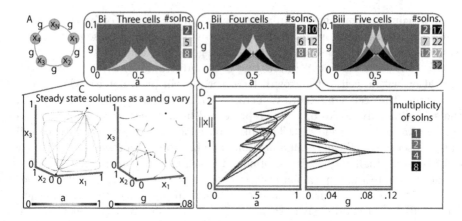

**Fig. 4.** Global region decomposition applied to coupled cell-network model. (A) Ring of cell-network. Each cell $x_i$ has bistable dynamics at stable states at 0 and 1, and unstable state given by $a$. The cells are coupled to neighbour cells by a coupling strength parameter $g$. (B) Region decomposition with parameters $a, g$, similar to a classification diagram, with the number of real stable steady-stages denoted by different colors. The number of stable real steady-states is given for each network where $N = 3, 4$, and 5. (C) Steady-state values in state space for $N = 3$ cell-network are plotted as parameter $a$ is varied between $[\varepsilon, 1]$ or $g$ is varied. The two black dots are stable steady-states, the red is an unstable steady-state, these steady-states are independent of parameter values $a$ and $g$. (D) Bifurcation diagram showing $\|x\|_2$ as the parameters are varied. (Color figure online)

| region | # stable steady-state solns |
|---|---|
| $[0, 0.0197)$ | 16 |
| $(0.0197, 0.0206)$ | 12 |
| $(0.0206, 0.0411)$ | 10 |
| $(0.0411, 0.0533)$ | 6 |
| $(0.0533, \infty)$ | 2 |

We now interpret the results in the context of cell-cell communication. We notice that intermediate values of $a$ generate the largest number of real stable steady-states; small and high values of $a$ yield fewer real stable steady-states. Interestingly, all cells synchronize for intermediate to strong values of the coupling parameter $g$ (two stable states in blue region in Fig. 4B). We conclude that strong cell-cell communication reduces the number of stable steady-state configurations that a population of cells can adopt and, thus, cell-cell communication could be used robustly to drive the cells to a small number of specific states. When coupling is weak ($0 < g < 0.1$), the interacting cells have more flexibility in terms of their final states, with the 5-cell network admitting up to 32 stable steady-states (Fig. 4B(iii)). Weaker cell-cell communication allows more patterns to emerge and may be appropriate when it is less important that neighboring cells share the same phenotype. We find that the regions of $(a, g)$ parameter space that give rise to more than two (synchronized) steady-states also increase in size as the number of cells increases.

## 3.3   Chain of Cells

We next consider a chain of cells rather than a ring. We demonstrate the ability of the local region decomposition method to analyze a generalization of Eq. 3 in which the coupling strength between two cells is not constant. For this caricature stem-cell model, we have classified the number of stable steady-states as the cell-cell coupling parameters are varied, and this increases the dimension of the parameter space. Visualization of the decomposition of higher dimensional parameter spaces is difficult, therefore we created a movie which we describe below. This movie along with code dependent on Bertini [6] and MATLAB used to generate it is available at the repository http://dx.doi.org/10.7274/R0P848V0.

To that end, we consider a generalization of the model given in Eq. 3 which uses coupling strengths $g_{i,i+1} = g_{i+1,i} \geq 0$ between cell $i$ and $i + 1$ with cyclic ordering $(N + 1 \equiv 1)$, namely

$$\frac{dx_i}{dt} = q(x_i) + \sum_{j=i-1}^{i+1} g_{i,j} \cdot (x_j - x_i), \text{ for all } i = 1, \dots, N. \tag{4}$$

With $N = 4$, the classical discriminant locus for this generalized model has degree 486. We consider the case with $\varepsilon = 0.01$, $a = 0.4$, and $g_{4,1} = g_{1,4} = 0$ leaving three free parameters: $g_{1,2}$, $g_{2,3}$, and $g_{3,4}$.

First, we consider the perturbed sweeping approach along the ray defined by $g_{i,i+1} = i \cdot \mu$ for $i = 1, 2, 3$ and $\mu \geq 0$. This decomposes the space $\mu \geq 0$ into 15 regions, approximately

| regions | # stable steady-state solns |
|---|---|
| $[0, 0.0079)$ | 16 |
| $(0.0079, 0.0080)$ | 15 |
| $(0.0080, 0.0132)$ | 14 |
| $(0.0132, 0.0136)$ | 13 |
| $(0.0136, 0.0137)$ | 12 |
| $(0.0137, 0.0142)$ | 11 |
| $(0.0142, 0.0153)$ | 10 |
| $(0.0153, 0.0161)$ | 9 |
| $(0.0161, 0.0171)$ | 8 |
| $(0.0171, 0.0237)$ | 7 |
| $(0.0237, 0.0352]$ | 6 |
| $(0.0352, 0.0360)$ | 5 |
| $(0.0360, 0.0407)$ | 4 |
| $(0.0407, 0.1264)$ | 3 |
| $(0.1264, \infty)$ | 2 |

As a comparison, the classical discriminant with respect to $\mu \in \mathbb{C}$ consists of 312 distinct points of which 84 are real. Of these, 42 are positive with 14 corresponding to a change in the number stable steady-state solutions.

Next, we vary $g_{1,2}$ and $g_{3,4}$ for a fixed $g_{2,3}$ (see Fig. 5). As this is a chain of cells, we observe a natural symmetry as $g_{1,2}$ and $g_{3,4}$ vary for fixed values of $g_{2,3}$. Additionally, as the strength of the cell-cell coupling is increased, the number of stable steady-states decreases monotonically from a maximum value of 16 when coupling is weak down to the minimum value of 2 when coupling is strong where the rate of decline depending on the choice of parameter values.

Finally, we determine how the how the number of stable steady-state solutions change as we move through the three-dimensional parameter using a movie. Each frame of the movie is based on a fixed value of $g_{1,2}$ and shows the decomposition of the plane involving $g_{2,3}$ and $g_{3,4}$. Figure 6 contains four frames from

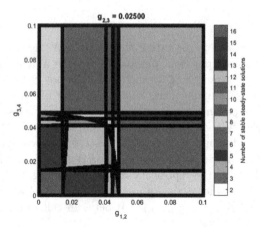

**Fig. 5.** Frame showing how, when $N = 4$ and $g_{2,3} = 0.025$, the number of stable steady-state solutions for Eq. 4 changes as $g_{1,2}$ and $g_{3,4}$ vary.

this movie. In particular, as the coupling $g_{1,2}$ increases, the maximum number of stable steady-states decreases (from 16 when $g_{1,2} = 0$ to 8 when $g_{1,2} = 0.1$). We note that when $g_{1,2} > 0$, the frames in the parameter space $(g_{2,3}, g_{3,4})$ are no longer symmetric. We note also the appearance of regions of parameter space in which, for fixed values of $g_{1,2}$, the number of stable steady-states no longer decreases monotonically with $g_{2,3}$ or $g_{3,4}$. These results, which would not easily be accessible using standard analytical tools, highlight the rich structure of the cell-network model and the power of the local region decomposition method for identifying these solutions.

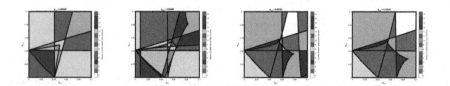

**Fig. 6.** Series of frames showing how, for fixed values of the parameter $g_{1,2} = 0.0$, 0.0325, 0.0675, 0.1 and $N = 4$, the number of stable steady-state solutions for Eq. 4 changes as $g_{2,3}$ and $g_{3,4}$ are varied.

## 4   Conclusion

We have presented a suite of numerical algebraic geometric methods for decomposing the parameter space associated with a dynamical system into distinct regions based on the multiplicity and stability of its steady-state solutions. The methods enable us to understand the parameter landscape of high-dimensional, ordinary differential models with large numbers of parameters. These methods have considerable potential: they could be used to analyze differential equation models associated with a wide range of real-world problems in biology, science, and engineering which cannot easily be tackled with existing approaches.

We have demonstrated that coupling the dynamics of cells, which individually exhibit bistable internal dynamics, can increase markedly the number of real stable steady-states that population exhibits. We considered different network topologies (rings and chains of cells) as well as heterogeneity of cell-cell coupling strengths using the local region decomposition. This methodology may help us to understand how stem cells within the intestinal crypt are able to generate differentiated cells with an array of absorptive and secretory phenotypes, just by considering the interaction of the cells as a network.

**Acknowledgement.** We thank J. Byrne and G. Moroz for helpful discussions. JDH was supported in part by NSF ACI 1460032 and CCF 1812746, Sloan Research Fellowship, and Army Young Investigator Program (YIP). HAH acknowledges funding from EPSRC Fellowship EP/K041096/1, Royal Society University Research Fellowship, and AMS Simons Travel Grant.

# References

1. Arnol'd, V.I., Goryunov, V.V., Lyashko, O.V., Vasil'ev, V.A.: Singularity theory II classification and applications. In: Arnol'd, V.I. (ed.) Dynamical Systems VIII. Encyclopaedia of Mathematical Sciences, vol. 39, pp. 1–235. Springer, Heidelberg (1993). https://doi.org/10.1007/978-3-662-06798-7_1
2. Bates, D.J., Brake, D.A., Hauenstein, J.D., Sommese, A.J., Wampler, C.W.: On computing a cell decomposition of a real surface containing infinitely many singularities. In: Hong, H., Yap, C. (eds.) ICMS 2014. LNCS, vol. 8592, pp. 246–252. Springer, Heidelberg (2014). https://doi.org/10.1007/978-3-662-44199-2_39
3. Bates, D.J., Brake, D.A., Niemerg, M.E.: Paramotopy: parameter homotopies in parallel (2015). http://paramotopy.com/
4. Bates, D.J., Hauenstein, J.D., Sommese, A.J., Wampler, C.W.: Adaptive multi-precision pathtracking. SIAM J. Numer. Anal. **46**(2), 722–746 (2008)
5. Bates, D.J., Hauenstein, J.D., Sommese, A.J., Wampler, C.W.: Numerically Solving Polynomial Systems with Bertini, vol. 25. SIAM, Philadelphia (2013)
6. Bates, D.J., Hauenstein, J.D., Sommese, A.J., Wampler II, C.W.: Software for numerical algebraic geometry: a paradigm and progress towards its implementation. In: Stillman, M., Verschelde, J., Takayama, N. (eds.) Software for Algebraic Geometry, vol. 148, pp. 1–14. Springer, New York (2008). https://doi.org/10.1007/978-0-387-78133-4_1
7. Besana, G.M., Di Rocco, S., Hauenstein, J.D., Sommese, A.J., Wampler, C.W.: Cell decomposition of almost smooth real algebraic surfaces. Numer. Algorithms **63**(4), 645–678 (2013)
8. Broer, H.W., Golubitsky, M., Vegter, G.: The geometry of resonance tongues: a singularity theory approach. Nonlinearity **16**(4), 1511–1538 (2003)
9. Brown, C.W.: QEPCAD B: a program for computing with semi-algebraic sets using CADs. ACM SIGSAM Bull. **37**(4), 97–108 (2003)

10. Chen, T., Lee, T.-L., Li, T.-Y.: Hom4PS-3: a parallel numerical solver for systems of polynomial equations based on polyhedral homotopy continuation methods. In: Hong, H., Yap, C. (eds.) ICMS 2014. LNCS, vol. 8592, pp. 183–190. Springer, Heidelberg (2014). https://doi.org/10.1007/978-3-662-44199-2_30

11. Clevers, H.: Stem Cells. What is an adult stem cell? Science **350**(6266), 1319–1320 (2015)

12. Collins, G.E.: Quantifier elimination for real closed fields by cylindrical algebraic decompostion. In: Brakhage, H. (ed.) GI-Fachtagung 1975. LNCS, vol. 33, pp. 134–183. Springer, Heidelberg (1975). https://doi.org/10.1007/3-540-07407-4_17

13. Cox, D.A., Little, J., O'Shea, D.: Using Algebraic Geometry. Graduate Texts in Mathematics, vol. 185, 2nd edn. Springer, New York (2005). https://doi.org/10.1007/b138611

14. Dhooge, A., Govaerts, W., Kuznetsov, Y.A.: MATCONT: a Matlab package for numerical bifurcation analysis of ODEs. ACM Trans. Math. Softw. (TOMS) **29**(2), 141–164 (2003)

15. Doedel, E.J.: AUTO: a program for the automatic bifurcation analysis of autonomous systems. Congr. Numer **30**, 265–284 (1981)

16. Farrell, P.E., Beentjes, C.H.L., Birkisson, A.: The computation of disconnected bifurcation diagrams (2016). arXiv:1603.00809

17. Fre, S., Huyghe, M., Mourikis, P., Robine, S., Louvard, D., Artavanis-Tsakonas, S.: Notch signals control the fate of immature progenitor cells in the intestine. Nature **435**(7044), 964–968 (2005)

18. Gel'fand, I.M., Kapranov, M.M., Zelevinsky, A.V.: Discriminants, Resultants, and Multidimensional Determinants. Mathematics: Theory & Applications. Birkhäuser, Boston (1994). https://doi.org/10.1007/978-0-8176-4771-1

19. Glassner, A.S. (ed.): An Introduction to Ray Tracing. Academic Press Ltd., London (1989)

20. Glendinning, P.: Stability, Instability and Chaos: An Introduction to the Theory of Nonlinear Differential Equations. Cambridge University Press, Cambridge (1994)

21. Golubitsky, M., Schaeffer, D.G.: Singularities and Groups in Bifurcation Theory, Volume I. Applied Mathematical Sciences, vol. 51. Springer, New York (1985). https://doi.org/10.1007/978-1-4612-5034-0

22. Golubitsky, M., Stewart, I., Schaeffer, D.G.: Singularities and Groups in Bifurcation Theory, Volume II. Applied Mathematical Sciences, vol. 69. Springer, New York (1988). https://doi.org/10.1007/978-1-4612-4574-2

23. Griffin, Z.A., Hauenstein, J.D.: Real solutions to systems of polynomial equations and parameter continuation. Adv. Geom. **15**(2), 173–187 (2015)

24. Gross, E., Harrington, H.A., Rosen, Z., Sturmfels, B.: Algebraic systems biology: a case study for the Wnt pathway. Bull. Math. Biol. **78**(1), 21–51 (2016)

25. Grün, D., et al.: Single-cell messenger RNA sequencing reveals rare intestinal cell types. Nature **525**(7568), 251–255 (2015)

26. Hanan, W., Mehta, D., Moroz, G., Pouryahya, S.: Stability and bifurcation analysis of coupled Fitzhugh-Nagumo oscillators. Extended Abstract Published in the Joint Conference of ASCM 2009 and MACIS 2009, Japan (2009). arXiv:1001.5420 (2010)

27. Hao, W., Hauenstein, J.D., Hu, B., Sommese, A.J.: A three-dimensional steady-state tumor system. Appl. Math. Comput. **218**(6), 2661–2669 (2011)

28. Hauenstein, J.D., Sommese, A.J.: Witness sets of projections. Appl. Math. Comput. **217**(7), 3349–3354 (2010)

29. Hauenstein, J.D., Sottile, F.: Algorithm 921: alphaCertified: certifying solutions to polynomial systems. ACM Trans. Math. Softw. **38**(4), 28 (2012)

30. Hauenstein, J.D., Wampler, C.W.: Isosingular sets and deflation. Found. Comput. Math. **13**(3), 371–403 (2013)
31. Hernandez-Vargas, E.A., Mehta, D., Middleton, R.H.: Towards modeling HIV long term behavior. IFAC Proc. Vol. **44**(1), 581–586 (2011)
32. Lazard, D., Rouillier, F.: Solving parametric polynomial systems. J. Symb. Comput. **42**(6), 636–667 (2007)
33. Lu, Y., Bates, D.J., Sommese, A.J., Wampler, C.W.: Finding all real points of a complex curve. Contemp. Math. **448**, 183–205 (2007). Algebra, Geometry and Their Interactions
34. Montaldi, J.: The path formulation of bifurcation theory. In: Chossat, P. (ed.) Dynamics, Bifurcation and Symmetry. NATO ASI Series (Series C: Mathematical and Physical Sciences), vol. 437, pp. 259–278. Kluwer, Dordrecht (1994). https://doi.org/10.1007/978-94-011-0956-7_21
35. Niu, W., Wang, D.: Algebraic approaches to stability analysis of biological systems. Math. Comput. Sci. **1**(3), 507–539 (2008)
36. Pettigrew, D.B., Smolen, P., Baxter, D.A., Byrne, J.H.: Dynamic properties of regulatory motifs associated with induction of three temporal domains of memory in Aplysia. J. Comput. Neuro. **18**(2), 163–181 (2005)
37. Piret, K., Verschelde, J.: Sweeping algebraic curves for singular solutions. J. Comput. Appl. Math. **234**(4), 1228–1237 (2010)
38. Simon, P.L., Farkas, H., Wittmann, M.: Constructing global bifurcation diagrams by the parametric representation method. J. Comput. Appl. Math. **108**(1–2), 157–176 (1999)
39. Sommese, A.J., Wampler II, C.W.: The Numerical Solution of Systems of Polynomials Arising in Engineering and Science. World Scientific Publishing Co., Pte. Ltd., Hackensack (2005)
40. Song, H., Smolen, P., Av-Ron, E., Baxter, D.A., Byrne, J.H.: Bifurcation and singularity analysis of a molecular network for the induction of long-term memory. Biophys. J. **90**(7), 2309–2325 (2006)
41. Sprinzak, D., et al.: Cis-interactions between Notch and Delta generate mutually exclusive signalling states. Nature **465**(7294), 86–90 (2010)
42. Sturmfels, B.: Solving Systems of Polynomial Equations. CBMS Regional Conference Series in Mathematics, vol. 97. Published for the Conference Board of the Mathematical Sciences, Washington, DC. American Mathematical Society, Providence (2002)
43. Verschelde, J.: Algorithm 795: PHCpack: a general-purpose solver for polynomial systems by homotopy continuation. ACM Trans. Math. Softw. (TOMS) **25**(2), 251–276 (1999)
44. Visvader, J.E., Clevers, H.: Tissue-specific designs of stem cell hierarchies. Nat. Cell Biol. **18**(4), 349–355 (2016)
45. Wampler, C., Sommese, A., Morgan, A.: Numerical continuation methods for solving polynomial systems arising in kinematics. J. Mech. Des. **112**(1), 59–68 (1990)
46. Xia, B.: Discoverer: a tool for solving semi-algebraic systems. ACM Commun. Comput. Algebra **41**(3), 102–103 (2007)
47. Yeung, T.M., Chia, L.A., Kosinski, C.M., Kuo, C.J.: Regulation of self-renewal and differentiation by the intestinal stem cell niche. Cell. Mol. Life Sci. **68**(15), 2513–2523 (2011)

# The Z-Polyhedra Library in MAPLE

Rui-Juan Jing and Marc Moreno Maza[(⊠)]

University of Western Ontario, London, Canada
rjing8@uwo.ca, moreno@csd.uwo.ca

**Abstract.** The Z-Polyhedra is a library written in MAPLE and dedicated to solving problems dealing with the integer points of polyhedral sets. Those problems include decomposing the integer points of polyhedral sets, solving parametric integer programs, performing dependence analysis in for-loop nests and determining the validity of certain Presburger formulas. This article discusses the design of the Z-Polyhedra library and provides numerous illustrations of its usage.

## 1 Introduction

Solving systems of linear equations is a well-studied and fundamental problem in mathematical sciences. When the input system includes equations as well as inequalities, the algebraic complexity of this problem increases from polynomial time to single exponential time with respect to the number of variables. When, in addition, the solution points with integer coordinates are the *only* ones of interest, the problem becomes even harder and is still actively investigated.

The integer points of polyhedral sets are, indeed, of interest in many areas of mathematical sciences, see for instance the landmark textbooks of Schrijver [13] and Barvinok [3], as well as the compilation of articles [4]. One of these areas is the analysis and transformation of computer programs. For instance, integer programming [5] is used by Feautrier in the scheduling of for-loop nests [6] while Barvinok's algorithm [2] (for counting integer points in polyhedra) is adapted by Köppe and Verdoolaege in [10] to answer questions like how many memory locations are touched by a for-loop nest. In [11], Pugh proposes an algorithm, called the *Omega Test*, for testing whether a polyhedron has integer points. In the same paper, Pugh shows how to use the Omega Test for performing dependence analysis [11] in for-loop nests.

In [12], Pugh also suggests, without stating a formal algorithm, that the Omega Test could be used for quantifier elimination on Presburger formulas. This observation has motivated our papers [7,8], where we propose a new approach for computing the integer points of systems of linear equations and inequalities. Here, solving means decomposing the solution set into geometrically meaningful components and providing compact representations of those components. Moreover, the proposed algorithm runs in polynomial time when the dimension of the ambient space is fixed and the input system satisfies mild assumptions. This work produced a MAPLE library, originally called `Polyhedra` and presented at

© Springer Nature Switzerland AG 2020
J. Gerhard and I. Kotsireas (Eds.): MC 2019, CCIS 1125, pp. 132–144, 2020.
https://doi.org/10.1007/978-3-030-41258-6_10

ISSAC 2017 as a software demonstration. To emphasize the fact that this library is primarily dedicated to the integer points of polyhedral sets, we re-baptized it the Z-polyhedra library. Improved design, new features and a faster core-solver (the command IntegerSolve) leads us to the present paper.

Section 2.1 discusses the implementation of the mathematical concepts involved in the manipulation of Z-polyhedra. Section 3 gives an overview of the user-interface and the main solvers implemented in the Z-polyhedra library. Section 4 illustrates the usage of the library through examples taken from the literature. We note that the new algorithm (to be reported in a soon coming article) behind the command IntegerSolve has reduced the execution of some problems from minutes to fractions of a second.

The Z-library is publicly available from the web site of the RegularChains library at www.regularchains.org. A comparison with related software can be found in the last section of [9].

## 2   Mathematical Concepts and Their Implementation

In this section, we review the basic concepts of polyhedral geometry that are involved in the specifications of the commands of our Z_Polyhedra library. We also discuss the implementation of those concepts, in particular their adaptation to the context of effective computations. Section 2.1 is dedicated to the notion of a polyhedral set while Sects. 2.2, 2.3 and 2.4 focus on lattices, Z-Polyhedra and parametric Z-Polyhedra.

**Notation 1.** We use bold letters, e.g. $\mathbf{v}$, to denote vectors and we use capital letters, e.g. $A$, to denote matrices. Also, we assume that vectors are column vectors. For row vectors, we use the transposition notation, that is, $A^t$ for the transposition of a matrix $A$. As usual, we denote by $\mathbb{Z}$, $\mathbb{Q}$ and $\mathbb{R}$ the ring of integers, the field of rational numbers and the field of complex numbers. Unless specified otherwise, all matrices and vectors have their coefficients in $\mathbb{Z}$.

### 2.1   Polyhedra

A subset $P \subseteq \mathbb{Q}^n$ is called a *convex polyhedron* (or simply a *polyhedron*) if $P = \{\mathbf{x} \mid A\mathbf{x} \leq \mathbf{b}\}$ holds, for a matrix $A \in \mathbb{Q}^{m \times n}$ and a vector $\mathbf{b} \in \mathbb{Q}^m$, where $n, m$ are positive integers; we call the linear system $\{A\mathbf{x} \leq \mathbf{b}\}$ a *representation* of $P$. Hence, a polyhedron is the intersection of finitely many half-spaces.

An inequality of the system $A\mathbf{x} \leq \mathbf{b}$ is *redundant* whenever it is implied by all the other inequalities in $A\mathbf{x} \leq \mathbf{b}$. A representation of a polyhedron is *minimal* if no inequality of that representation is redundant.

An inequality $\mathbf{a}^t\mathbf{x} \leq b$ (with $\mathbf{a} \in \mathbb{Q}^n$ and $b \in \mathbb{Q}$) is an implicit equation of the inequality system $A\mathbf{x} \leq \mathbf{b}$ if $\mathbf{a}^t\mathbf{x} = b$ holds for all $\mathbf{x} \in P$. The *dimension* of the polyhedron $P$, denoted by $\dim(P)$, is $n - r$, where $n$ is dimension[1] of the ambient

---

[1] Of course, this notion of dimension coincides with the topological one, that is, the maximum dimension of a ball contained in $P$.

space (that is, $\mathbb{Q}^n$) and $r$ is the maximum number of implicit equations defined by linearly independent vectors. We say that $P$ is *full-dimensional* whenever $\dim(P) = n$ holds. In other words, $P$ is full-dimensional if and only if it does not have any implicit equations.

The article [9] presents an efficient algorithm for computing a minimal representation of the polyhedron $P$ from any representation of $P$. This algorithm builds upon ideas proposed by Balas in [1]; it is implemented in the Z_Polyhedra library by the command `MinimalRepresentation` of the module `PolyhedraTools`.

Let $p, q$ be two positive integers such that $p + q = n$ holds. We rank the coordinates $(x_1, \ldots, x_n)$ of an arbitrary point $\mathbf{x}$ as $x_1 > \cdots > x_n$ and we denote by $\mathbf{u}$ (resp. $\mathbf{v}$) the first $p$ (last $q$) coordinates of $\mathbf{x}$. We denote by $\mathsf{proj}(P; \mathbf{v})$ the *projection of $P$ on $\mathbf{v}$*, that is, the subset of $\mathbb{Q}^q$ defined by:

$$\mathsf{proj}(P; \mathbf{v}) = \{\mathbf{v} \in \mathbb{Q}^q \mid \exists\, \mathbf{u} \in \mathbb{Q}^p, \ (\mathbf{u}, \mathbf{v}) \in P\}.$$

Fourier-Motzkin elimination (FME for short) is an algorithm computing the projection $\mathsf{proj}(P; \mathbf{v})$ of the polyhedron of $P$ by successively eliminating the $\mathbf{u}$-variables from a representation of $P$.

Consider a representation $R$ of $P$ and a positive integer $i$ such that $1 \leq i \leq n$. Denote by $R^{(x_i)}$ the inequalities in $R$ whose largest variable is $x_i$. A *projected representation* of $P$ induced by $R$ is a representation of $P$ consisting of

1. $R^{(x_1)}$, if $n = 1$,
2. $R^{(x_1)}$ and a projected representation of $\mathsf{proj}(P; (x_2, \ldots, x_n))$, otherwise.

The article [9] presents an efficient algorithm for computing a minimal projected representation of the polyhedron $P$ from any $R$ representation of $P$. This algorithm is implemented in the Z_Polyhedra library by the command `MinimalProjectedRepresentation` of the module `PolyhedraTools`. In particular, this command provides a much more efficient way of performing FME than the command `Project` of the `PolyhedralSets` library in MAPLE, as illustrated by the comparative implementation reported in [9].

## 2.2 Lattices

The n-dimensional *integer lattice*, namely $\mathbb{Z}^n$, is the lattice in the Euclidean space $\mathbb{R}^n$ whose lattice points are all $n$-tuples of integers. More generally, a lattice of $\mathbb{R}^n$ consists of all linear combinations with integer coefficients of a basis of $\mathbb{R}^n$ (as a vector space). The data-type `Lattice` of the Z_Polyhedra library implements lattices in that latter sense, with some adaptation to support our purpose of studying the integer points of polyhedra. This adaptation is actually taken from the article [14]. To be precise,

1. we restrict the basis vectors, given by $n \times n$ matrix $A$, to have integer coefficients,
2. we allow a shift of the origin by a vector $\mathbf{b} \in \mathbb{Z}^n$.

Therefore, we call an *integer lattice* of $\mathbb{Z}^n$ any set of the form

$$\{A\mathbf{x} + \mathbf{b} \mid \mathbf{x} \in \mathbb{Z}^n\}$$

where $A \in \mathbb{Z}^{n \times n}$ is a full-rank matrix and $\mathbf{b} \in \mathbb{Z}^n$ is a vector; such a set is denoted by $\mathcal{L}(A, \mathbf{b})$.

## 2.3 Z-Polyhedra

Following [14] here again, we call a Z-*Polyhedron* the intersection of a polyhedron with an integer lattice. The purpose of this notion is, for us, to support the description of the integer points of a polyhedron $P \subseteq \mathbb{Q}^n$, that is, the description of the set $P \cap \mathbb{Z}^n$. This leads us to some preliminary remarks.

Consider first the problem of solving a Diophantine equation over $\mathbb{Z}$, say in 2 variables $x$ and $y$. For instance, consider $3x - 4y = 7$; its solutions, as computed by MAPLE, are of the form $x = 5 + 4\_Z1, y = 2 + 3\_Z1$, the description of which requires the use of the auxiliary variable $\_Z1$. In his Omega test [11,12] Pugh extended that idea for solving arbitrary systems of linear equations of $\mathbb{Z}$. For instance, for the system

$$\begin{cases} 7x + 12y + 31z = 17 \\ 3x + 5y + 14z = 7 \end{cases}$$

our implementation of the Omega test produces

$$\begin{cases} z = -t_0 - 1 \\ y = -5t_0 - 3 \\ x = 13t_0 + 12 \end{cases}$$

Of course, the introduction of the parameter $t_0$ can be avoided by re-writing $x$ and $z$ as a function of $z$, leading to:

$$\begin{cases} x = -1 - 13z \\ y = 2 + 5z \end{cases}$$

Consider now this other polyhedron $P$ of $\mathbb{Q}^3$:

$$\begin{cases} x = 19 \\ y = 25 + (1/2)z \\ z \leq 18 \\ z \leq 0 \end{cases}$$

Because of the presence of the rational number $1/2$, the above input system cannot be considered as a description of the set $P \cap \mathbb{Z}^3$. Using our algorithm [7,8] inspired by the Omega test, we obtain the following:

$$\begin{cases} x = 19 \\ y = 25 + t_0 \\ z = 2t_0 \\ -t_0 \leq 0 \\ t_0 \leq 9 \end{cases}$$

Inspired by the work [14], we have substantially improved our algorithm in terms of efficiency and in terms of output conciseness. In particular, for the above example, we obtain in a Maple session, the result below: On the left-hand side of Fig. 1, we retrieve our original polyhedron $P$ and on the right-hand side, we have the lattice $\mathcal{L}$ of $\mathbb{Z}^n$ consisting of the points $(x, y, z)$ where $z/2$ is integer. The intersection $P \cap \mathcal{L}$ is exactly $P \cap \mathbb{Z}^3$. More generally, encoding the integer points of a polyhedron using the above format, that we call the *PL format*, and thus using lattices, allows us to totally avoid the recourse to auxiliary variables. In addition, it is easy to convert any set of the form $P \cap \mathbb{Z}^n$ (where $P \subseteq \mathbb{Q}^n$ is a polyhedron) from PL format, say $P \cap \mathcal{L}(C, \mathbf{d})$, to the Omega test format, simply by substituting $\mathbf{x}$ with $C\mathbf{t} + \mathbf{d}$ into the representation of $P$, say $A\mathbf{x} \leq \mathbf{b}$.

$$
\left[ \left[ \begin{array}{c} x = 19 \\ y = 25 + \dfrac{z}{2} \\ z \leq 18 \\ -z \leq 0 \end{array} \right], \left[ \begin{array}{c} x \\ y \\ z \end{array} \right], \left[ \left[ \begin{array}{ccc} 1 & 0 & 0 \\ 0 & 1 & 0 \\ 0 & 0 & 2 \end{array} \right], \left[ \begin{array}{c} 0 \\ 0 \\ 0 \end{array} \right] \right] \right]
$$

**Fig. 1.** A $\mathbb{Z}$-Polyhedron in PL format.

## 2.4   Parametric $\mathbb{Z}$-Polyhedra

A *parametric $\mathbb{Z}$-Polyhedron* is a family of $\mathbb{Z}$-polyhedra

- given by the representation of a $\mathbb{Z}$-polyhedron where,
- the defining matrix or the defining vector depend linearly on parameters.

This notion is particularly useful in application problems, like parametric integer linear programming, where the feasible region, and thus optimal solutions, depend on the values of parameters.

```
> with(Z_Polyhedra);
[EnumerateIntegerPoints, IntegerSolve, Lattice, LexicographicalMinimum, ParametricIntegerSolve, Parametric_Z_polyhedron, PlotIntegerPoints3d,
   PolyhedraTools, Z_polyhedron, hasIntegerPoints]
> ineqs := [x_1 + 3* x_2 <= 9 - 2 * theta_1 + theta_2, 2 * x_1 + x_2 <= 8 + theta_1 - 2* theta_2, x_1 <= 4 + theta_1 +
   theta_2, -x_1 <= 0, -x_2 <=0]: eqs := []: vars := [x_1, x_2]: paras := [theta_1, theta_2]:
> P := Parametric_Z_polyhedron[new](eqs, ineqs, vars, parameters = paras);
                                P := Parametric_Z_polyhedron
> Parametric_Z_polyhedron[Display](P);
```

$$
\left[ \left[ \left[ \begin{array}{c} x\_1 + 3\,x\_2 \leq 9 - 2\,theta\_1 + theta\_2 \\ 2\,x\_1 + x\_2 \leq 8 + theta\_1 - 2\,theta\_2 \\ x\_1 \leq 4 + theta\_1 + theta\_2 \\ -x\_1 \leq 0 \\ -x\_2 \leq 0 \end{array} \right], \left[ \begin{array}{c} x\_1 \\ x\_2 \end{array} \right], \left[ \left[ \begin{array}{cccc} 1 & 0 & 0 & 0 \\ 0 & 1 & 0 & 0 \end{array} \right], \left[ \begin{array}{c} 0 \\ 0 \end{array} \right] \right] \right], \left[ [\,], \left[ \begin{array}{c} theta\_1 \\ theta\_2 \end{array} \right], \left[ \left[ \begin{array}{cc} 1 & 0 \\ 0 & 1 \end{array} \right], \left[ \begin{array}{c} 0 \\ 0 \end{array} \right] \right] \right] \right]
$$

**Fig. 2.** Making a new object of type `Parametric_Z_polyhedron`.

The MAPLE session on Fig. 2 shows how to create (command `Parametric_Z_polyhedron[new]`) and display (command `Parametric_Z_polyhedron[Display]`) a parametric Z-Polyhedron from a list of equations, a list of inequalities, a list of the involved variables and a list of the involved parameters. In this example, the parameters are `theta_1` and `theta_2` while the unknowns are `x_1` and `x_2`. On can see that an object of type `Parametric_Z_polyhedron` is represented by a pair of Z-polyhedra in PL format:

- one in the parameter space,
- one the whole ambient space.

# 3 Core Algorithms and Their Implementation

The Z_Polyhedra library in MAPLE implements commands to manipulate Z-polyhedra and in particular the integer points of polyhedra defined over $\mathbb{Q}$. To this end, the Z_Polyhedra library offers

1. 3 data-types in the form MAPLE modules: `Z_polyhedron`, `Lattice` and `Parametric_Z_polyhedron`,
2. a collection of solvers to compute the integer points of (parametric) Z-polyhedra,
3. a fourth module gathering commands to operate on polyhedra and their rational points.

Data-Types and solvers are further discussed below.

## 3.1 Data-Types

An object of the data-type `Z_polyhedron`, encodes the integer points of a polyhedron, using the PL format, specified in Sect. 2.3. An object of the data-type `Lattice` encodes a lattice as defined in Sect. 2.2. Finally, an object of the data-type `Parametric_Z_polyhedron` encodes a parametric Z-polyhedron as defined in Sect. 2.4.

Each of these data-types is implemented in an "object-oriented" fashion using the MAPLE language construct of a module. Each of these three modules offers "get" methods to access the different attributes of an object, see Fig. 3.

## 3.2 Solvers

The most commonly used solver is `IntegerSolve`. It takes as input a system of linear equations and inequalities, that is, a representation of some polyhedron $P \subseteq \mathbb{Q}^n$. It returns finitely many Z-polyhedra

- either in PL format $P_1 \cap \mathcal{L}_1, \ldots, P_e \cap \mathcal{L}_e$ such that

$$P \cap \mathbb{Z}^n = (P_1 \cap \mathcal{L}_1) \cup \cdots \cup P_e \cap \mathcal{L}_e,$$

```
> with(Z_Polyhedra);
[EnumerateIntegerPoints, IntegerSolve, Lattice, LexicographicalMinimum,

   ParametricIntegerSolve, Parametric_Z_polyhedron, PlotIntegerPoints3d,

   PolyhedraTools, Z_polyhedron, hasIntegerPoints]

> with(Lattice);
          [DefiningMatrix, DefiningVector, Display, IsPointInLattice]

> with(Z_polyhedron);
            [Display, Equations, Inequalities, IsContained, Unknowns]

> with(Parametric_Z_polyhedron);
   [ConstraintsOnParameters, ConstraintsOnUnknowns, Display, Parameters, Unknowns]

> with(PolyhedraTools);
[IsNegative, IsNonNegative, IsNonPositive, IsPositive, IsRedundant, IsZero,

   MinimalProjectedRepresentation, MinimalRepresentation, hasRationalPoints]
```

**Fig. 3.** MAPLE session showing the commands and modules available to an end-user of the Z-Polyhedra library.

```
> eqs := [];
  ineqs := [3*x-2*y+z<= 7, -2*x+2*y-z <= 12, -4*x+y+3*z <= 15, -y <= -25];
  vars := [op(indets(ineqs))];
```
$$eqs := [\,]$$
$$ineqs := [3\,x - 2\,y + z \le 7, -2\,x + 2\,y - z \le 12, -4\,x + y + 3\,z \le 15, -y \le -25]$$
$$vars := [x, y, z]$$

```
> L5 := IntegerSolve(eqs, ineqs, vars);map(Z_polyhedron:-Display, %);
```
$$L5 := [\,Z\_polyhedron, Z\_polyhedron, Z\_polyhedron, Z\_polyhedron, Z\_polyhedron\,]$$

$$
\begin{bmatrix}
x = 19 \\
y = 25 + \_Z1 \\
z = 2\,\_Z1 \\
\_Z1 \le 9 \\
-\_Z1 \le 0
\end{bmatrix},
\begin{bmatrix}
x = 15 \\
y = 27 \\
z = 16
\end{bmatrix},
\begin{bmatrix}
x = 18 \\
y = 33 \\
z = 18
\end{bmatrix},
\begin{bmatrix}
x = 14 \\
y = 25 \\
z = 15
\end{bmatrix},
\begin{bmatrix}
x = \_Z1 \\
y = \_Z2 \\
z = \_Z3 \\
3\,\_Z1 - 2\,\_Z2 + \_Z3 \le 7 \\
-2\,\_Z1 + 2\,\_Z2 - \_Z3 \le 12 \\
-4\,\_Z1 + \_Z2 + 3\,\_Z3 \le 15 \\
2\,\_Z2 - \_Z3 \le 48 \\
-5\,\_Z2 + 13\,\_Z3 \le 67 \\
-\_Z2 \le -25 \\
\_Z3 \le 17 \\
-\_Z3 \le -2
\end{bmatrix}
$$

**Fig. 4.** Using `IntegerSolve` for decomposing the integer points of a polyhedron.

– or in Omega test format (thus using auxiliary variables) as on the example shown on Fig. 4.

The core solver of the $\mathbb{Z}$-Polyhedra library is `ParametricIntegerSolve`. Its specifications are similar to those of `IntegerSolve` but using parametric $\mathbb{Z}$-polyhedra instead of $\mathbb{Z}$-polyhedra. In fact, `IntegerSolve` is derived from `ParametricIntegerSolve` by letting the list of parameters be empty. Figure 5 shows what `ParametricIntegerSolve` does on the example of Sect. 2.4, that is, computing its minimal projected representation.

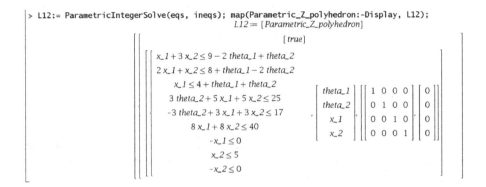

**Fig. 5.** Using `ParametricIntegerSolve`.

While solving a system of constraints does not mean enumerating its solutions, enumeration is sometimes what the user needs, in particular when plotting is involved Figure 6 shows how the command `EnumerateIntegerPoints` enumerates the integer points of a polyhedron, after computing a minimal projected representation of that polyhedron. This is used by the command `PlotIntegerPoints3d` for plotting the same polyhedron.

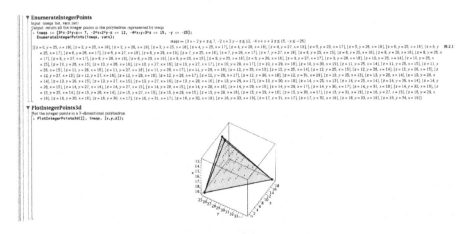

**Fig. 6.** Using `EnumerateIntegerPoints`.

# 4   Applications

## 4.1   Dependence Analysis

Consider the following for-loop nest:

$$\text{for } i \; = \; 1 \text{ to } 5 \text{ do}$$
$$\text{for } j \; = \; i \text{ to } 5 \text{ do}$$
$$A[i, \; j+1] \; = \; A[5, \; j]$$

It is natural to ask whether two different iterations of the above for-loop nest can access the same coefficient in the array A, with at least one of those iterations writing that coefficient.

Consider first the case where one iteration $(i, j)$ writes the same coefficient in A that another iteration $(i', j')$ reads. If such a couple of iterations exists then the system below must have solutions for $i', j', i, j \in \mathbb{Z}$

$$\begin{cases} 1 \leq i \leq j \leq 5 \\ 1 \leq i' \leq j' \leq 5 \\ \quad i = 5 \\ j + 1 = j' \end{cases}$$

```
> with(Z_Polyhedra):
> eqs := [i=5, j+1=jj]; ineqs := [1 <= i, i <= j, j <= 5, 1 <= ii, ii<= jj, jj<= 5];
                              eqs := [i = 5, j + 1 = jj]
                    ineqs := [1 ≤ i, i ≤ j, j ≤ 5, 1 ≤ ii, ii ≤ jj, jj ≤ 5]
> IntegerSolve(eqs, ineqs);
                              [ ]
```

**Fig. 7.** Using `IntegerSolve` for dependence analysis.

The MAPLE session on Fig. 7 shows that the polyhedron defined by `eqs` and `ineqs` has no integer points. Similarly, it can be shown that no two iterations access the same coefficient in A both in writing. Consequently, each for-loop in the above for-loop nest can be executed in a parallel fashion without any race conditions.

We consider now a more difficult example used by Paul Feautrier in his lectures, see Fig. 8. Here again, the `IntegerSolve` command can be used to prove that none of these 3 statements yield dependence, see Fig. 9.

## 4.2   Cache Lines Accessed by a For-Loop Nest

The question considered here is counting the total number of cache lines accessed by a for-loop nest (a 5-point stencil computation code) see Example 5 in [12].

```
     for(i=1; i<=n; i++){
1:       x = a[i][i];
         for(k=1; k<i; k++)
2:           x = x - a[i][k]*a[i][k];
3:       p[i] = 1.0/sqrt(x);
         for(j=i+1; j<=n; j++){
4:           x = a[i][j];
             for(k=1; k<i; k++)
5:               x = x - a[j][k]*a[i][k];
6:           a[j][i] = x * p[i];
         }
```

| $(5, i, j, k) : A[j][k]$ | | $(6, i', j') : A[j'][i']$ |
|---|---|---|
| $0 \le i \le n$ | | $0 \le i' \le n$ |
| $i + 1 \le j \le n$ | | $i' + 1 \le j' \le n$ |
| $1 \le k \le i - 1$ | | |
| | $j = j', k = i'$ | |
| $i < i'$ | $i = i', j < j'$ | $i = i', j = j'$ |

**Fig. 8.** On the right: pseudo-code for Choleski LU. on the left: dependence analysis of 3 statements of this pseudo-code.

The polyhedron studied in the MAPLE session on Fig. 10 is more general than the one of William Pugh. Indeed, we have replaced the loop bound 500 by a parameter N. What `IntegerSolve` computes in this case is a minimal projected representation of that polyhedron. This essentially provides an enumeration of its integer points. Focusing on the variables i and j leads to the desired answer.

### 4.3  Parametric Linear Programming

This last application presents work in progress. Consider the following for-loop nest:

$$\text{for } i = 0 \text{ to } m \text{ do}$$
$$\text{for } j = 0 \text{ to } n \text{ do}$$
$$A[2 * i + j] = i + j$$

```
> with(Z_Polyhedra): t := time();
                                    t := 1.043
> equations :=[j = j1, k=i1];
  inequalities := [i >=0, i <= n, j >= i+1, j<= n, k >= 1, k <= i-1, i1 >= 0, i1 <= n, i1 + 1 <= j1, j1 <= n, i<= i1 -1];
  vars :=[i, j, k, i1, j1,n];
  res := IntegerSolve(equations, inequalities, vars);
                      equations := [j = j1, k = i1]
       inequalities := [0 ≤ i i ≤ n, i+1 ≤ j, j ≤ n, 1 ≤ k, k ≤ i-1, 0 ≤ i1, i1 ≤ n, i1+1 ≤ j1, j1 ≤ n, i ≤ i1-1]
                      vars := [i, j, k, i1, j1, n]
                             res := [ ]
> equations := [j = j1, k=i1, i=i1];
  inequalities := [i >=0, i <= n, j >= i+1, j<= n, k >= 1, k <= i-1, i1 >= 0, i1 <= n, i1 + 1 <= j1, j1 <= n, j <= j1-1];
  vars :=[i, j, k, i1, j1,n];
  res := IntegerSolve(equations, inequalities, vars);
                      equations := [j = j1, k = i1, i = i1]
       inequalities := [0 ≤ i i ≤ n, i+1 ≤ j, j ≤ n, 1 ≤ k, k ≤ i-1, 0 ≤ i1, i1 ≤ n, i1+1 ≤ j1, j1 ≤ n, j ≤ j1-1]
                      vars := [i, j, k, i1, j1, n]
                             res := [ ]
> equations := [j = j1, k=i1, i=i1, j=j1];
  inequalities :=[ i >=0, i <= n, j >= i+1, j<= n, k >= 1, k <= i-1, i1 >= 0, i1 <= n, i1 + 1 <= j1, j1 <= n ];
  vars :=[i, j, k, i1, j1,n];
  res := IntegerSolve(equations, inequalities, vars); time() -t;
                      equations := [j = j1, k = i1, i = i1, j = j1]
       inequalities := [0 ≤ i i ≤ n, i+1 ≤ j, j ≤ n, 1 ≤ k, k ≤ i-1, 0 ≤ i1, i1 ≤ n, i1+1 ≤ j1, j1 ≤ n]
                      vars := [i, j, k, i1, j1, n]
                             res := [ ]
                                   0.131
```

**Fig. 9.** Using `IntegerSolve` for dependence analysis on Choleski LU.

**Fig. 10.** Using `IntegerSolve` for cache line accesses.

A natural question of concurrency is to determine, for a given $k$, what is the last iteration $(i, j)$ at which $A[2 * i + j]$ receives the value $k$, see the driving problem in [5] by Feautrier.

The command `LexicographicalMinimum` addresses a similar question with *maximum* replaced by *minimum*. Hence, to answer the original question, we need a natural change of coordinates where $(i, j)$ is mapped to $(m - i, n - j)$. We are now looking at the following parametric $\mathbb{Z}$-polyhedron $P(k, m, n)$:

$$
\begin{cases}
0 \leq i \\
i \leq m \\
0 \leq j \\
j \leq n \\
2i + j - k \leq -k + 2m + n \\
-2i - j + k \leq -k + 2m + n
\end{cases}
$$

In its current version, `LexicographicalMinimum` finds the lexicographical minimum of $(i, j)$ within $P(k, m, n)$, viewing $i, j, k, m, n$ as rational numbers (instead of integers) which yields the solution shown on Fig. 11. The output consists of 4 pairs; each pair gives a lexicographical minimum together with the corresponding conditions on $k, m, n$ under which this minimum is reached.

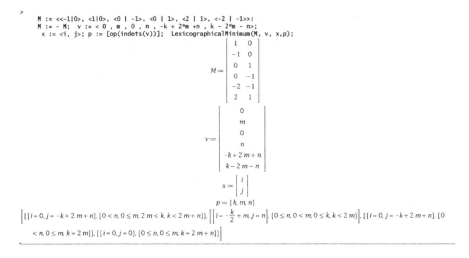

**Fig. 11.** Using `LexicographicalMinimum` for parametric linear programming.

**Acknowledgements.** The authors would like to thank IBM Canada Ltd. (CAS project 880) and NSERC of Canada (CRD grant CRDPJ500717-16), as well as the Fields Institute of Canada for supporting their work.

# References

1. Balas, E.: Projection with a minimal system of inequalities. Comput. Optim. Appl. **10**(2), 189–193 (1998)
2. Barvinok, A.I.: A polynomial time algorithm for counting integral points in polyhedra when the dimension is fixed. Math. Oper. Res. **19**(4), 769–779 (1994)
3. Barvinok, A.I.: Integer points in polyhedra. Contemporary mathematics. European Mathematical Society, Zurich (2008)
4. Beck, M.: Integer Points in Polyhedra-Geometry, Number Theory, Representation Theory, Algebra, Optimization, Statistics: AMS-IMS-SIAM Joint Summer Research Conference, 11–15 June 2006, Snowbird, Utah. Contemporary mathematics - American Mathematical Society, American Mathematical Society (2008)
5. Feautrier, P.: Parametric integer programming. RAIRO Recherche Opérationnelle **22**, 243–268 (1988). http://www.numdam.org/article/RO_1988__22_3_243_0.pdf
6. Feautrier, P.: Automatic parallelization in the polytope model. In: Perrin, G.-R., Darte, A. (eds.) The Data Parallel Programming Model. LNCS, vol. 1132, pp. 79–103. Springer, Heidelberg (1996). https://doi.org/10.1007/3-540-61736-1_44
7. Jing, R.-J., Moreno Maza, M.: Computing the integer points of a polyhedron, I: algorithm. In: Gerdt, V.P., Koepf, W., Seiler, W.M., Vorozhtsov, E.V. (eds.) CASC 2017. LNCS, vol. 10490, pp. 225–241. Springer, Cham (2017). https://doi.org/10.1007/978-3-319-66320-3_17
8. Jing, R.-J., Moreno Maza, M.: Computing the integer points of a polyhedron, II: complexity estimates. In: Gerdt, V.P., Koepf, W., Seiler, W.M., Vorozhtsov, E.V. (eds.) CASC 2017. LNCS, vol. 10490, pp. 242–256. Springer, Cham (2017). https://doi.org/10.1007/978-3-319-66320-3_18

9.  Jing, R.-J., Moreno Maza, M., Talaashrafi, D.: Complexity estimates for fourier-motzkin elimination (2018). CoRR, abs/1811.01510
10. Köppe, M., Verdoolaege, S.: Computing parametric rational generating functions with a primal barvinok algorithm. Electr. J. Comb. **15**(1), R16 (2008)
11. Pugh, W.: The omega test: a fast and practical integer programming algorithm for dependence analysis. In: Martin, J.L. (ed.) Proceedings Supercomputing 1991, Albuquerque, NM, USA, 18–22 November 1991, pp. 4–13. ACM (1991)
12. Pugh, W.: Counting solutions to presburger formulas: how and why. In: Sarkar, V., Ryder, B.G., Lou Soffa, M., (eds.) Proceedings of the ACM SIGPLAN 1994 Conference on Programming Language Design and Implementation (PLDI), Orlando, Florida, USA, 20–24 June 1994, pp. 121–134. ACM (1994)
13. Schrijver, A.: Theory of Linear and Integer Programming. John Wiley, New York (1986)
14. Seghir, R., Loechner, V., Meister, B.: Integer affine transformations of parametric $\mathbb{Z}$-polytopes and applications to loop nest optimization. TACO **9**(2), 8:1–8:27 (2012)

# Detecting Singularities Using the PowerSeries Library

Mahsa Kazemi$^{(\boxtimes)}$ and Marc Moreno Maza

Department of Computer Science, University of Western Ontario,
London, ON, Canada
mkazemin@uwo.ca, moreno@csd.uwo.ca

**Abstract.** Local bifurcation analysis of singular smooth maps plays a fundamental role in understanding the dynamics of real world problems. This analysis is accomplished in two steps: first performing the Lyapunov-Schmidt reduction to reduce the dimension of the state variables in the original smooth map and then applying singularity theory techniques to the resulting reduced smooth map. In this paper, we address an important application of the so-called Extended Hensel Construction (EHC) for computing the aforementioned reduced smooth map, which, consequently, leads to detecting the type of singularities of the original smooth map. Our approach is illustrated via two examples displaying pitchfork and winged cusp bifurcations.

**Keywords:** Singularities · Smooth maps · Lyapunov-Schmidt reduction · Extended Hensel Construction · PowerSeries library

## 1 Introduction

Consider the smooth map

$$\Phi : \mathbb{R}^n \times \mathbb{R}^m \longrightarrow \mathbb{R}^n, \quad \Phi_i(\mathbf{x}, \boldsymbol{\alpha}) = 0, \ i = 0, \ldots, n \tag{1}$$

where the vectors $\mathbf{x} = (x_1, \ldots, x_n)$ and $\boldsymbol{\alpha} = (\alpha_1, \ldots, \alpha_m)$ represent state variables and parameters, respectively. We assume that $\Phi_i(\mathbf{0}, \mathbf{0}) = 0$. The smooth map $\Phi$ is called *singular* when $\det(\mathrm{d}\Phi)_{(\mathbf{0},\mathbf{0})} = 0$. The local zeros of a singular map may experience *qualitative* changes when small perturbations are applied to the parameters $\boldsymbol{\alpha}$. These changes are called *bifurcation*. Local bifurcation analysis of zeros of the singular smooth map (1) plays a pivotal role in exploring the behaviour of many real world problems [4,7,8,10]. *Lyapunov-Schmidt reduction* is a fundamental tool converting the singular map (1) into $g : \mathbb{R}^p \times \mathbb{R}^m \longrightarrow \mathbb{R}^p$ with $p = n - \mathrm{rank}(\mathrm{d}\Phi_{0,0})$. The reduction is achieved through producing an equivalent map to (1) made up of a pair of equations and making use of the Implicit Function Theorem. This solves the $n-1$ variables of $\mathbf{x}$ in the first equation; thereafter, substituting the result into the second one gives an equation for the remaining variable. It is proved that the local zeros of the map $g$ are

© Springer Nature Switzerland AG 2020
J. Gerhard and I. Kotsireas (Eds.): MC 2019, CCIS 1125, pp. 145–155, 2020.
https://doi.org/10.1007/978-3-030-41258-6_11

in one-to-one correspondence with the local zeros of $\Phi$; for more details see [8, pp. 25–34]. Hence, the study of local zeros of (1) is facilitated throught treating their counterparts in $g$. Singularity theory is an approach providing a comprehensive framework equipped with effective tools for this study. The pioneering work of René Thom established the original ideas of the theory which was then extensively developed by John Mather and V. I. Arnold. The book series [8] written by Marty Golubitsky, Ian Stewart and David G. Schaeffer is a collection of significant contributions of the authors in dealing with a wide range of real world problems using singularity theory techniques as well as explaining the underlying ideas of the theory in ways accessible to applied scientists and mathematicians particularly those dealing with bifurcation problems in the presence of parameters and symmetries. The singularity theory tools are applied to the problems that have emerged as an output of the Lyapunov-Schmidt reduction. Following [8, p. 25], we focus on the reduction when $\mathrm{rank}(\mathrm{d}\Phi_{0,0}) = n - 1$, $m = 1$ and refer to $\alpha_1 = \lambda$ as the bifurcation parameter. In other words, we consider the following map

$$g : \mathbb{R} \times \mathbb{R} \longrightarrow \mathbb{R} \quad g(x, \lambda) = 0. \tag{2}$$

Two smooth maps are regarded as *germ-equivalent* when they are identical on some neighborhood of the origin. In fact, a *germ-equivalence* class of a smooth map is called a *germ*. We denote by $\mathscr{E}_{x,\lambda}$ the space of all scalar smooth germs which is a local ring with $\mathcal{M}_{\mathscr{E}_{x,\lambda}} = \langle x, \lambda \rangle_{\mathscr{E}_{x,\lambda}}$ as the unique maximal ideal; see also [8, p. 56] and [4, p. 3]. Due to the existence of germs with infinite Taylor series and flat germs (whose Taylor series is zero), there does not exist a computational tool to automatically study local bifurcations in $\mathscr{E}_{x,\lambda}$. This has motivated the authors of [4] to propose circumstances under which the computations supporting the bifurcation analysis in $\mathscr{E}_{x,\lambda}$ are transferred to smaller local rings and verify that the corresponding results are valid in $\mathscr{E}_{x,\lambda}$. For instance, the following theorem permits the use of formal power series $K[[x, \lambda]]$ ring as a smaller computational ring in computation of algebraic objects involved in the analysis of bifurcation.

**Theorem 1** *( [4, Theorem 4.3]). Suppose that $\{f_i\}_{i=1}^m \in \mathscr{E}_{x,\lambda}$. For $k, N \in \mathbb{N}$ with $k \leq N$,*

$$\mathcal{M}^k_{K[[x,\lambda]]} \subseteq \langle J^N f_1, \ldots, J^N f_m \rangle_{K[[x,\lambda]]} \quad \text{iff} \quad \mathcal{M}^k_{\mathscr{E}_{x,\lambda}} \subseteq \langle f_1, \ldots, f_m \rangle_{\mathscr{E}_{x,\lambda}}$$

*where $\mathcal{M}^k = \langle x^{\alpha_1} \lambda^{\alpha_2} : \alpha_1 + \alpha_2 = k \rangle$ and $J^N f_i$ is the sum of terms of degree $N$ or less in the Taylor series of $f_i$.*

This, along with other criteria in [4,6], highlights the importance of alternative rings in performing automatic local bifurcation analysis of scalar and $\mathbb{Z}_2$-equivariant singularities.

The work presented here addresses one of the applications of the so-called Extended Hensel Construction (EHC) invented by Sasaki and Kako, see [12]. We show that the EHC can be used in computing the reduced system $g \in \mathscr{E}_{x,\lambda}$, which, as a result, leads to determining the type of singularity hidden in system (1).

This EHC has been studied and improved by many authors. In particular, the papers [1–3] present algorithmic improvements (where the EHC relies only linear algebra techniques and univariate polynomial arithmetic) together with applications of the EHC in deriving real branches of space curves and consecuently computing limitis of real multivariate rational functions. The same authors implemented their version of the EHC as the `ExtendedHenselConstruction` command of the `PowerSeries` library[1],

The EHC comes into two flavors. In the case of bivariate polynomials it behaves as Newton-Puiseux algorithm while with multivariate polynomials it acts as an effective version of Jung-Abhyankar Theorem. In both cases, it provides a factorization of the input object in the vicinity of the origin. We believe that this capabiliy makes the EHC a desirable tool for an automatic derivation of the zeros of a polynomial system locally near the origin. The rest of this paper is organized as follows. In Sect. 2, some of the ideas in singularity theory are reviewed. We then discuss the EHC procedure followed by an overview on the `PowerSeries` Library. Finally, our proposed approach is illustrated through two examples revealing pitchfork and winged cusp bifurcations.

## 2   Background

### 2.1   Concepts from Singularity Theory

In this section we explain the materials required for defining recognition problem of a singular germ. These concepts are accompanied by examples. We skip the technical details of singularity theory-related concepts as they are beyond the scope of this paper. The interested readers are referred to [4,5,8] for the principal ideas, algebraic formulations and automatic computation of the following objects.

**Contact Equivalence.** We say that two smooth germs $f, g \in \mathscr{E}_{x,\lambda}$ are *contact equivalent* when

$$g(x, \lambda) = S(x, \lambda) f(X(x, \lambda), \Lambda(\lambda)) \tag{3}$$

is held for a smooth germ $S(x, \lambda) \in \mathscr{E}_{x,\lambda}$ and local diffeomorphisms $((x, \lambda) \longrightarrow (X(x, \lambda), \Lambda(\lambda))) : \mathbb{R}^2 \longrightarrow \mathbb{R}^2$ satisfying

$$S(x, \lambda), X_x(x, \lambda), \Lambda(\lambda) > 0$$

**Normal Form.** Bifurcation analysis of local zeros of $g$ in (2) requires computing a contact equivalent germ to $g$ which has simpler structure and makes the analysis efficient. Indeed, each step of this analysis, for instance recognition problem, involves normal form computation. To be more precise, the simplest representative of the class of $g \in \mathscr{E}_{x,\lambda}$ under contact equivalence is called a *normal form* of $g$.

---

[1] http://www.regularchains.org/downloads.html.

*Example 1.* Consider the smooth germ $g(x, \lambda) = \sin(x^3) - \lambda x + \exp(\lambda^3) - 1 \in \mathscr{E}_{x,\lambda}$. Note that $g(0,0) = \frac{\partial}{\partial x}g(0,0) = 0$; therefore, the origin is the singular point of $g$. The procedure in [4, Sect. 6] returns $x^3 - \lambda x$ as the normal form of $g$ denoted by $\mathrm{NF}(g)$. The equation $x^3 - \lambda x = 0$ is called the *pitchfork bifurcation problem* and the *bifurcation diagram* for pitchfork is defined by the local variety $\{(x, \lambda) \mid x^3 - x\lambda = 0\}$. When $\lambda$ smoothly varies around the origin, the number of solutions of the pitchfork bifurcation problem changes from one to three; see Fig. 1.

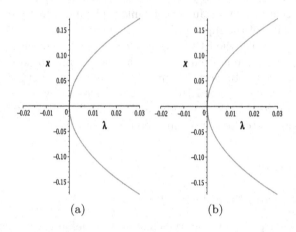

(a)                                        (b)

**Fig. 1.** Figures (*a*) and (*b*) depict the bifurcation diagrams of $g$ and $\mathrm{NF}(g)$, respectively.

Now, modulo monomials of degree $\geq 5$, we compute the transformation $(X(x, \lambda), S(x, \lambda), \Lambda(\lambda))$ through which $g$ is converted into NF(g) in (3).

$$X(x, \lambda) := x + \lambda^2 + \lambda x + \lambda^2 x + \lambda x^2 + x^3,$$
$$S(x, \lambda) := 1 - \lambda + 2\lambda x + x^2,$$
$$\Lambda(\lambda) := \lambda.$$

**Recognition Problem.** Let $g \in \mathscr{E}_{x,\lambda}$ be a singular germ. *Recognition problem* for a normal form of $g$ computes a list of zero and non-zero conditions on derivatives of a singular germ $f \in \mathscr{E}_{x,\lambda}$ under which $f$ is contact-equivalent to $g$. The proposed algorithm in [8, pp. 86–93 ], divides monomials (in $\mathscr{E}_{x,\lambda}$) into three categories; *low, intermediate* and *high order terms. Low order terms* refer to the monomials of the form $x^{\alpha_1}\lambda^{\alpha_2}$ that do not participate in the representation of any germ equivalent to $g$. The *high order terms* consist of the monomials $x^{\alpha_1}\lambda^{\alpha_2}$ which do not change the structure of the local zeros of $g$ when they are present; that is, adding $x^{\alpha_1}\lambda^{\alpha_2}$ to $g$ creates a germ contact equivalent to $g$. Due to the sophisticated structure of intermediate order terms we skip defining them here and instead introduce *intrinsic* generators $x^{\alpha_1}\lambda^{\alpha_2}$ which contribute to every equivalent germ and provide information about intermediate order terms.

Low order terms and intrinsic generators are identified through the following theorem.

**Theorem 2** [8, Theorems 8.3 and 8.4, p. 88]. *Suppose that $f, g \in \mathscr{E}_{x,\lambda}$ and there exists a positive integer $k$ such that $\mathcal{M}^k_{\mathscr{E}_{x,\lambda}} \subset \langle g, x\frac{\partial}{\partial x}g, \lambda\frac{\partial}{\partial x}g \rangle_{\mathscr{E}_{x,\lambda}}$.*

*(a) if $f$ is equivalent to $g$ and $x^{\alpha_1}\lambda^{\alpha_2}$ belongs to low order terms of $g$ then $\frac{\partial^{\alpha_1}}{\partial x^{\alpha_1}}\frac{\partial^{\alpha_2}}{\partial \lambda^{\alpha_2}}f(0,0) = 0$.*

*(b) furthermore, assume that $x^{\alpha_1}\lambda^{\alpha_2}$ belongs to intrinsic generators of $g$. If $f$ is equivalent to $g$ then $\frac{\partial^{\alpha_1}}{\partial x^{\alpha_1}}\frac{\partial^{\alpha_2}}{\partial \lambda^{\alpha_2}}f(0,0) \neq 0$.*

*Example 2.* For the smooth germ $g$ given by Example 1, we deduce the vector space $\mathbb{R}\{1, \lambda, x, x^2\}$ as low order terms. It follows from Theorem 2(a) that any germ $f$ equivalent to $g$ satisfies

$$f(0,0) = \frac{\partial}{\partial \lambda}f(0,0) = \frac{\partial}{\partial x}f(0,0) = \frac{\partial^2}{\partial x^2}f(0,0) = 0 \tag{4}$$

Moreover, the higher order terms of $g$ are determined by the ideal

$$\langle x^4, \lambda^4, x^3\lambda, x\lambda^3, x^2\lambda^2 \rangle_{\mathscr{E}_{x,\lambda}} + \langle x^2\lambda, \lambda^3, x\lambda^2 \rangle_{\mathscr{E}_{x,\lambda}} + \langle \lambda^2 \rangle_{\mathscr{E}_{x,\lambda}}$$

which means that adding/removing any monomial, taken from this ideal, to/from $g$ gives a new germ equivalent to $g$. Finally, the corresponding intrinsic generators of $g$ are described via $\{x^3, \lambda x\}$ verified by Theorem 2(b) that for any germ $f$ equivalent to $g$ the following is valid

$$\frac{\partial^3}{\partial x^3}f(0,0) \neq 0, \quad \frac{\partial}{\partial \lambda}\frac{\partial}{\partial x}f(0,0) \neq 0 \tag{5}$$

To sum up, the recognition problem for a normal form of $g$ is characterized by (4) and (5).

## 2.2   The Extended Hensel Construction

This part is summarized from [1].

**Notation 1.** Suppose that $\mathbb{K}$ is an algebraic number field whose algebraic closure is denoted by $\overline{\mathbb{K}}$. Assume that $F(X,Y) \in \mathbb{K}[X,Y]$ is a bivariate polynomial with complex number coefficients. Let also $F$ be a univariate polynomial in $X$ which is monic and square-free. The partial degree of $F$ w.r.t. $X$ is represented by $d$. We denote by $\mathbb{K}[[U^*]] = \bigcup_{\ell=1}^{\infty} \mathbb{K}[[U^{\frac{1}{\ell}}]]$ the ring of *formal Puiseux series*. Hence, given $\varphi \in \mathbb{K}[[U^*]]$, there exists $\ell \in \mathbb{N}_{>0}$ such that $\varphi \in \mathbb{K}[[U^{\frac{1}{\ell}}]]$ holds. Thus, we can write $\varphi = \sum_{m=0}^{\infty} a_m U^{\frac{m}{\ell}}$, for some $a_0, \ldots, a_m, \ldots \in \mathbb{K}$. We denote by $\mathbb{K}((U^*))$ the quotient field of $\mathbb{K}[[U^*]]$. Let $\varphi \in \mathbb{K}[[U^*]]$ and $\ell \in \mathbb{N}$ such that $\varphi = f(U^{\frac{1}{\ell}})$ holds for some $f \in \mathbb{K}[[U]]$. We say that the Puiseux series $\varphi$ is *convergent* if we have $f \in \mathbb{K}\langle U \rangle$. We recall Puiseux's theorem: if $\mathbb{K}$ is an algebraically closed field of characteristic zero, the field $\mathbb{K}((U^*))$ of formal Puiseux series over $\mathbb{K}$ is the algebraic closure of the field of formal Laurent series over $\mathbb{K}$; moreover, if $\mathbb{K} = \mathbb{C}$, then the field $\mathbb{C}(\langle Y^* \rangle)$ of convergent Puiseux series over $\mathbb{C}$ is algebraically closed as well.

The purpose of the EHC is to factorize $F(X,Y)$ as $F(X,Y) = G_1(X,Y) \cdots$ $G_r(X,Y)$, with $G_i(X,Y) \in \overline{\mathbb{K}}(\langle Y^* \rangle)[X]$ and $\deg_X(G_i) = m_i$, for $1 \leq i \leq r$. Thus, the EHC factorizes $F(X,Y)$ over $\overline{\mathbb{K}}(\langle Y^* \rangle)$, thus over $\mathbb{C}(\langle Y^* \rangle)$.

**Newton Line.** We plot each non-zero term $cX^{e_x}Y^{e_y}$ of $F(X,Y)$ to the point of coordinates $(e_x, e_y)$ in the Euclidean plane equipped with Cartesian coordinates. We call *Newton Line* the straight line $L$ passing through the point $(d, 0)$ and another point, such that no other points lie below $L$. The equation of $L$ is $e_x/d + e_y/\delta = 1$ for some $\delta \in \mathbb{Q}$. We define $\widehat{\delta}, \widehat{d} \in \mathbb{Z}^{>0}$ such that $\widehat{\delta}/\widehat{d} = \delta/d$ and $\gcd \widehat{\delta}, \widehat{d} = 1$ both hold.

**Newton Polynomial.** The sum of all the terms of $F(X,Y)$ which are plotted on the Newton line of $F$ is called the *Newton polynomial* of $F$. We denote it by $F^{(0)}$. Observe that the Newton polynomial is a homogeneous polynomial in $(X, Y^{\delta/d})$. Let $\zeta_1, \ldots, \zeta_r \in \overline{\mathbb{K}}$ be the distinct roots of $F^{(0)}(X, 1)$, for some $r \geq 2$. Hence we have $\zeta_i \neq \zeta_j$ for all $1 \leq i < j \leq r$ and there exist positive integers $m_1 \leq m_2 \leq \cdots \leq m_r$ such that, using the homogeneity of $F^{(0)}(X, Y)$, we have

$$F^{(0)}(X,Y) = (X - \zeta_1 Y^{\delta/d})^{m_1} \cdots (X - \zeta_r Y^{\delta/d})^{m_r}.$$

The *initial factors* of $F^{(0)}(X,Y)$ are $G_i^{(0)}(X,Y) := (X - \zeta_i Y^{\delta/d})^{m_i}$, for $1 \leq i \leq r$. For simplicity, we put $\widehat{Y} = Y^{\widehat{\delta}/\widehat{d}}$.

**Theorem 3 (Extended Hensel Construction).** *We define the ideal*

$$S_k = \langle X^d Y^{(k+0)/\widehat{d}}, X^{d-1}Y^{(k+\widehat{\delta})/\widehat{d}}, \ldots, X^0 Y^{(k+d\widehat{\delta})/\widehat{d}} \rangle, \tag{6}$$

*for* $k = 1, 2, \ldots$ *Then, for all integer* $k > 0$, *we can construct* $G_i^{(k)}(X,Y) \in$ $\mathbb{C}\langle Y^{1/\widehat{d}} \rangle[X]$, *for* $i = 1, \ldots, r$, *satisfying*

$$F(X,Y) = G_1^{(k)}(X,Y) \cdots G_r^{(k)}(X,Y) \mod S_{k+1}, \tag{7}$$

*and* $G_i^{(k)}(X,Y) \equiv G_i^{(0)}(X,Y) \mod S_1$, *for all* $i = 1, \ldots, r$.

### 2.3  The PowerSeries Library

The PowerSeries library consists of two modules, dedicated respectively to multivariate power series over the algebraic closure of $\mathbb{Q}$, and univariate polynomials with multivariate power series coefficients. Figure 2 illustrates *Weiertrass Preparation Factorization*. The command PolynomialPart displays all the terms of a power series (or a univariate polynomial over power series) up to a specified degree. In fact, each power series is represented by its terms that have been computed so far together with a program for computing the next ones. A command like WeiertrassPreparation computes the terms of the factors $p$ and $\alpha$ up to the specified degree; moreover, the encoding of $p$ and $\alpha$ contains a program for computing their terms in higher degree. Figures 3 and 4 illustrate the *Extended Hensel Construction* (EHC)[2] For the case of an input bivariate polynomial, see

---

[2] The factorization based on Hensel Lemma is in fact a weaker construction since: (1) the input polynomial must be monic and (2) the output factors may not be linear.

```
> PS := PowerSeries([X, Y]) :
  with(PS) :
  UPoPS := UnivariatePolynomialOverPowerSeries([X, Y], Z) :
  with(UPoPS) :
  u := UPoPS:-FromListOfPolynomials([Y, 1, X + 1]);
  UPoPS:-PolynomialPart(u, 2);
  (p, alpha) := UPoPS:-WeierstrassPreparation(u, 2);
  UPoPS:-PolynomialPart(p, 2);
  UPoPS:-PolynomialPart(alpha, 2);
```

$$u := polynomial\_over\_power\_series$$

$$Y + Z + (X + 1) Z^2$$

$$p, \alpha := polynomial\_over\_power\_series, polynomial\_over\_power\_series$$

$$Y^2 + Y + Z$$

$$-X Y - Y^2 - Y + 1 + (X + 1) Z$$

**Fig. 2.** Weierstrass Preparation Factorization for a univariate polynomial with multivariate power series coefficients.

```
> P := PowerSeries([y, z]) :
  U := UnivariatePolynomialOverPowerSeries([y, z], x) :
  poly := y·x³ + (-2 · y + z + 1)·x + y :
  U:-ExtendedHenselConstruction(poly, [0, 0], 3):
```

$$\left[ \left[ x = \frac{-RootOf(\_Z^2 + y) + RootOf(\_Z^2 + y) y - \frac{1}{2} RootOf(\_Z^2 + y) z + \frac{1}{2} y^2}{y} \right], \right.$$

$$\left[ x = \frac{RootOf(\_Z^2 + y) - RootOf(\_Z^2 + y) y + \frac{1}{2} RootOf(\_Z^2 + y) z + \frac{1}{2} y^2}{y} \right],$$

$$\left. [x = -y] \right]$$

**Fig. 3.** Extended Hensel construction applied to a trivariate polynomial for computing its absolute factorization.

```
> P := PowerSeries([y]) :
> U := UnivariatePolynomialOverPowerSeries([y], x) :
> poly := y·x³ + (-2·y + 1)·x + y :
> OutputFlag :: name := 'parametric':
> parametricVar :: name := T :
> iter := 3 :
> verificationFlag :: boolean := true :
> U:-ExtendedHenselConstruction(poly, 0, iter, OutputFlag, parametricVar, verificationFlag);
```

$$\left[ [y = T^2, x = -T^3], \left[ y = T^2, x = \frac{RootOf(\_Z^2 + 1) T - T^3 RootOf(\_Z^2 + 1) + \frac{1}{2} T^4}{T} \right], \left[ y = T^2, x \right. \right.$$

$$= \frac{-RootOf(\_Z^2 + 1) T + T^3 RootOf(\_Z^2 + 1) + \frac{1}{2} T^4}{T} \right]$$

**Fig. 4.** Extended Hensel construction applied to a bivariate polynomial for computing its Puiseux parametrizations around the origin.

Fig. 4, this coincides with the Newton-Puiseux algorithm, thus computing the Puiseux parametrizations of a plane curve about a point; this functionality is at the core of the `LimitPoints` command. For the case of a univariate polynomial with multivariate polynomial coefficients, the EHC is a weak version of Jung-Abhyankar Theorem.

## 3    Applications

In this section we are concerned with two smooth maps $\Phi, \Psi : \mathbb{R}^2 \times \mathbb{R} \to \mathbb{R}^2$ whose state variables and bifurcation parameter are denoted by $(x, y)$ and $\lambda$, respectively. Since the Jacobian matrix of each map is not full rank at the origin, the Implicit Function Theorem fails at solving $(x, y)$ as a function of $\lambda$ locally around the origin. This causes bifurcations to reside in local zeros of each singular smooth map. We recall that these bifurcations are treated via first applying the Lyapunov-Schmidt reduction to a singular smooth map ending up with a reduced map of the form (2) and then passing the result through singularity theory techniques. Here, we follow the same approach except that we employ the `ExtendedHenselConstruction` command to compute the reduced map. The latter factorizes one of the equations around the origin and the resulting real branches that go through the origin are plugged in the other one to obtain the desired map (2). Once the map is computed we use the concept of recognition problem to identify the type of singularity.

### 3.1    The Pitchfork Bifurcation

In spite of simple structure, the pitchfork bifurcation is highly observed in physical phenomena mostly in the presence of symmetry breaking. For instance, [9] reports on *spontaneous mirror-symmetry breaking through a pitchfork bifurcation in a photonic molecule made up of two coupled photonic-crystal nanolasers.* Furthemore, authors in [11] study the pitchfork bifurcation arising in Lugiato–Lefever (LL) equation which is a model for a *passive Kerr resonator in an optical fiber ring cavity.* Finally, [6, Example 4.1] captures pitchfork bifurcation while analysing the local bifurcations of Chua's circuit. Here, we consider the exercise 3.2 on [8, p. 34]. Suppose that $\Phi : \mathbb{R}^2 \times \mathbb{R} \to \mathbb{R}^2$ is defined by $\begin{pmatrix} \Phi_1 \\ \Phi_2 \end{pmatrix}$ where

$$\Phi_1(x, y, \lambda) = 2x - 2y + 2x^2 + 2y^2 - \lambda x \tag{8}$$
$$\Phi_2(x, y, \lambda) = x - y + xy + y^2 - 3\lambda x.$$

To obtain the reduced system $g$ in (2) we pass $\Phi_1$ to the `ExtendedHensel` `Construction` giving rise to the branches in Fig. 5. Note that the second branch is not of interest as it does not pass the origin. Substituting the first branch into $\Phi_2$, modulo monomials of degree $\geq 4$, results in (Fig. 6)

$$g(y, \lambda) = 2y^3 - \frac{5}{2}y\lambda + \frac{9}{2}y^2\lambda - \frac{5}{4}y\lambda^2. \tag{9}$$

```
[> P := PowerSeries([y, lambda]) :
[> U := UnivariatePolynomialOverPowerSeries([y, lambda], x) :
[> poly := 2·x − 2·y + 2·x² + 2·y² − lambda·x :
[> U:-ExtendedHenselConstruction(poly, [0, 0], 4);
```

$$\left[\left[\left[y=0, \lambda=0, x=-2\,y^2+\frac{1}{2}\,y\lambda+y+4\,y^3-2\,y^2\lambda+\frac{1}{4}\,y\lambda^2+\frac{1}{8}\,y\lambda^3+8\,y^3\lambda-\frac{7}{4}\,y^2\lambda^2\right.\right.\right.$$
$$\left.-12\,y^4\right],\left[y=0, \lambda=0, x=2\,y^2-\frac{1}{2}\,y\lambda-y+\frac{1}{2}\,\lambda-1-4\,y^3+2\,y^2\lambda-\frac{1}{4}\,y\lambda^2-\frac{1}{8}\,y\lambda^3\right.$$
$$\left.\left.\left.-8\,y^3\lambda+\frac{7}{4}\,y^2\lambda^2+12\,y^4\right]\right]\right]$$

**Fig. 5.** EHC applied to $\Phi_1(x, y, \lambda)$.

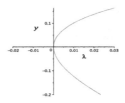

**Fig. 6.** Pitchfork bifurcation diagram associated with $g$ in Eq. (9).

Given $g$ in (9), the low order terms and the intrinsic generators are determined by $\mathbb{R}\{1, y, \lambda, y^2\}$ and $\{y\lambda, y^3\}$, respectively. Thus, Theorem 2 implies that $g$ satisfies the recognition problem for pitchfork

$$f(0,0) = \frac{\partial}{\partial y}f(0,0) = \frac{\partial}{\partial \lambda}f(0,0) = \frac{\partial^2}{\partial y^2}f(0,0) = 0$$

$$\frac{\partial}{\partial y}\frac{\partial}{\partial \lambda}f(0,0) \neq 0, \quad \frac{\partial^3}{\partial y^3}f(0,0) \neq 0$$

This proves that the original system $\Phi$ has pitchfork singularity located at the origin.

## 3.2 The Winged Cusp Bifurcation

The *winged cusp bifurcation problem* is defined by the equation $x^3 + \lambda^2 = 0$ and its corresponding bifurcation diagram $\{(x, \lambda) \mid x^3 + \lambda^2 = 0\}$ is exhibited via Fig. 7. Singularity theory tools have been utilized in the area of chemical engineering with the aim of studying the solutions of the continuous flow stirred tank reactor (CSTR) model. This study proves that the winged cusp bifurcation is the normal form for describing the *organizing center* of the bifurcation diagrams of the model produced by numerical methods. It, further, unravels more bifurcation diagrams that have not been reported through these numerical methods; see [7,8,13,14].

Now assume that $\Psi : \mathbb{R}^2 \times \mathbb{R} \rightarrow \mathbb{R}^2$ is given by $\begin{pmatrix} \Psi_1 \\ \Psi_2 \end{pmatrix}$ where

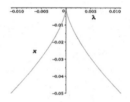

**Fig. 7.** The winged cusp bifurcation diagram.

$$\Psi_1(x, y, \lambda) = -2x + 3y + \lambda^2 + y^3 + x^4 \tag{10}$$
$$\Psi_2(x, y, \lambda) = 2x - 3y + y^2\lambda + x^3.$$

Applying the `ExtendedHenselConstruction` to $\Psi_2$ leads to the branches in Fig. 8.

```
> P := PowerSeries([y, lambda]) :
> U := UnivariatePolynomialOverPowerSeries([y, lambda], x) :
> poly := 2·x - 3·y + y²·lambda + x³ :
> U:-ExtendedHenselConstruction(poly, [0, 0], 3);
```
$$\left[\left[\left[y = 0, \lambda = 0, x = \frac{3}{2}y - \frac{1}{2}y^2\lambda - \frac{27}{16}y^3\right], \left[y = 0, \lambda = 0, x = -RootOf(\_Z^2 + 2) - \frac{3}{4}y\right.\right.\right.$$
$$\left.- \frac{27}{64}y^2\, RootOf(\_Z^2 + 2) + \frac{1}{4}y^2\lambda + \frac{27}{32}y^3\right], \left[y = 0, \lambda = 0, x = RootOf(\_Z^2 + 2) - \frac{3}{4}y\right.$$
$$\left.\left.\left.+ \frac{27}{64}y^2\, RootOf(\_Z^2 + 2) + \frac{1}{4}y^2\lambda + \frac{27}{32}y^3\right]\right]\right]$$

**Fig. 8.** EHC applied to $\Psi_2(x, y, \lambda)$.

Substituting the first branch into $\Psi_1$, modulo monomials of degree $\geq 4$, yields

$$g(y, \lambda) = \frac{35}{8}y^3 + \lambda^2 + y^2\lambda. \tag{11}$$

As $\{1, y, \lambda, y^2, y\lambda\}$ spans the space of low order terms and intrinsic generators are $\{\lambda^2, y^3\}$, Theorem 2 guarantees that $g$ satisfies the recognition problem for the winged cusp (Fig. 9)

$$f(0,0) = \frac{\partial}{\partial y}f(0,0) = \frac{\partial}{\partial \lambda}f(0,0) = \frac{\partial^2}{\partial y^2}f(0,0) = \frac{\partial}{\partial y}\frac{\partial}{\partial \lambda}f(0,0) = 0$$

$$\frac{\partial^2}{\partial \lambda^2}f(0,0) \neq 0, \quad \frac{\partial^3}{\partial y^3}f(0,0) \neq 0$$

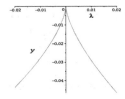

**Fig. 9.** Bifurcation diagram associated with $g$ in 11.

# References

1. Alvandi, P., Ataei, M., Kazemi, M., Moreno Maza, M.: On the extended hensel construction and its application to the computation of limit points. J. Symb. Comput. (2019, to appear)
2. Alvandi, P., Kazemi, M., Moreno Maza, M.: Computing limits of real multivariate rational functions. In: Proceedings of ISSAC 2016, pp. 39–46. ACM, New York (2016)
3. Alvandi, P., Ataei, M., Moreno Maza, M.: On the extended Hensel construction and its application to the computation of limit points. In: Proceedings of ISSAC 2017, pp. 13–20. ACM, New York (2017)
4. Gazor, M., Kazemi, M.: Symbolic local bifurcation analysis of scalar smooth maps. ArXiv:1507.06168 (2016)
5. Gazor, M., Kazemi, M.: A user guide for singularity. ArXiv:1601.00268 (2017)
6. Gazor, M., Kazemi, M.: Normal form analysis of $\mathbb{Z}_2$-equivariant singularities. Int. J. Bifurcat. Chaos **29**(2), 1950015-1–1950015-20 (2019)
7. Golubitsky, M., Keyfitz, B.L.: A qualitative study of the steady-state solutions for a continuous flow stirred tank chemical reactor. SIAM J. Math. Anal. **11**(2), 316–339 (1980)
8. Golubitsky, M., Stewart, I., Schaeffer, D. G.: Singularities and Groups in Bifurcation Theory, vol. 1–2. Springer, New York (1985 and 1988). https://doi.org/10.1007/978-1-4612-5034-0
9. Hamel, P., et al.: Spontaneous mirror-symmetry breaking in coupled photonic-crystal nanolasers. Nat. Photonics **9**, 311–315 (2015)
10. Labouriau, I.: Applications of singularity theory to neurobiology. Ph.D. thesis, Warwick University, (1984)
11. Rossi, J., Carretero-González, R., Kevrekidis, P.G., Haragus, M.: On the spontaneous time-reversal symmetry breaking in synchronously-pumped passive Kerr resonators. J. Phys. A: Math. Theor. **49**(45), 455201–455221 (2016)
12. Sasaki, T., Kako, F.: Solving multivariate algebraic equation by Hensel construction. Jpn. J. Ind. Appl. Math. **16**, 257–285 (1999)
13. Uppal, A., Ray, W.H., Poore, A.B.: The classification of the dynamic behavior of continuous stirred tank reactors-influence of reactor residence time. J. Chem. Eng. Sci. **31**(3), 205–214 (1976)
14. Zeldovich, Y.V., Zisin, U.A.: On the theory of thermal stress. Flow in an exothermic stirred reactor, II. Study of heat loss in a flow reactor. J. Tech. Phys. **11**(6), 501–508 (1941). (Russian)

# A Maple Package for the Symbolic Computation of Drazin Inverse Matrices with Multivariate Transcendental Functions Entries

Jorge Caravantes[1(⊠)], J. Rafael Sendra[1], and Juana Sendra[2]

[1] Dpto. de Física y Matemáticas, Universidad de Alcalá, Alcalá de Henares, Spain
jorge.caravantes@uah.es, rafael.sendra@uah.es
[2] Dpto. de Matemática Aplicada a las TIC, Universidad Politécnica de Madrid,
Madrid, Spain
juana.sendra@upm.es

**Abstract.** The study of Drazin inverses is an active research area that is developed, among others, in three directions: theory, applications and computation. This paper is framed in the computational part.

Many authors have addressed the problem of computing Drazin inverses of matrices whose entries belong to different domains: complex numbers, polynomial entries, rational functions, formal Laurent series, meromorphic functions. Furthermore, symbolic techniques have proven to be a suitable tools for this goal.

In general terms, the main contribution of this paper is the implementation, in a package, of the algorithmic ideas presented in [10,11]. Therefore, the package computes Drazin inverses of matrices whose entries are elements of a finite transcendental field extension of a computable field. The computation strategy consists in reducing the problem to the computation of Drazin inverses, via Gröbner bases, of matrices with rational functions entries.

More precisely, this paper presents a Maple computer algebra package, named **DrazinInverse**, that computes Drazin inverses of matrices whose entries are elements of a finite transcendental field extension of a computable field. In particular, the implemented algorithm can be applied to matrices over the field of meromorphic functions, in several complex variables, on a connected domain.

**Keywords:** Maple · Drazin inverse · Gröbner bases · Symbolic Computation · Meromorphic functions

---

The authors are partially supported by FEDER/Ministerio de Ciencia, Innovación y Universidades - Agencia Estatal de Investigación/MTM2017-88796-P (Symbolic Computation: new challenges in Algebra and Geometry together with its applications). J. R. Sendra and J. Caravantes are members of the research group ASYNACS (REF. CT-CE 2019/683).

J. Gerhard and I. Kotsireas (Eds.): MC 2019, CCIS 1125, pp. 156–170, 2020.
https://doi.org/10.1007/978-3-030-41258-6_12

# 1   Introduction

The study of Drazin inverses is an active research area that is developed, among others, in three directions: theory (see e.g. Chap. 4 in [2], Chap. 7 in [4,6,8]), applications (see e.g. Chapter 4 in [2], Chap. 7 in [4,6,8]) and computation (see e.g. [1,2,4,7–9,12]). This paper is framed in the computational part.

Many authors have addressed the problem of computing Drazin inverses of matrices whose entries belong to different domains: complex numbers, polynomial entries, rational functions, formal Laurent series, meromorphic functions (see [3,6,7,10,11,13]). Furthermore, symbolic techniques have proven to be a suitable tools for this goal.

In general terms, the main contribution of this paper is the implementation, in a package, of the algorithmic ideas presented in [10,11]. Therefore, the package computes Drazin inverses of matrices whose entries are elements of a finite transcendental field extension of a computable field. The computation strategy consists in reducing the problem to the computation of Drazin inverses, via Gröbner bases, of matrices with rational functions entries.

The structure of the paper is as follows. In Sect. 2 we fix the notation, we recall the basic notions on Drazin inverses, and we summarize the main ideas, results and algorithmic processes presented in [10,11]; for this purpose, the section is structured in several subsections. The algorithms in [10] and [11] are summarized here in Algorithms 1 and 3. In Sect. 3 we present the creation of a package in the computer algebra system Maple, that we call **DrazinInverse**. We also describe the different procedures in the package, and we illustrate them with a Maple Worksheet as well as with several examples.

# 2   Theoretical and Algorithmic Framework

In this section we fix the notation, and we recall the main notions, results and algorithmic processes that are used in our implementation; for further details we refer to [10] and [11].

## 2.1   The Notion of Drazin Inverse

Let $\mathbb{F}$ be a field, and $\mathbb{K}$ a computable subfield of $\mathbb{F}$. Let $\mathbf{w} = (w_1 \ldots, w_r)$ be a tuple of variables, and let $\mathbf{t} = (t_1, \ldots, t_r) \in (\mathbb{F} \setminus \mathbb{K})^r$ be a fixed tuple of different transcendental elements over $\mathbb{K}$. For instance, $\mathbb{K}$ could be $\mathbb{C}$ (the complex numbers field), $\mathbb{F}$ the meromorphic functions on a connected domain of $\mathbb{C}$, and $\mathbf{t} = (z, \sin(z), \mathrm{e}(z))$. We will also consider the following rings

$$\mathrm{R_w} = \mathcal{M}_{n \times n}(\mathbb{K}(\mathbf{w})), \text{ and } \mathrm{R_t} = \mathcal{M}_{n \times n}(\mathbb{K}(\mathbf{t}))$$

of $n \times n$ matrices with entries in $\mathbb{K}(\mathbf{w})$ and $\mathbb{K}(\mathbf{t})$, respectively. In this situation, the notions of Drazin index and Drazin inverse are defined as follows.

**Definition 1.** *Let $A$ be an $n \times n$ matrix with entries in a field; e.g. $A \in \mathrm{R_w}$ or $A \in \mathrm{R_t}$.*

1. *The* Drazin index *of $A$ is the smallest non-negative integer $k$ such that* $\text{rank}(A^k) = \text{rank}(A^{k+1})$; *we denote the index by* $\text{index}(A)$.
2. *The* Drazin inverse *of $A$ is the unique matrix satisfying the following matrix equations:*

$$\begin{cases} A^{\text{index}(A)+1} \cdot X = A^{\text{index}(A)} \\ A \cdot X = X \cdot A \\ X \cdot A \cdot X = X. \end{cases} \tag{1}$$

*We denote by $\mathfrak{D}(A)$ the Drazin inverse of $A$.*

Next, we introduce the notion of denominator of $A$.

**Definition 2.** *The denominator of a matrix $A \in \mathrm{R_w}$, denoted by* $\text{denom}(A)$, *is the least common multiple of the denominators of all entries, taken in reduced form, of $A$.*

## 2.2  Gröbner Basis Computation of Drazin Inverses

Using that Drazin inverse computation can be translated into an elimination theory question, in [10], we use Gröbner bases to determine it when the matrix belongs to $\mathrm{R_w}$. The strategy in [10] is as follows. We decompose the system of Eq. (1) in two subsystems, one of them carrying the linear equations, and the other the quadratic equations. More precisely, we consider the system

$$\mathcal{L} = \begin{cases} A^{\text{index}(A)+1} \cdot X - A^{\text{index}(A)} = \mathbf{O} \\ A \cdot X - X \cdot A = \mathbf{O} \end{cases}$$

that is linear, and the subsystem $\{X \cdot A \cdot X - X = \mathbf{O}\}$ that is an algebraic system of quadratic polynomials. Solving the compatible system $\mathcal{L}$, and substituting the solution in the quadratic system we get a new (in general) quadratic system, equivalent to (1), and having less variables. Let $\mathcal{Q}$ be the resulting system, and $\mathcal{F}$ be the set of polynomials defining $\mathcal{Q}$. The solution of a Gröbner basis of $\mathcal{F}$, w.r.t. a lexicographic order of its variables, jointly with the solution of $\mathcal{L}$, provides $\mathfrak{D}(A)$. These ideas yield to the following algorithm.

---

**Algorithm 1.** Drazin inverse via Gröbner basis (see Fig. 1)

---

Given $A \in \mathrm{R_w}$ the algorithm computes its Drazin inverse $\mathfrak{D}(A)$.

1: Compute $k := \text{index}(A)$.
2: Solve the linear system $\mathcal{L} = \{A^{k+1}\hat{X} - A^k = \mathbf{O}, A\hat{X} - \hat{X}A = \mathbf{O}\}$ and substitute its solution $\mathcal{S}$ in $\hat{X}A\hat{X} - \hat{X} = \mathbf{O}$. Let $XAX - X = \mathbf{O}$ be the resulting system and $V$ the set of variables.
3: Compute a Gröbner basis $\mathcal{G}$ of the polynomials defining $XAX - X = \mathbf{O}$ with respect to a lexicographic order of $V$.
4: Substitute the solution provided by $\mathcal{G}$ and $\mathcal{S}$ in $\hat{X}$ to get $\mathfrak{D}(A)$.

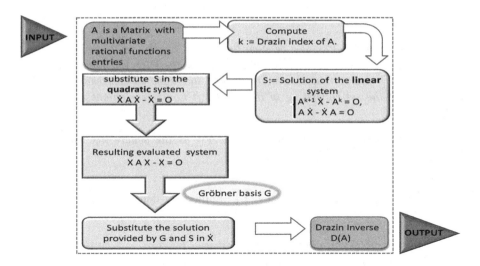

**Fig. 1.** Scheme of Algorithm 1

### 2.3  Drazin Inverses Under Specializations

In this subsection we summarize the ideas in [11] to compute the Drazin inverse of matrices in $R_t$. The strategy consists in reducing the computation in $R_t$ to the computation in $R_w$. More precisely, we proved the existence, and actual computation, of a multivariate polynomial in $\mathbb{C}[\mathbf{w}]$ (that we call evaluation polynomial, see Definition 4), such that if it does not vanish at $\mathbf{t}$ the computation of the Drazin inverse over $\mathbb{C}(\mathbf{t})$ is reduced to $\mathbb{C}(\mathbf{w})$.

In this context, we define $\operatorname{denom}(A)$ to be the least common multiple of all denominators in the entries of $A$. Then, we consider the map

$$\Phi_\mathbf{t} : \{A \in R_\mathbf{w} \mid \operatorname{denom}(A)(\mathbf{t}) \neq 0\} \longrightarrow R_\mathbf{t}$$
$$A(\mathbf{w}) = (a_{i,j}(\mathbf{w}))_{1 \leq i,j \leq n} \longmapsto A(\mathbf{t}) = (a_{i,j}(\mathbf{t}))_{1 \leq i,j \leq n}$$

and we analyze the behavior of the Drazin inverse of matrices in $R_\mathbf{t}$ under the action of $\Phi_\mathbf{t}$. More precisely, if $A \in R_\mathbf{t}$, we study when there exists $A^*$ such that

$$\Phi_\mathbf{t}\left(\mathfrak{D}(A^*)\right) = \mathfrak{D}(\Phi_\mathbf{t}\left(A^*\right)),$$

that is, when

$$\mathfrak{D}(A^*)(\mathbf{t}) = \mathfrak{D}(A(\mathbf{t})).$$

In this case, we compute the Drazin inverse $\mathfrak{D}(A)$ specializing $\mathfrak{D}(A^*)$. This motivates the next definition.

**Definition 3.** *We say that $A \in R_\mathbf{t}$ behaves properly under specialization if there exists $A^* \in R_\mathbf{w}$ such that $\Phi_\mathbf{t}\left(\mathfrak{D}(A^*)\right) = \mathfrak{D}(A)$.*

In order to certify that the proper behavior, under specialization, of a matrix in $R_\mathbf{t}$, we will use a polynomial in $\mathbb{K}[\mathbf{w}]$.

**Definition 4**

1. Let $A \in \mathrm{R_t}$ and let $A^* \in \mathrm{R_w}$ be such that $\Phi_\mathbf{t}(A^*) = A$.
2. For $j \in \{1, \dots, \mathrm{index}(A^*)\}$, let $P_j A_j^* = L_j U_j$ be the PA=LU factorization of $A_j^*$, where $P_j$ are permutation matrices and the diagonal entries of $L_j$ are 1.
3. Let $D_{U_j}$ be the set consisting in the most left non-zero element in each row of $U_j$

We define the evaluation polynomial of $(A, A^*)$ as the square-free part of the polynomial

$$\prod_{j=1}^{\mathrm{index}(A)} \mathrm{denom}(L_j)\,\mathrm{denom}(U_j) \cdot \prod_{g \in D_{U_j}} \mathrm{numer}(g) \cdot \mathrm{denom}(\mathfrak{D}(A^*))(\mathbf{w})$$

where numer denotes the numerator of a rational function in $\mathbb{K}(\mathbf{w})$. We denote the evaluation polynomial by $\mathrm{EvalPol}_{A,A^*}(\mathbf{w})$.

The following result, proved in [11], establishes the main property of the evaluation polynomial.

**Theorem 1.** Let $A \in \mathrm{R_t}$ and let $A^* \in \mathrm{R_w}$ be such that $\Phi_\mathbf{t}(A^*) = A$. If $\mathrm{EvalPol}_{A,A^*}(\mathbf{t}) \neq 0$, then $A$ specializes properly at $(A^*, \mathbf{t})$.

In the sequel, using the previous ideas, we derive an algorithm. First of all, we need to design a method that certifies the correctness of the output; this is done in Algorithm 2. For this purpose we will use the evaluation polynomial. Let $A \in \mathrm{R_t}$ and $A^* \in \mathrm{R_w}$ such that $\Phi_\mathbf{t}(A^*) = A$. We have to check whether $\mathrm{EvalPol}_{A,A^*}(\mathbf{t})$ is zero. So, we need to check whether an algebraic expression involving several transcendental elements is zero or not. This implies to simplify expressions relating transcendental elements; this, in general, may be a complicated task (see e.g. [5]). Algorithmically, we will proceed as follows; see Fig. 2. First, we try to simplify $\mathrm{EvalPol}_{A,A^*}(\mathbf{t})$. If we get zero, then the answer is clear. If we get a non-zero element, it still may happen that a further simplification yields to zero. In this situation, we iteratively evaluate the expression at different, randomly chosen, real numbers, till either we get a non-zero quantity, in which case we can ensure that $\mathrm{EvalPol}_{A,A^*}(\mathbf{t}) \neq 0$, or till 200 evaluations have been performed, in which case, we cannot ensure that $\mathrm{EvalPol}_{A,A^*}(\mathbf{t}) \neq 0$. In the later case, we will need to check whether the answer exists (see Remark 1) and is correct by substituting in the Eq. (1). The next algorithm is derived from the previous ideas.

**Algorithm 2.** Evaluation Test (See Fig. 2)

Given $A \in R_t$ and $A^* \in R_w$ such that $\Phi_t(A^*) = A$ the algorithm checks whether $A$ specializes properly at $(A^*, \mathbf{t})$.

1: Compute $\text{EvalPol}_{A,A^*}(\mathbf{w})$.
2: $T(\mathbf{t}(\mathbf{z})) := \text{EvalPol}_{A,A^*}(\mathbf{t})$                     ▷ Let $\mathbf{t}$ depend on the variables $\mathbf{z}$
3: **if** $T(\mathbf{t}(\mathbf{z})) = 0$ **then return 0.**
4: **else**
5:     $k = 0$.
6:     $K := T(\mathbf{t}(\mathbf{z}_k))$                     ▷ Let $\mathbf{z}_k$ be a tuple of random numbers.
7:     **while** $k < 200$ and $K = 0$ **do**
8:         $k = k + 1$.
9:         $K := T(\mathbf{t}(\mathbf{z}_k))$              ▷ Let $\mathbf{z}_k$ be a tuple of random numbers.
10:     **end while**
11: **end if**
12: **if** $K = 0$ **then return 0**
13: **else**
14: **return 1**
15: **end if**

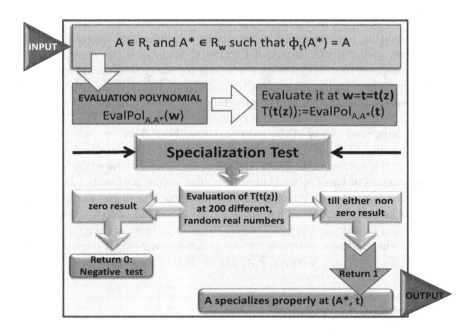

**Fig. 2.** Scheme of Algorithm 2

The next algorithm is the main one, and summarizes all previous ideas.

---

**Algorithm 3.** Drazin inverse via Specialization (see Fig. 3)

---

Given $A \in R_t$ the algorithm computes its Drazin inverse $\mathfrak{D}(A)$.

1: [Specialization] Replace $\mathbf{t}$ by $\mathbf{w}$ in $A$ to get $A^* \in R_{\mathbf{w}}$.     ▷ note that $\Phi_t(A^*) = A$
2: [Inverse computation] Apply Algorithm 1 to $A^*$ to get $\mathfrak{D}(A^*)$.
3: [Specialization test] Apply Algorithm 2 to get $\mathbf{c}$.
4: [Specialization of the inverse] $M := \Phi_t(\mathfrak{D}(A^*))$.
5: **if** $\mathbf{c} = 1$ **then return** $M$
6: **else**
7:     check whether $M$ exists and the pair $M, A$ satisfy the equations (1)
8:     **if yes then return** $M$
9:     **else**
10:         **return** The method fails.
11:     **end if**
12: **end if**

---

*Remark 1*

1. The evaluating polynomial contains all the denominators appearing in the algorithm, included the output. If EvalPol($\mathbf{t}$) vanishes by an specialization, it could happen that $M$ contains a non existing entry (because of a division by 0), but the simplification algorithms did not detect it. In this case, we specialize the matrix at several real numbers to check whether the resulting matrix is real; if not, we cannot guarantee the existence of $M$. This, however, has not happened in any of the tests that have been done.
2. The vanishing of EvalPol($\mathbf{t}$) proves the existence of an algebraic relation involving the entries of $\mathbf{t}$. Let $I$ the ideal in $\mathbb{K}[\mathbf{w}]$ generated by the irreducible factors of EvalPol($\mathbf{w}$) that vanish after specialization. If $I$ is a prime ideal, then $\mathbb{K}[\mathbf{w}]/I$ is an integral domain and then we can run the algorithm again considering the field of fractions of $\mathbb{K}[\mathbf{w}]/I$ instead of $\mathbb{K}(\mathbf{w})$. Otherwise, there would be zero divisors in $\mathbb{K}[\mathbf{w}]/I$, so perhaps there are zero divisors in $\mathbb{K}[\mathbf{t}]$ and the existence of Drazin inverse is not guaranteed. If the entries of $\mathbf{t}$ are analytic functions, however, $\mathbb{K}[\mathbf{t}]$ has no zero divisors.

## 2.4   An Illustrating Example

Let $\mathbf{t} = (\cos(z), e^z)$ and $\mathbf{w} = (w_1, w_2)$. Let

$$A(\mathbf{t}) = \begin{pmatrix} 0 & 0 & 2\,e^z \cos(z) \\ 2\cos(z)\,e^{-z} & 2\,e^{-z} & 2 - e^{-z} \\ 3\cos(z)\,e^{-z} & 3\,e^{-z} & 6\,e^{-z} \end{pmatrix} \in R_t.$$

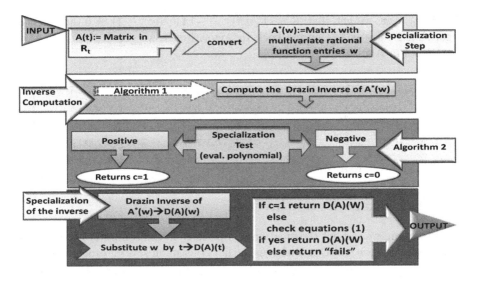

**Fig. 3.** Scheme of Algorithm 3

We want to compute the Drazin inverse $\mathfrak{D}(A)$ of $A$. For this purpose, we first associate to $A$ a matrix $A^* \in R_{\mathbf{w}} = \mathcal{M}_{3\times 3}(\mathbb{C}(\mathbf{w}))$. In the following, we describe the computations of the heart of the algorithm.

Step 1 in Algorithm 3. By replacing $w_1 := \cos(z), w_2 := e^z$ we get

$$
A^*(\mathbf{w}) = \begin{pmatrix}
0 & 0 & 2\,w_2 w_1 \\
\dfrac{2\,w_1}{w_2} & \dfrac{2}{w_2} & \dfrac{2}{w_2} \\
\dfrac{3\,w_1}{w_2} & \dfrac{3}{w_2} & \dfrac{6}{w_2}
\end{pmatrix} \in R_{\mathbf{w}}.
$$

Step 2 in Algorithm 3. Applying Algorithm 1 we get

$$
\mathfrak{D}(A^*)(\mathbf{w}) = \begin{pmatrix}
\dfrac{-4\,w_2{}^3 w_1{}^2}{3\left(w_1{}^4 w_2{}^4 - 2\,w_1{}^2 w_2{}^2 + 1\right)} & \dfrac{-4\,w_2{}^3 w_1}{3\left(w_1{}^2 w_2{}^2 - 1\right)^2} & \dfrac{3\left(w_1{}^2 w_2{}^2 + 5\right) w_2{}^3 w_1}{9\left(w_1{}^2 w_2{}^2 - 1\right)^2} \\
\dfrac{\left(w_1{}^2 w_2{}^2 + 3\right) w_2 w_1}{3\left(w_1{}^2 w_2{}^2 - 1\right)^2} & \dfrac{\left(w_1{}^2 w_2{}^2 + 3\right) w_2}{3\left(w_1{}^2 w_2{}^2 - 1\right)^2} & \dfrac{-\left(5\,w_1{}^2 w_2{}^2 + 3\right) w_2}{9\left(w_1{}^2 w_2{}^2 - 1\right)^2} \\
\dfrac{w_2 w_1}{2\left(w_1{}^2 w_2{}^2 - 1\right)} & \dfrac{w_2}{2\left(w_1{}^2 w_2{}^2 - 1\right)} & \dfrac{-w_2}{3\left(w_1{}^2 w_2{}^2 - 1\right)}
\end{pmatrix}.
$$

Step 3 in Algorithm 3. We execute Algorithm 2. First we get the evaluation polynomial (see Step 1 in Algorithm 2)

$$
\text{EvalPol}_{A,A^*}(\mathbf{w}) = 144 w_2 w_1 (w_2 w_1 - 1)(w_1 w_2 + 1) \in \mathbb{C}[\mathbf{w}].
$$

and we evaluate it at $\mathbf{t}$ (see Step 2 in 2) to get

$$
T((\mathbf{z})) = 144 e^z \cos(\mathbf{z})(e^z \cos(\mathbf{z}) - 1)(e^z \cos(\mathbf{z}) + 1).
$$

Taking $z_0 = \pi$, we get that $K = T(\mathbf{t}(\pi)) = -144e^{\pi}(e^{2\pi} - 1) \neq 0$, and hence Algorithm 2 returns $\mathbf{1}$. Coming back to Algorithm 3, we get $M = \Phi_{\mathbf{t}}(\mathfrak{D}(A^*))$ (see Step 4 in Algorithm 3)

$$M = \begin{pmatrix} \dfrac{-4\,(e^z)^3\cos(z)^2}{3\,(\cos(z)^4(e^z)^4 - 2\cos(z)^2(e^z)^2 + 1)} & \dfrac{-4\,(e^z)^3\cos(z)}{3\,(\cos(z)^2(e^z)^2 - 1)^2} & \dfrac{3\,(\cos(z)^2(e^z)^2 + 5)(e^z)^3\cos(z)}{9\,(\cos(z)^2(e^z)^2 - 1)^2} \\[2ex] \dfrac{(\cos(z)^2(e^z)^2 + 3)e^z\cos(z)}{3\,(\cos(z)^2(e^z)^2 - 1)^2} & \dfrac{(\cos(z)^2(e^z)^2 + 3)e^z}{3\,(\cos(z)^2(e^z)^2 - 1)^2} & \dfrac{-(5\cos(z)^2(e^z)^2 + 3)e^z}{9\,(\cos(z)^2(e^z)^2 - 1)^2} \\[2ex] \dfrac{e^z\cos(z)}{2\,(\cos(z)^2(e^z)^2 - 1)} & \dfrac{e^z}{2\,(\cos(z)^2(e^z)^2 - 1)} & \dfrac{-e^z}{3\,(\cos(z)^2(e^z)^2 - 1)} \end{pmatrix}$$

Finally, since $\mathbf{c} = \mathbf{1}$, Algorithm 3 returns $\mathfrak{D}(A) = M$     □.

# 3   The Package DrazinInverse

In this section, we present the creation of a package in the computer algebra system Maple, that we call **DrazinInverse** that consists in several procedures that are based and implement Algorithms 1, 2 and 3, described above. This package computes the Drazin inverse of matrices whose entries are elements of a finite transcendental field extension of a computable field. More precisely, matrices whose coefficients are rational functions of finitely transcendental elements over a computable subfield of the field $\mathbb{C}$ of the complex numbers.

## 3.1   Summary: Overview of the Software Structure

The created Maple package **DrazinInverse** is initialized by the command:

> with(DrazinInverse):

The main procedure in the package is **DrazinInv**, which determines the Drazin inverse of a matrix. This command implements Algorithm 3 and invokes the following sub-procedures based on the definitions and results introduces by the authors in [10] and [11] and summarized in Sect. 2:

- **TypeMatrixAlg:** Checks whether a given matrix is in $R_{\mathbf{w}}$.
- **DrazinIndex:** Determines the Drazin Index of an square matrix (see Definition 1).
- **RationalDrazin:** Computes the Drazin inverse of a matrix in $R_{\mathbf{w}}$.
- **MatrixDenom:** Determines the denominator of a matrix (see Definition 2).
- **FirstPivot:** Determines the first non zero coefficient of a specific row of a given matrix.
- **EvaluationPol:** Determines the evaluation polynomial of a matrix (see Definition 4).
- **SpecializationTest:** Checks whether a multivariate polynomial does not vanish for random values of the variables.
- **CheckDarzinInverse:** Checks whether two given matrices are the Drazin inverse of each other.

In the following, we give a brief description of the procedures. The package is available at http://www3.uah.es/rsendra/software.html.

## 3.2   Description of the Individual Software Components

> **TypeMatrixAlg**

(i) *Feature:* This procedure checks the algebraic character of a given matrix. Briefly, the input and output of the procedure can be stated as follows:
  ▷ INPUT: Given a matrix A.
  ▷ OUTPUT: "true" if A has rational function entries otherwise "false".
(ii) *Calling Sequence:* > TypeMatrixAlg(A);
(iii) *Parameters: A* is a matrix.
(iv) *Synopsis:* The procedure checks whether the input matrix $A$ has rational function entries, it is whether $A$ is in the ring of matrices $\mathcal{M}_{n \times n}(\mathbb{K}(\mathbf{w}))$. The procedure outputs the message "true" or "false".

> **DrazinIndex**

(i) *Feature:* This procedure determines the Drazin Index of an square matrix. Briefly, the input and output of the procedure can be stated as follows:
  ▷ INPUT: Given an square matrix $A$.
  ▷ OUTPUT: Compute the Drazin Index of $A$.
(ii) *Calling Sequence:* > DrazinIndex(A);
(iii) *Parameters: A* is an square matrix.
(iv) *Synopsis:* The procedure computes the Drazin Index of $A$ (Definition 1).

> **RationalDrazin**

(i) *Feature:* This procedure computes the Drazin inverse of a matrix with multivariate rational functions as entries. Briefly, the input and output of the procedure can be stated as follows:
  ▷ INPUT: Given an square matrix $A$ with multivariate rational functions as entries.
  ▷ OUTPUT: Compute the Drazin inverse of $A$.
(ii) *Calling Sequence:* > RationalDrazin(A);
(iii) *Parameters: $A \in \mathcal{M}_{n \times n}(\mathbb{K}(\mathbf{w}))$.*
(iv) *Synopsis:* Basically, the procedure computes the Drazin inverse of a matrix with multivariate rational functions as entries using Gröbner basis. See Algorithm 1. First, the procedure checks whether $A \in \mathcal{M}_{n \times n}(\mathbb{K}(\mathbf{w}))$ with the above command TypeMatrixAlg. In the affirmative case, it outputs the Drazin inverse and the Drazin Index of $A$. Otherwise, the procedure outputs the message "$A$ is not a Matrix with rational function elements".

> **MatrixDenom**

(i) *Feature:* This procedure determines the denominator of a matrix $A$ (see Definition 2). Briefly, the input and output of the procedure can be stated as follows:

▷ INPUT: Given a matrix $A$.

▷ OUTPUT: Compute the least common multiple of the denominators of all entries, taken in reduced form, of $A$.

(ii) *Calling Sequence:* > MatrixDenom(A);

(iii) *Parameters:* $A$ is a matrix.

(iv) *Synopsis:* First, the procedure checks whether $A \in \mathcal{M}_{n \times n}(\mathbb{K}(\mathbf{w}))$ with the above command TypeMatrixAlg. If it succeeds, it computes the denominator of $A$, denom($A$) (Definition 2). Otherwise, the procedure outputs the message "$A$ is not a Matrix with rational function elements".

## > FirstPivot

(i) *Feature:* This procedure determines the pivot of an specific row of a matrix. Briefly, the input and output of the procedure can be stated as follows:

▷ INPUT: Given a matrix $A$ and a natural number $i$.

▷ OUTPUT: Compute the first non zero coefficient of the $i$-th row of $A$.

(ii) *Calling Sequence:* > FirstPivot(A,i);

(iii) *Parameters:* $A$ is a matrix, $i \in \mathbb{N}$.

(iv) *Synopsis:* The procedure determines the first non zero coefficient of a specific row of a matrix.

## > EvaluationPol

(i) *Feature:* This procedure determines the evaluation polynomial of a matrix $A$ (Definition 4). Briefly, the input and output of the procedure can be stated as follows:

▷ INPUT: Given an square matrix $A$ and $k = \text{index} A$.

▷ OUTPUT: Compute the evaluation polynomial of $A$.

(ii) *Calling Sequence:* > EvaluationPol(A,B,k);

(iii) *Parameters:* $A \in \mathcal{M}_{n \times n}(\mathbb{K}(\mathbf{w}))$, $B$ is the Drazin inverse of $A$, and $k \in \mathbb{N}$.

(iv) *Synopsis:* First, the procedure checks whether the input matrices $A, B$ belong to $\mathcal{M}_{n \times n}(\mathbb{K}(\mathbf{w}))$ with the above command TypeMatrixAlg. If it succeeds, it computes the evaluation polynomial of $A$ (Definition 4) invoking the command MatrixDenom.

## > SpecializationTest

(i) *Feature:* This procedure checks whether a multivariate polynomial does not vanish for random values of the variables. Briefly, the input and output of the procedure can be stated as follows:

▷ INPUT: Given a multivariate polynomial $P$.

▷ OUTPUT: Check if $P(\alpha) = 0$ for randomly chosen 200 values of $\alpha$.

(ii) *Calling Sequence:* > SpecializationTest(P);

(iii) *Parameters:* $P$ is a multivariate polynomial.

(iv) *Synopsis:* Basically, the procedure tests whether the multivariate polynomial $P$ does not vanish for at most random 200 values, stoping either when a non-zero value is reached or when the 200 specializations have been performed. It outputs the messages "positive test" or "negative test". This procedure will be applied to check the specialization criterium described in Subsect. 2.3.

## > DrazinInv

(i) *Feature:* This procedure determines the Drazin inverse of a matrix whose entries are elements of a finite transcendental field extension of a computable field. Briefly, the input and output of the procedure can be stated as follows:
  ▷ INPUT: Given an square martix $A$.
  ▷ OUTPUT: Compute the Drazin inverse, $\mathfrak{D}(A)$, of $A$.
(ii) *Calling Sequence:* > DrazinInv(A);
(iii) *Parameters:* $A \in \mathcal{M}_{n \times n}(\mathbb{K}(\mathbf{t}))$.
(iv) *Synopsis:* Basically, the procedure computes Drazin inverse $\mathfrak{D}(A)$ of $A$ using Algorithm 3. More precisely, first the procedure associates to $A$ a matrix $A^*$, whose entries are rational functions in several variables; secondly it computes the Drazin inverse of $A^*$ using the RationalDrazin command. Following it checks the correctness of the method using the SpecializationTest and CheckDrazinInverse commands.

## > CheckDarzinInverse

(i) *Feature:* This procedure checks if two given matrices are the Drazin inverse of each other. Briefly, the input and output of the procedure can be stated as follows:
  ▷ INPUT: Given, two matrices $A, B$ and the Drazin Index $k$ of $A$.
  ▷ OUTPUT: Check if $B$ is the Drazin inverse of $A$.
(ii) *Calling Sequence:* > CheckDarzinInverse(A,B,k);
(iii) *Parameters:* $A, B \in \mathcal{M}_{n \times n}(\mathbb{K}(\mathbf{t}))$ and $k \in \mathbb{N}$ ($k = \text{index}(A)$).
(iv) *Synopsis:* The procedure checks whether $B$ is the Drazin inverse of $A$ substituting the matrices in the system of Eq. (1).

### 3.3  Illustrative Examples of the Usage of the Package Commands

In order to use the package, download the file DrazinInverse.m from http://www3.uah.es/rsendra/software.html and save it as your local folder. After starting Maple you redefine the variable libname as
  > libname:=libname,'path of user local folder';
    In this situation, after executing the command with(DrazinInverse), the package is ready to be used. In Fig. 4 we provide a Maple Worksheet illustrating the usage of our package for the computation of an inverse Drazin matrix over the field of meromorphic functions.
    We also provide, at http://www3.uah.es/rsendra/software.html, several sample files with the results obtained applying the package to matrices with different meromorphic functions entries. To read the sample files we only have to invoke:
  > read('C:/user root saved file/sample.txt');

```
> with(DrazinInverse);
```
$[$ *CheckDrazinInverse, DrazinIndex, DrazinInv, FirstPivot, MatrixDenom, RationalDrazin,*     (1)
   *SpecializationTest, TypeMatrixAlg*$]$

```
> A := Matrix(3, 3, {(1, 1) = 0, (1, 2) = 0, (1, 3) = 2*exp(z)*cos
  (z), (2, 1) = 2*cos(z)/exp(z), (2, 2) = 2/exp(z), (2, 3) = 2/exp
  (z), (3, 1) = 3*cos(z)/exp(z), (3, 2) = 3/exp(z), (3, 3) = 6/exp
  (z)});
```

$$A := \begin{bmatrix} 0 & 0 & 2\,e^{z}\cos(z) \\ \dfrac{2\cos(z)}{e^{z}} & \dfrac{2}{e^{z}} & \dfrac{2}{e^{z}} \\ \dfrac{3\cos(z)}{e^{z}} & \dfrac{3}{e^{z}} & \dfrac{6}{e^{z}} \end{bmatrix} \qquad (2)$$

```
> DA:=DrazinInv(A);
```

*The Drazin Index is*

1

*Positive test*

*The Inverse Drazin Matrix is*

$$DA := \left[\left[ -\frac{4}{3}\,\frac{\left(e^{z}\right)^{3}\cos(z)^{2}}{\cos(z)^{4}\left(e^{z}\right)^{4} - 2\cos(z)^{2}\left(e^{z}\right)^{2} + 1}, \ -\frac{4}{3}\,\frac{\left(e^{z}\right)^{3}\cos(z)}{\left(\cos(z)^{2}\left(e^{z}\right)^{2} - 1\right)^{2}}, \right.\right. \qquad (3)$$

$$\left. \frac{1}{9}\,\frac{\left(3\cos(z)^{2}\left(e^{z}\right)^{2} + 5\right)\left(e^{z}\right)^{3}\cos(z)}{\left(\cos(z)^{2}\left(e^{z}\right)^{2} - 1\right)^{2}} \right],$$

$$\left[ \frac{1}{3}\,\frac{\left(\cos(z)^{2}\left(e^{z}\right)^{2} + 3\right)e^{z}\cos(z)}{\left(\cos(z)^{2}\left(e^{z}\right)^{2} - 1\right)^{2}}, \ \frac{1}{3}\,\frac{\left(\cos(z)^{2}\left(e^{z}\right)^{2} + 3\right)e^{z}}{\left(\cos(z)^{2}\left(e^{z}\right)^{2} - 1\right)^{2}}, \right.$$

$$\left. -\frac{1}{9}\,\frac{\left(5\cos(z)^{2}\left(e^{z}\right)^{2} + 3\right)e^{z}}{\left(\cos(z)^{2}\left(e^{z}\right)^{2} - 1\right)^{2}} \right],$$

$$\left.\left[ \frac{1}{2}\,\frac{e^{z}\cos(z)}{\cos(z)^{2}\left(e^{z}\right)^{2} - 1}, \ \frac{1}{2}\,\frac{e^{z}}{\cos(z)^{2}\left(e^{z}\right)^{2} - 1}, \ -\frac{1}{3}\,\frac{e^{z}}{\cos(z)^{2}\left(e^{z}\right)^{2} - 1} \right]\right]$$

```
> CheckDrazinInverse(A,DA,DrazinIndex(A));
```

*YES*     (4)

**Fig. 4.** Maple worksheet

In the following we describe the data of the files. Each file contains the result of applying the package to a matrix over the field of rational functions $\mathbb{C}(\mathbf{t})$ where $\mathbf{t} = (t_1, t_2)$ for different selections of the transcendental elements $t_1, t_2$. Moreover, In the five first cases the matrix is $3 \times 3$ while in the second to the last is $5 \times 5$ and the last is $10 \times 10$.

1. In the sample file "sample1.text" we take $\mathbf{t} = (\Gamma(z), \zeta(z))$ with $z$ independent complex variable, where $\Gamma(z)$ is the *Gamma function*, that is

$$\Gamma(z) = \int_0^\infty e^{-t} t^{z-1} dt,$$

and $\zeta(z)$ is the *Riemann Zeta* function defined for $Re(z) > 1$ by

$$\zeta(z) = \sum_{i=1}^\infty \frac{1}{i^z}.$$

2. In the sample file "sample2.text" we take $\mathbf{t} = (\Gamma(z), \mathsf{B}(z_1, z_2))$ with $z, z_1, z_2$ independent complex variables, where $\Gamma(z)$ is the *Gamma function* and $\mathsf{B}(z_1, z_2)$ is the *Beta function*, that is

$$\mathsf{B}(z_1, z_2) = \frac{\Gamma(z_1)\Gamma(z_2)}{\Gamma(z_1 + z_2)}.$$

3. In the sample file "sample3.text" we take $\mathbf{t} = (\Gamma(z), \text{HankelH2}(v_1, v_2))$ with $z, v_1, v_2$ independent complex variables, where $\Gamma(z)$ is the *Gamma function* and HankelH2 is the *Hankel* function, also known as the Bessel function of the third kind.

4. In the sample file "sample4.text" we take $\mathbf{t} = (\Gamma(z), \text{Si}(x))$ with $z, x$ independent complex variables, where $\Gamma(z)$ is the *Gamma function* and $\text{Si}(x)$ is the *Sine Integral*, that is

$$\text{Si}(z) = \int_0^x \frac{\sin(t)}{t} dt.$$

5. In the sample file "sample5.text" we take $\mathbf{t} = (\text{FresnelC}(x), \text{Shi}(z))$ with $x, z$ independent complex variables, where $\text{FresnelC}(x)$ is the *Fresnel Cosine Integral* defined by:

$$\text{FresnelC}(x) = \int_0^x \cos(\frac{\pi t^2}{2}) dt,$$

and $\text{Shi}(z)$ is the *Hyperbolic Sine Integral* defined as

$$\text{Si}(z) = \int_0^z \frac{\sinh(t)}{t} dt.$$

6. In the sample file "sample6.text" we take $\mathbf{t} = (\ln(z), \text{CoulombF}(L, n, p))$ with $z, L, n, p$ independent complex variables, where $\text{CoulombF}(L, n, p)$ is the *Regular Coulomb wave function* satisfying the differential equation:

$$y''(x) + (1 - \frac{2n}{x} - \frac{L(L+1)}{x^2})y(x) = 0.$$

7. In the sample file "sample7.text" we take $(t_1, t_2) = (\cos(z)^2, e(z))$.

# References

1. Avrachenkov, K.E., Filar, J.A., Howlett, P.G.: Analytic Perturbation Theory and Its Applications. SIAM (2013)
2. Ben-Israel, A., Greville, T.N.E.: Generalized Inverses: Theory and Applications, 2nd edn. Springer, Heidelberg (2003). https://doi.org/10.1007/b97366
3. Bu, F., Wei, Y.: The algorithm for computing the Drazin inverses of two-variable polynomial matrices. Appl. Math. Comput. **147**, 805–836 (2004)
4. Campbell S.L., Meyer, C.D.: Generalized inverses of linear transformations series: classics in applied mathematics. SIAM (2009)
5. Corless, R.M., Davenport, J.H., Jeffrey, D.J., Litt, G., Watt, S.M.: Reasoning about the elementary functions of complex analysis. In: Campbell, J.A., Roanes-Lozano, E. (eds.) AISC 2000. LNCS (LNAI), vol. 1930, pp. 115–126. Springer, Heidelberg (2001). https://doi.org/10.1007/3-540-44990-6_9
6. Diao, H., Wei, Y., Qiao, S.: Displacement rank of the Drazin inverse. J. Comput. Appl. Math. **167**, 147–161 (2004)
7. Ji, J.: A finite algorithm for the Drazin inverse of a polynomial matrix. Appl. Math. Comput. **130**, 243–251 (2002)
8. Ljubisavljević, J., Cvetković-Ilić, D.S.: Additive results for the Drazin inverse of block matrices and applications. J. Comput. Appl. Math. **235**, 3683–3690 (2011)
9. Miljković, S., Miladinović, M., Stanimirović, P.S., Weib, Y.: Gradient methods for computing the Drazin-inverse solution. J. Comput. Appl. Math. **253**, 255–263 (2013)
10. Sendra, J., Sendra, J.R.: Gröbner basis computation of drazin inverses with multivariate rational function entries. Appl. Math. Comput. **259**, 450–459 (2015)
11. Sendra, J.R., Sendra, J.: Symbolic computation of Drazin inverses by specializations. J. Comput. Appl. Math. **301**, 201–212 (2016)
12. Stanimirović, P.S., Cvetković-Ilić, D.S.: Successive matrix squaring algorithm for computing outer inverses. Appl. Math. Comput. **203**, 19–29 (2008)
13. Stanimirović, P.S., Tasić, M.B., Vu, K.M.: Extensions of Faddeev's algorithms to polynomial matrices. Appl. Math. Comput. **214**, 246–258 (2009)

# A Poly-algorithmic Quantifier Elimination Package in Maple

Zak Tonks$^{(\boxtimes)}$ (iD)

University of Bath, Bath, Somerset BA2 7AY, UK
z.p.tonks@bath.ac.uk

**Abstract.** The problem of Quantifier Elimination (QE) in Computer Algebra is that of eliminating all quantifiers from a statement featuring polynomial constraints. This problem is known to be worst case time complexity worst case doubly exponential in the number of variables. As such implementations are sometimes seen as undesirable to use, despite problems arising in algebraic geometry and even economics lending themselves to formulations as QE problems. This paper largely concerns discussion of current progress of a package `QuantifierElimination` written using Maple that uses a poly-algorithm between two well known algorithms to solve QE: Virtual Term Substitution (VTS), and Cylindrical Algebraic Decomposition (CAD). While mitigation of efficiency concerns is the main aim of the implementation, said implementation being built in Maple reconciles with an aim of providing rich output to users to make use of algorithms to solve QE valuable. We explore the challenges and scope such an implementation gives in terms of the desires of the Satisfiability Modulo Theory (SMT) community, and other frequent uses of QE, noting Maple's status as a Mathematical toolbox.

**Keywords:** Quantifier Elimination · Virtual Term Substitution · Cylindrical Algebraic Decomposition · Symbolic computation

## 1 Introduction

Quantifier Elimination (QE) over the real numbers is a problem in Computer Algebra with origins in Logic that concerns the elimination of quantifiers from a boolean formula of polynomial constraints. We concisely define such formulae.

**Definition 1.** *A* Tarski Formula *is a polynomial constraint, $f \rho 0$, where $\rho \in \{<, \leq, \neq, =\}$, $f \in \mathbb{Q}[x_1, \ldots, x_n]$, $n \in \mathbb{N}$, or a boolean formula of Tarski formulae, where allowable boolean operators may include $\wedge, \vee, \Rightarrow, \neg, \veebar$.*

*A* Quantified Tarski Formula *is a Tarski formula which allows quantifiers ($\exists, \forall$), and as such quantified variables preceding any subformula.*

---

Thanks to my supervisor James Davenport.

J. Gerhard and I. Kotsireas (Eds.): MC 2019, CCIS 1125, pp. 171–186, 2020.
https://doi.org/10.1007/978-3-030-41258-6_13

A reader may be aware of the concept of a "Prenex" Tarski formula, where there is precedence for all quantifiers to exist only at the beginning of such a formula, and then in the remaining unquantified Tarski formula, only $\vee, \wedge$ are allowable boolean operators. This includes pushing of $\neg$ to the leaves of a formula such that relational operators get inverted ($=$ to $\neq$, $<$ to $\geq$, etc.). We note that any quantified Tarski formula can be converted to such a prenex form, and so we can obtain

$$Q_{n-m+1}x_{n-m+1} \cdots Q_n x_n \; \Phi(x_1, \ldots, x_n) \tag{1}$$

as a "Quantified Prenex Tarski Formula", where each $Q_i \in \{\forall, \exists\}, i = n - m + 1 \ldots n$, $n \geq m \in \mathbb{N}$, and $\Phi$ is a quantifier free Tarski formula featuring at most boolean operators from $\{\wedge, \vee\}$. Note that this means there are $m$ quantified variables. Further, we will assume the input formula for QE is a prenex quantified Tarski formula, given the available conversion. As such, the goal is to eliminate $Q_{n-m+1}x_{n-m+1} \cdots Q_n x_n$ from (1), and obtain the quantifier free equivalent formula $\Psi(x_1, \ldots, x_{n-m})$. A variable $x_i$ where $i \in \{1, \ldots, n-m\}$ is described as a "free variable". When $m = n$, i.e. there are no free variables, the formula is fully quantified, and as such (1) is certainly equivalent to $\top$ or $\bot$, which in Maple are represented by *true* and *false* respectively. Considering the context of Maple, we will use *true* and *false* in this paper to mean $\top$ or $\bot$ respectively.

## 2    Background of Techniques and Other Software

Tarski (after whom Definition 1 is named) first suggested an algorithm to solve QE problems in 1951 [16], but his method was essentially completely infeasible for implementation. Later algorithms were feasible, but deduced to be doubly exponential worst case complexity [7], hence the discussion surrounding mitigation of running times of such algorithms.

The Cylindrical Algebraic Decomposition (CAD) method for QE started with Collins in 1975 [6]. This algorithm became a popular technique to solve QE problems, with improvements such as "Partial CAD" [9] completely tailored to QE, resulting in the software QEPCAD B [2]. Meanwhile, most research has focused on optimisation of the projection operator intrinsic to the usual first stage of the method, to which essentially all of the complexity can be attributed [4].

Weispfenning suggested the Virtual Term Substitution (VTS) method (sometimes known as just "Virtual Substitution" (VS)) in 1988 [17], at the time allowing for elimination of quantified variables appearing at most quadratically in $\Phi$. A plethora of improvements to VTS can be attributed to Košta, who provided an extension of the method to variables appearing in degree up to 3 in $\Phi$ [10].

We acknowledge the existence of packages engaging with QE publicly available for Maple thus far. RegularChains [1] provides an implementation of QE via CAD, and is included in current editions of Maple. Similarly SyNRAC [18] offers an implementation of QE via VTS. ProjectionCAD [8] is yet another CAD based solution to QE. With the exception of RegularChains, the aforementioned are publicly available online, but not included natively with Maple. Outside of Maple, an implementation of QE exists in Mathematica [15], with this implementation at least similar to this work in the sense that VTS (written as "Virtual Substitution" in `RealPolynomialSystems`) and CAD cohabit (amongst other Computer Algebra tools), and the implementation can use either of these algorithms for QE depending on the degree of the variables existing in the QE problem passed. However, it is not poly-algorithmic in the sense that there is any intention to examine subproblems produced by VTS, but could use CAD on the entire quantified result of VTS eliminating a subset of the original quantified variables. For the purposes of this work, subproblems produced by VTS are meaningful to examine, and can be solved by incremental CAD.

## 3    Virtual Term Substitution

VTS offers a direct, but degree limited solution to QE. Referring to (1), to eliminate $Q_i x_i$, $i \in \{n - m + 1, \ldots, n\}$ VTS will attempt to substitute a point from every interval of the real line according to those intervals formed by the real roots of all polynomials contained in $\Phi$ in $x_i$. Given that such polynomials are multivariate (and indeed here we view them as elements of $\mathbb{Q}[x_1, \ldots, x_{i-1}, x_{i+1}, \ldots, x_n][x_i]$), one may immediately realise that the roots of such polynomials in $x_m$ will not, in general, be real numbers at this stage. Indeed, this is one of the ways in which the term "virtual" arises, as substitutions of these root expressions requires assertions on sign conditions of their coefficients, for example at the very least to ensure that such a root will exist. These "guards", that contain such assertions are Tarski formulae (over all other variables) in themselves, and the results of virtual substitution are similarly further Tarski formulae.

Further problems give rise to the use of the term "virtual" here. One is that (at least some) of the above mentioned intervals must be open if we are to cover the real line. Hence, presence of other variables means that one cannot use an exact root expression to represent substitution of a point from intervals where at least one end is open. Another is that the end intervals must feature $\pm\infty$, again such that we can cover the entire real line. Lastly, substitution of terms arising exactly from the real root of a polynomial implies dealing with potentially irrational expressions, potentially including other as yet uneliminated variables. Recalling the claim that VTS is degree limited, usage of the latest techniques from [10] allows us usage of test points arising from polynomials of up to degree 3, which implies the usage of irrational numbers at least on paper.

Solutions to these challenges are via the following:

- Technology to allow test points (those are the expressions used for substitution of any one variable) featuring the infinitesimals $\epsilon$, $\infty$.
- Especially owing to [10], where one wishes to substitute a root of $f$ in $x$ into $g \rho 0$, usage of pseudoremainder of $g$ by $f$ to ensure that one only requires knowledge of how to substitute a root of a polynomial of degree $d$ into a polynomial of degree at most $d - 1$ (as if $\deg(g, x) = d$ then $\deg(\text{prem}(g, f, x), x) \leq d - 1$). In this case, this is done via enumeration of formulae schemes, which describe the resulting logic that ensures substitution

of what would be an otherwise irrational root from $f$ results in formulae over polynomials in $\mathbb{Q}[x_1,\ldots,x_n]$. In other words, from data including the degrees of $f$ and $g$, the real index of the root to be substituted, and the real type of $f$, one can directly look up the resulting formula for the virtual substitution. If $h := \mathrm{prem}(g, f, x)$ is of degree at most 0 in $x$, then the virtual substitution $(g\,\rho\,0)[x/\!/\mathrm{RootOf}(f)]$ (ie. (virtual) substitution of a root of $f$ in $x$ of some description into $g\,\rho\,0$) returns $h\,\rho\,0$. When virtually substituting into a formula with boolean structure (that is a non atomic formula) the process is defined recursively, with each constraint replaced with its virtual substitution (which may expand as a non atomic formula). Note that the use of pseudoremainder is justified by $f$ being 0 at a root of $f$.

VTS can be seen as looking for valid "examples" (via test points) to solve an existential question, and similarly valid counterexamples to solve a universal one. As such, VTS aims to acquire truth out of a virtual substitution for an existential quantifier, and likewise falsity out of a substitution for a universal quantifier. In particular existential VTS acquires:

$$\exists x\ \varPhi = G(t_1) \wedge \varPhi[x/\!/t_1] \vee \cdots \vee G(t_k) \wedge \varPhi[x/\!/t_k] \tag{2}$$

where $t_1,\ldots,t_k$ are the test points acquired from polynomials in $\varPhi$, and $G$ is a map from test points to their relevant guards. Given that VTS is by invention a tool to tackle existential questions, one really acquires technology to deal with universal quantifiers by the following equivalence:

$$\forall x\ \varPhi \equiv \neg\exists x\ \neg\varPhi \tag{3}$$

leading to:

$$\forall x\ \varPhi \equiv \neg(G(t_1) \wedge \neg\varPhi[x/\!/t_1]) \wedge \cdots \wedge \neg(G(t_k) \wedge \neg\varPhi[x/\!/t_k]) \tag{4}$$

Inspection of (2) or (4), and recollection of the prenex input (1) will lead one to the understanding that using VTS to eliminate the innermost quantifier will lead to a string of quantifiers preceding either a disjunction or conjunction of results of virtual substitution on the input formula. Via distribution of such quantifiers into this disjunction or conjunction, one receives (a choice!) of further QE problems, of which one can use VTS or CAD to tackle. Beyond the choice of which operand to propagate VTS on further, one may even have a choice of which quantified variable to distribute in, due to commutativity of quantifiers of the same type. That is, if $Q_j = Q_{n-m+1}$ for some $j < n - m + 1$, then one can distribute in any $Q_i x_i, i \in \{j \ldots n{-}m{+}1\}$, albeit this requires some commitment, as one has to distribute this into all operands of the disjunction/conjunction, and so ideally one must be sure this choice of variable is a close to optimal choice for all such QE problems formed further. Most improvements thus far for VTS have focused on the univariate case, with the implications of the multivariate case, and discussion of implementation left somewhat alone.

## 3.1    Implementation

Multivariate VTS defines a canonical tree structure (see Fig. 2), with edges of the tree being structural test points from VTS, and nodes being formulae that are the results of (perhaps successive) substitution of test points on $\Phi$. Such formulae exist within the implicit disjunction/conjunction formed by VTS. True finished tree leaves are those that hold the formula *true* or *false*, and meaningful leaves are those that imply termination of VTS in some way (due to the implicit disjunction/conjunction evaluating to *true* or *false*). Such a tree structure is immediately amenable to an object based implementation of the VTS nodes. Objects are supported in Maple using a special option for modules [11], which enables inheritance. Given that such formulae for each node are really QE problems in themselves, these objects are called IQERs (Intermediate Quantifier Elimination Results), and past storage of the associated formula for the node, offer support for storage of the preceding tree edge (in essence a VTS test point), parent node, child nodes, and other information including information to allow evolutionary computation.

As far as QuantifierElimination is concerned, QE input in Maple is defined by usage of the inert operators forall, exists, Implies, And, Or, Xor, and Not. QuantifierElimination accepts quantified input that is non prenex (it is converted to prenex form upon input). forall and exists in Maple allow for a string of similarly quantified variables via passing a list of variables to either operator as their first operand, else any Maple name is allowable. Figure 1 shows an example of some QE input and output via QuantifierElimination in Maple.

Maple's inert boolean operators are ideal for usage as QE input, being inert and typesetting well in the interface. However such operators make for formulae that are poorly mutable, where if one were to add an extra operand to a conjunction/disjunction with, say, $n \in \mathbb{N}$ operands, then appending the extra operand actually requires rebuilding the entire formula, hence $n + 1 \in \mathcal{O}(n)$ operations due to $n$ intermediate objects to be discarded in garbage collection. Meanwhile, Maple also provides the Array construct, a mutable storage solution that allows for circumvention of this issue. As building of arbitrary formulae is vital to operation of VTS, usage of Arrays under the hood for intermediate Tarski formulae arising in VTS is an obvious choice. Conversion back to formulae using Maple's inert constructs occurs just before output for the purpose of aesthetics.

## 4    Cylindrical Algebraic Decomposition

The perhaps more well known approach to QE, CAD, takes a more basic approach to substitution of values to deduce a quantifier free equivalent to (1). That being said, its methodology for finding meaningful polynomials defining intervals to substitute from at each level requires far more computation than that of VTS. While VTS does not directly attempt "back substitution" to solve the problem, thanks to the virtual methodology, CAD can be seen as more analogous to back substitution. The usual first stage, *projection*, attempts to form bases of polynomials (starting from the polynomials in $\Phi$) in progressively fewer variables, until one receives a basis of univariate polynomials. From here, one can begin the second stage, *lifting*, where from the intervals defined by the univariate basis, one need only perform real root isolation to be able to define "cells" on which all polynomials in $\Phi$ will be sign invariant (with respect to the variable the cells are created over). The sign invariance of all polynomials on a cell implies $\Phi$ is truth invariant on that cell. Sample points (some candidate point from each cell) can be substituted into polynomials in the basis above to propagate the process further, forming "stacks" of cells which are arranged cylindrically. Cylindricity implies that projections of any pair of cells onto some space corresponding to $x_1, \ldots, x_k$ for some $k < n - M$ (where $M$ is the maximum of the levels of these two cells) will either completely coincide, or be entirely disjoint. This condition is not strictly necessary for unquantified variables, but is included algorithmically in practice for simplicity.

In particular, the projection stage was stated to be the most time consuming process here - indeed this stage requires taking operations such as resultants between polynomials, and discriminants on each polynomial. Such objects may be relevant to deducing critical points in real space to solve the QE problem. However the nature of degree bloat in taking operations such as resultants implies exponentially increasing time must be invested to perform this process. This is why much research has taken place in optimising the projection process, including a "reduced" projection operator in the presence of equational constraints in $\Phi$ [12,13], which is of interest for the author's QE implementation.

### 4.1  Implementation

In a similar manner to VTS, CAD lends itself to a tree structure, where nodes are CAD cells, and node parenting is equivalent to the implication that a cell is in the stack over another given cell. As such, the CAD implementation in `QuantifierElimination` is also object based. With respect to ongoing research on increasingly efficient projection operators, `QuantifierElimination` uses the most contemporary operator currently available for CAD, the Lazard projection operator [14]. The author keenly awaits ongoing research by colleagues into the equivalent of [12] for the Lazard operator - equational constraints for CAD would be of most use, as is discussed in Sect. 5.1. There are few other interesting nuances to speak of for the implementation here - Maple's `RootFinding` package offers the real root isolation that CAD relies upon heavily, and Maple 2019's update to this isolator allows for usage of algebraic numbers (and in general, non rational) coefficients in polynomials for isolation.

## 5  The QuantifierElimination Package

`QuantifierElimination` is a package being written using Maple in collaboration with Maplesoft, that is intended to be the first implementation of a poly-algorithm between VTS and CAD in Maple to tackle QE problems. `QuantifierEliminate` (Fig. 1) is the main procedure implementing the poly-algorithm to achieve QE.

## 5.1   VTS and CAD

VTS attempts to make meaningful (virtual) substitutions for variables based on the polynomials appearing in any one intermediate formula, while CAD will make substitutions based off of polynomials obtained via repeated projection on the polynomials from the initial input. In the latter case, polynomials in the bases at intermediate levels may have roots meaningless to the original problem, or even to real space. As such, VTS can loosely be seen to be "more concise", where meaningless substitutions only occur as a result of the boolean structure of the input problem. One may expect a better average case for VTS as a result, which is in itself an investigation, together with further complexity analysis of VTS.

If this is the case, the main motivation behind using both VTS and CAD consecutively is as a result of the limitations of VTS. As stated, VTS is entirely degree limited, where with presently available techniques VTS will only be able to eliminate quantifiers for variables appearing at most as cubics in $\Phi$. In particular, there are two main conveniences if one is to consider a poly-algorithm between VTS and CAD.

The first convenience is that, generically, one propagation of VTS produces additional QE problems. Propagation of VTS on a node of the VTS tree implicitly forms a conjunction or disjunction (from a universal or existential quantifier respectively) of the results of virtual substitution via each applicable test point from the node. More concisely, propagation of VTS on $Q_{n-m+1}x_{n-m+1}\ldots Q_n x_n\ \Phi(x_1,\ldots,x_n)$ (noting that we claim any node to be a true prenex QE problem in itself) forms $Q_{n-m+1}x_{n-m+1}\ldots Q_{n-1}x_{n-1}$ $B(\Phi[x_n//t_1],\ldots,\Phi[x_n//t_k])$, where using makeshift prefix notation $B$ is either `And` or `Or` depending on the quantifier. Henceforth, one can distribute $Q_{n-1}x_{n-1}$ (or a choice of quantified variable, considering commutativity of quantifiers to be discussed below) into all operands of $B$, obtaining a choice of potential QE problems. As long as we have such a choice, and we prefer to propagate VTS, we can do whatever necessary to keep receiving intermediate QE problems amenable to VTS, and then use CAD as a "last resort" when there is no choice due to excessive degree of all quantified variables.

An immediate consequence that one may find concerning is that this, as standard, implies that one could build a (potentially exponential!) number of CADs to solve ensuing intermediate QE problems that arrive, all of which of excessive degree for VTS to traverse. Hence, we arrive at another potential convenience. Note that a CAD for a set of polynomials arising from one QE problem can be used to solve any QE problem featuring the same order of quantified variables, and a subset of polynomials from the original problem. In the event that several VTS nodes of excessive degree feature formulae involving similar polynomials (in terms of set theoretics), then one need only build one CAD from scratch for the first non VTS-amenable problem. Further, one uses CAD incrementality to add to this CAD to obtain a CAD that solves each successive IQER of excessive degree for VTS encountered. Hence a "Master CAD" arises for any one QE problem, which is empty for any problem solvable completely by VTS.

One notes that this assumes that the situation of "similar" VTS nodes arises frequently, else in the worst case of completely disjoint sets of polynomials for successive intermediate formulae to be traversed by CAD, we are essentially building a CAD for each from scratch. The author aims to specify a definition of "similar" for VTS nodes in terms of their associated formulae, and investigate how prevalent examples are (or are not) that lend themselves to such a "Master CAD" approach in the future.

The second convenience is that quantifiers of the same type commute. At the very least this provides some freedom for choice-of-variable strategy, as above. Most importantly for now, this provides opportunities to eliminate a quantifier of a variable appearing of, say, lowest degree, before one appearing in excessive degree. In terms of VTS, this excessive degree is currently degree 4. To be precise, if we have $Q_{n-m+1}x_{n-m+1} \ldots Q_n x_n \; \Phi(x_1, \ldots, x_n)$ where $Q_n = Q_{n-1}$, and $x_n$ appears as a quartic in $\Phi$, then VTS is at present unsuitable to work with the formula, but one can swap $Q_n x_n$ and $Q_{n-1}x_{n-1}$, and hence the innermost problem is amenable to VTS.

In the best case, the poly-algorithm aims to avoid usage of CAD completely, instead receiving a quantifier free answer purely via VTS, bypassing non VTS-amenable nodes. This has implications with respect to output depending on what the user requests - discussion of this can be found in Sect. 6. There is significant scope for strategy in terms of VTS propagation, to attempt to avoid high degree cases, and instead find a satisfying VTS leaf for QE (especially in the fully quantified case) by doing the least work possible, in terms of selection of test points to use, and selection of node to propagate upon. Indeed strategy (in particular with respect to this implementation) is something the author is highly interested in, but must engage with later.

A last potential convenience is that the "guards" mentioned for VTS - Tarski formulae that assert that a substitution of a test point in VTS is valid, may include equational constraints by nature. This could arise in the case where we assume the leading coefficient of a quadratic vanishes in order to substitute the linear root implied by the reductum, for example. Additionally, at least in the case of regular test points not featuring $\epsilon$ or $\infty$, equational constraints from the original quantifier free $\Phi$ will be preserved by VTS in further IQERs. As a result, optimisations in projection for CAD implied by existence of equational constraints are certainly of use in the poly-algorithm this work suggests. In particular, we may be able to guide strategy to attempt to happen upon IQERs as dense with equational constraints as possible. This, however, assumes that the usage of such optimisations are valid for the Lazard operator which QuantifierElimination currently uses.

The author is also in the process of developing evolutionary techniques for QE. "Evolutionary" in this context refers to techniques that take data structures from a previously computed QE problem, and recompute the same QE for a modification of the problem, such as addition or subtraction of a subformula from the input (referred to as "incrementality" and "decrementality" further). The intention is that such recomputation should be fast - at the very least, take significantly less work than recomputation of the new QE problem from scratch. While full discussion of such techniques is out of the scope of this paper for the purposes of brevity, such evolutionary operations included by `QuantifierElimination`, `InsertFormula` and `DeleteFormula` allow incrementality and decrementality for QE respectively. These procedures require a `QEData` object as an argument, which can be requested as an output argument from `QuantifierEliminate` if future evolutionary operations are desired. A `QEData` object contains various data (including VTS and CAD results) from the previously computed problem amenable to modification to enable incrementality or decrementality.

## 5.2   QuantifierTools

The `QuantifierTools` package is due to be included as a subpackage of `QuantifierElimination`. `QuantifierTools` is designed to be a package that will allow users to manipulate Tarski formulae in various ways, to enable Maple as a toolbox for all mathematics. Additionally, `QuantifierTools` facilitates some understanding of Tarski formulae, and hence attempts to make such formulae more tractable. Current notable included functions include:

– A procedure to "alpha-convert" a Tarski formula, removing potential conflicts between variables in subformulae of the given expression if it is non prenex
– A procedure to convert a Tarski formula to prenex form (and hence performs alpha-conversion)
– A procedure to negate a Tarski formula
– A procedure to convert a formula with rational functions (of polynomials) in constraints to a proper Tarski formula
– A procedure to get the set of all polynomials appearing in a Tarski formula, possibly for the intent of performing full CAD on such a set.

# 6   Aims for QE Output

A main goal of the implementation is to mitigate the high running times of calls to QE on large problems. One way that this can be achieved is by avoiding unnecessary computation. In particular, as mentioned previously in this work, avoiding usage of CAD completely on intermediate problems of high degree if VTS can find a satisfactory answer alone. In terms of the author's implementation, this corresponds to unfinished VTS tree leaves being left unused by termination of the algorithm. However, to meet the goal of providing richest output and to enable understanding of how the algorithm was used, such unfinished leaves should be presented to the user if requested. Recall that an unfinished VTS tree leaf represents a QE problem in itself, via appropriate distribution of quantifiers amongst (at most) $Q_{n-m+1}x_{n-m+1} \ldots Q_n x_n$ into the implicit disjunction/conjunction formed by operation of VTS.

Hence such unevaluated parts of the problem, while immutable in terms of the *quantifier free equivalent* of (1) which is the main part of the output of a call to QE, should be presented to the user as inert calls to an appropriate `QuantifierElimination` procedure. Figure 1 is an example of such inertized calls that could be presented to the user. The obvious precedent is that if they are amenable to usage of VTS as per the variable ordering as was chosen at the time by QE, then they are presented as inert calls to the poly-algorithmic `QuantiferEliminate` procedure. Otherwise, in terms of the author's package they are only amenable to quantifier elimination as a result of CAD, and as such they are presented as inert calls to `CylindricalAlgebraicDecompose`. The purpose of evaluating such an inert call may be to understand a QE problem further, else potentially acquiring a surplus of satisfying witnesses for the problem (see Sect. 6.1).

## 6.1   Production of Meaningful Witnesses

While the main goal of quantifier elimination is to produce a quantifier free equivalent to the input formula (1), such an output is completely without proof - there is no substitution that can be done from the quantifier free equivalent that easily proves the output is valid. Software for the not too distant Satisfiability problem (SAT) will produce a satisfying assignment in the case that the input problem is satisfiable, hence providing a proof of the satisfiability in some sense.

We are able to discuss an analogous concept to a satisfying assignment, "witnesses". These are equations featuring quantified variables that provide proof of the equivalence of the input formula to some node of the VTS tree. Note that operation of VTS is entirely based around finding valid examples for an existential quantifier, else valid counterexamples for a universal quantifier. The test points used for this purpose entirely lend themselves to the purpose of witnesses, except for the presence of non-standard infinitesimal symbols: $\infty$ and $\epsilon$. Ignoring this for a moment, we note that in the case of a fully quantified formula, at least one set of assignments for all existentially quantified variables that makes the input formula equivalent to *true* suffices as a proof of such an equivalence (and by the nature of the universal quantifiers, any and every assignment of universal quantifiers should do). Similarly, at least one set of assignments for all universally quantified variables can suffice as a proof of the quantifier free equivalent *false*.

The nature of multivariate VTS is to eliminate one quantifier at a time, receiving a set of prenex quantified Tarski formulae with one less quantifier, and in particular one less variable than before. As such, considering VTS test points may include expressions with uneliminated quantified variables, one can envisage a "back-substitution" process here, where for any one VTS node, one follows the path to the root, using previous test points to substitute into test points from

> *QuantifierEliminate* $(\exists\,([x,y,z],\ 0 < xyz)\,,'\,mode\ =\ breadth')$

$[[true, z = \frac{1}{2}, y = \frac{1}{2}, x = \frac{1}{2}], [\text{`\%QuantifierEliminate'}\ (\exists\,(x,\ -x < 0)), z = -\infty, y =$
$-\infty], [\text{`\%QuantifierEliminate'}\ (\exists\,(x,\ x < 0)), z = -\infty, y =$
$\epsilon], [\text{`\%QuantifierEliminate'}\ (\exists\,(x,\ x < 0)), z = \frac{1}{2}, y = -\infty]]$

**Fig. 1.** Example output of QE in Maple where unevaluated QE problems belonging to the VTS tree are presented to the user. Note for this rather trivial example, one needs to traverse the tree breadthwise rather than depthwise (as would be default) to get this situation to occur, but results in something closer to a comprehensive set of (pre)witnesses that describe all possible proofs of equivalence of the input to *true* (given that the inertized QE calls here are all equivalent to *true*).

lower levels, in order to receive equations featuring only real numbers, which will fit the definition of "witnesses". For a fully quantified input formula, there is the potential to receive univariate problems one level below the leaves (which will necessarily hold the formula *true* or *false*).

As a result, Algorithm 1 is the detailing of a back-substitution process to produce a set of witnesses for any node of the VTS tree (assuming one started with a maximum level leaf, i.e. full elimination on a fully quantified formula occured). The difficulty of such a process is the handling of test points involving $\epsilon$ and $\infty$, which so far we have neglected. The main ideas for these originate with [10], but are made rigorous here. Further, we will refer to an unprocessed witness expression, which may include $\epsilon$ or $\infty$, and other free/quantified variables as a "prewitness". See Fig. 1 for a coincidental example of both witnesses and prewitnesses, where the first operand of the list corresponds to a maximum level leaf which terminated further usage of VTS due to being a satisfactory answer for the whole QE problem. That maximum level leaf was amenable to witness processing, while the rest are presented as prewitnesses featuring $\epsilon$ and $\infty$.

**Theorem 1.** *Algorithm* GetWitnesses *successfully produces a set of witnesses for a maximum level leaf of the VTS tree representing full elimination of a fully quantified problem (1), and in particular such witnesses are all equations featuring real numbers.*

*Proof.* First, due to assertion that the input formula is fully quantified, and recollection of the nature of the VTS tree where every node is implicitly a QE problem with quantifiers commensurate with its level, a true max level leaf node of the VTS tree resulted from a univariate QE problem. Therefore as long as we can find an appropriate witness for this QE problem, substitution of this witness into appropriate objects the level above will result in univariate objects, hence receiving an inductive claim on the back substitution's validity.

– The handling of $\infty$ test points on line 9. If a test point used $x = \pm\infty$, then this really implies that $x$ should be large or small enough. Hence, it suffices that we choose $x$ such that $x$ exceeds all real roots of $\Phi$, in any variable.

---

**Algorithm 1.** GetWitnesses($L, Q_1x_1 \ldots Q_nx_n\ \Phi(x_1, \ldots, x_n)$)

---

**Data:** $L$, a leaf `IQER`, $Q_1x_1 \ldots Q_nx_n\ \Phi(x_1, \ldots, x_n)$ the original fully quantified prenex quantified Tarski formula given to QE

**Result:** A list of processed witnesses $x_1 = r_1, \ldots, x_n = r_n$, where $r_i \in \mathbb{R}$, $i = 1, \ldots, n$, and $\Psi$ evaluated at $x_1 = r_1, \ldots, x_n = r_n$ is equivalent to the quantifier free formula associated with *leaf*

1  set *temp* $= L$;

2  set *witnesses* to be the empty list;

3  **while** temp *has a parent* `IQER` **do**

4  　　let $F$ be the quantifier free formula associated to *temp*, and $i$ be its level wrt the VTS tree;

5  　　construct the prewitness associated to the `IQER` *temp* as $x_{n-i+1} = t$ from the test point of *temp* (essentially the edge above *temp*), where $t$ may feature $\epsilon$ or $\infty$;

6  　　$t \leftarrow t$ evaluated at all current *witnesses*;

7  　　set *temp* as the parent `IQER` of *temp*;

8  　　**if** $t = \pm\infty$ **then**

9  　　　　add $x = \mathbf{sgn}(t)\,M$ to *witnesses*, where $M$ is the maxima amongst Cauchy root bounds of all polynomials appearing in $\Phi$;

10  　　**else if** $t$ *contains* $\epsilon$ **then**

11  　　　　Let $F_{parent}$ be the quantifier free formula associated to *temp*, evaluated at all current *witnesses* (and substitute 0 for any variables not used in *witnesses*);

12  　　　　let *rootList* be a complete ordered list of sample points (via real root isolation) at which the multiplication of all polynomials appearing in $F_{parent}$ is non zero;

13  　　　　let $s$ be the sign of $\epsilon$ in $t$;

14  　　　　let $r$ be $t - se$;

15  　　　　**if** $s = -1$ **then**

16  　　　　　　reverse *rootList*;

17  　　　　**for** $y$ *in* rootList **do**

18  　　　　　　**if** $y < r$ **then**

19  　　　　　　　　$r \leftarrow \frac{y+r}{2}$;

20  　　　　　　　　break;

21

22  　　　　add $x = r$ to *witnesses*;

23  　　**else**

24  　　　　add $x = t$ to *witnesses*;

25

26  **return** *witnesses*;

---

The Cauchy root bound will achieve such a purpose - the maxima among all Cauchy root bounds of all polynomials appearing in $\Phi$ (as usual, ignoring boolean structure, in practice implemented recursively) exceeds all real roots of every polynomial in $\Phi$ in all variables. This is prepended with $\pm$ depending on the sign of $\infty$ used in the test point.

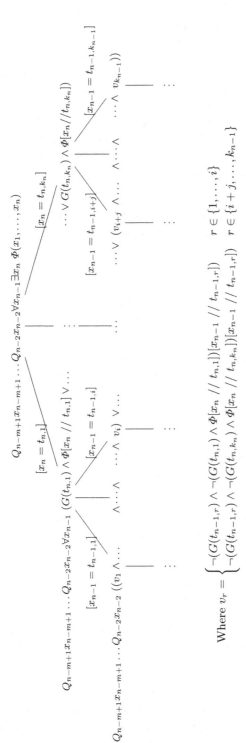

**Fig. 2.** An example of the VTS tree, where here we assert the last two quantifiers are $\forall$ and $\exists$ to demonstrate a conjunction nested within a disjunction formed by VTS. This Tarski formula formed by the tree necessarily eventually forms the quantifier free output of QE by VTS.

- The handling of test points involving $\epsilon$, the "elseif" associated to line 10. Recollection of the motive that the usage of the test point $t \pm \epsilon$ represents usage of a term "just less/more than $t$", and as such it suffices to find a real number less/more than $t$, but more/less than the next root of the quantifier free formula it arose from (after back substitution). Thanks to back substitution, real root isolation can be performed on a univariate formula to find a suitable point. Presently, this algorithm implies usage of root isolation on a polynomial that is the multiple of all polynomials in the quantifier free formula. This is sufficient such that one can isolate $t$ from its relevant next nearest root, but perhaps not necessary, and may be able to be optimised.
- Else (line 22), the test point used no such infinitesimals, and the valid complete back substitution implies that all indeterminates are eliminated from the term used for the right hand side of the test point, resulting in a real number (possibly involving irrational terms).

## 7    Conclusions

### 7.1    Comparison to Other QE Implementations

The `QuantifierElimination` package is the first attempt to have an active poly-algorithm between VTS and CAD. In contrast to QE software already publicly available for Maple, the `QuantifierElimination` package is the first to encapsulate both VTS and CAD in one package. Furthermore, the implementation of CAD in `QuantifierElimination` is the first of its kind in Maple to use the Lazard projection, instead of the McCallum, Collins, or Brown projection operators. Tentatively, support for the reduced Lazard projection operator in the presence of equational constraint(s) in $\Phi$ is included, in the event of the appearance of the equivalent of [12,13] for the Lazard projection. Superficially, `QuantifierElimination` is the first package implementing VTS to be included with Maple as standard, the second to include native QE, and likewise the second native implementation of CAD. Inclusion of `QuantifierTools` is intended at least for the purposes of pedagogy - making the package even more user friendly, and explore quantified Tarski formulae further.

### 7.2    Future and Further Aims

As a collaboration with Maplesoft, the aim is to include the `QuantifierElimination` package in a future official Maple release. Full testing is underway to ensure the implementation is suitable for inclusion. At the time of writing, new implementations of both VTS and CAD exist in Maple, together with new evolutionary methods for VTS. At present the interface between them is rudimentary—an implementation of incremental CAD is required next to realise the "Master CAD" idea in light of Sect. 5.1, and also enables a full evolutionary QE system (extensive discussion of which is omitted here).

With a view to providing valuable output for users, one notes that the output quantifier free formula is generally the most important part of output of any QE call, but VTS and CAD alike do not necessarily form a "simple" formula by default for output due to their natures. There has been some work done in this area thanks to [3,5], but the nature of the implementation implies that such a simplifier for `QuantifierElimination` could already feature a convolution of techniques, given quantifier free output could come from either algorithm. In contrast to CAD, VTS uses Tarski formulae extensively throughout. As such, it is additionally of interest to be able to simplify intermediate formulae, but the key here should be that one doesn't expend so much computational effort simplifying formulae that one doesn't see the performance benefit of doing so, or even sees a performance loss. An appropriate Tarski formula simplifier would be appropriate for inclusion in the `QuantifierTools` package, while also being used under the hood in `QuantifierElimination`.

The notion of UNSAT cores from the SAT & SMT communities extends to the first order logic of Tarski formulae, and in particular quantified ones. It is therefore of interest to be able to produce for any QE problem where the answer obtained was *false*, the subformula(e) that were most pertinent in obtaining such a false answer. A method to do so would not be out of place in the `QuantifierElimination` package, presumably working from a `QEData` object.

# References

1. Alvandi, P., Chen, C., Lemaire, F., Maza, M., Xie, Y.: The RegularChains Library. http://www.regularchains.org/
2. Brown, C.W.: QEPCAD B: a program for computing with semi-algebraic sets using CADs. SIGSAM Bull. **37**(4), 97–108 (2003). https://doi.org/10.1145/968708. 968710
3. Brown, C.W.: Fast simplifications for Tarski formulas based on monomial inequalities. J. Symb. Comput. **47**(7), 859–882 (2012). https://doi.org/10.1016/j.jsc.2011. 12.012
4. Brown, C.W., Davenport, J.H.: The complexity of quantifier elimination and cylindrical algebraic decomposition. In: Proceedings of the 2007 International Symposium on Symbolic and Algebraic Computation, ISSAC 2007, pp. 54–60. ACM, New York, NY, USA (2007). https://doi.org/10.1145/1277548.1277557
5. Chen, C., Maza, M.M.: Simplification of cylindrical algebraic formulas. In: Gerdt, V.P., Koepf, W., Seiler, W.M., Vorozhtsov, E.V. (eds.) CASC 2015. LNCS, vol. 9301, pp. 119–134. Springer, Cham (2015). https://doi.org/10.1007/978-3-319-24021-3_9
6. Collins, G.E.: Quantifier elimination for real closed fields by cylindrical algebraic decompostion. In: Brakhage, H. (ed.) GI-Fachtagung 1975. LNCS, vol. 33, pp. 134–183. Springer, Heidelberg (1975). https://doi.org/10.1007/3-540-07407-4_17
7. Davenport, J., Heintz, J.: Real quantifier elimination is doubly exponential. J. Symb. Comput. **5**(1), 29–35 (1988). https://doi.org/10.1016/S0747-7171(88)80004-X
8. England, M., Wilson, D., Bradford, R., Davenport, J.H.: Using the regular chains library to build cylindrical algebraic decompositions by projecting and lifting. In: Hong, H., Yap, C. (eds.) ICMS 2014. LNCS, vol. 8592, pp. 458–465. Springer, Heidelberg (2014). https://doi.org/10.1007/978-3-662-44199-2_69

9. Hong, H., Collins, G.: Partial cylindrical algebraic decomposition for quantifier elimination. J. Symb. Comput. 299–328 (1991). https://doi.org/10.1016/S0747-7171(08)80152-6

10. Košta, M.: New concepts for real quantifier elimination by virtual substitution. Ph.D. thesis, Universität des Saarlandes (2016). https://doi.org/10.22028/D291-26679

11. Maplesoft: Maple Programming Guide, pp. 360–372. https://www.maplesoft.com/documentation_center/maple2019/ProgrammingGuide.pdf

12. McCallum, S.: On projection in CAD-based quantifier elimination with equational constraint. In: Proceedings ISSAC 1999, pp. 145–149 (1999). https://doi.org/10.1145/309831.309892

13. McCallum, S.: On propagation of equational constraints in CAD-based quantifier elimination. In: Proceedings ISSAC 2001, pp. 223–231 (2001). https://doi.org/10.1145/384101.384132

14. McCallum, S., Parusiński, A., Paunescu, L.: Validity proof of Lazard's method for CAD construction. J. Symb. Comput. **92**, 52–69 (2019). https://doi.org/10.1016/j.jsc.2017.12.002

15. Strzebonski, A.: Real Polynomial Systems, Wolfram Mathematica. https://reference.wolfram.com/language/tutorial/RealPolynomialSystems.html

16. Tarski, A.: A Decision Method for Elementary Algebra and Geometry, 2nd edn. Univ. Cal. Press (1951). Reprinted in Quantifier Elimination and Cylindrical Algebraic Decomposition (ed. B.F. Caviness & J.R. Johnson), pp. 24–84. Springer, Wein-New York (1998). https://doi.org/10.1007/978-3-7091-9459-1_3

17. Weispfenning, V.: The complexity of linear problems in fields. J. Symb. Comput. **5**(1), 3–27 (1988). https://doi.org/10.1016/S0747-7171(88)80003-8

18. Yanami, H., Anai, H.: SyNRAC: a maple toolbox for solving real algebraic constraints. ACM Commun. Comput. Algebra **41**(3), 112–113 (2007). https://doi.org/10.1145/1358190.1358205

# Full Papers – Education/Applications Stream

# The Creation of Animated Graphs to Develop Computational Thinking and Support STEM Education

Alice Barana[1] , Alberto Conte[1], Cecilia Fissore[1] ,
Francesco Floris[1] , Marina Marchisio[2] ,
and Matteo Sacchet[1(✉)]

[1] Department of Mathematics, University of Turin, Turin, Italy
{alice.barana, alberto.conte, cecilia.fissore,
francesco.floris, matteo.sacchet}@unito.it
[2] Department of Molecular Biotechnology and Health Sciences,
University of Turin, Turin, Italy
marina.marchisio@unito.it

**Abstract.** Problem solving and computational thinking are the key competences that all individuals need for professional fulfillment, personal development, active citizenship, social inclusion and employment. In mathematics, during contextualized problem solving using Maple, the differences between these two skills become thinner. A very important feature of Maple for problem solving is the programming of animated graphs: an animation obtained by generalizing a static graph, choosing the parameter to be varied and its interval of variation. The first objective of this research is to analyze the computational thinking processes behind the creation of animated graphs for the resolution of a contextualized problem. To this end, we selected and analyzed some resolutions of problems carried out by fourth-grade students of upper secondary schools in Italy (grade 12). The paper shows some examples in which different processes of computational thinking have emerged, which reflect resolutive strategies and different generalization processes. From the analysis it emerged that all the processes underlying the mental strategies of the computational thought useful for solving problems are activated in the creation of animated graphs. In the second part of the article we discuss examples of animations created during training activities with secondary school teachers, and how animations can support the learning of scientific concepts. It is very important to train the teachers in this regard, both to understand the processes that the students would activate during the creation of animated graphics and to enrich the theoretical or practical explanations with animated representations.

**Keywords:** Advanced computing environment · Animated graphs ·
Computational thinking · Problem solving · Animations · STEM education

© Springer Nature Switzerland AG 2020
J. Gerhard and I. Kotsireas (Eds.): MC 2019, CCIS 1125, pp. 189–204, 2020.
https://doi.org/10.1007/978-3-030-41258-6_14

# 1  Introduction

According to the recommendations of the Council of the European Union of 22 May 2018 [9] among the eight key competences for lifelong learning to achieve progress and success are the mathematical competence, defined as "the ability to develop and apply mathematical thinking and insight in order to solve a range of problems in everyday situations" and the digital competence that "involves the confident, critical and responsible use of, and engagement with, digital technologies for learning, at work, and for participation in society. It includes information and data literacy, communication and collaboration, media literacy, digital content creation (including programming), safety (including digital well-being and competences related to cybersecurity), intellectual property related questions, problem solving and critical thinking". The eight key competences are linked to the development of skills such as: problem solving, critical thinking, ability to cooperate, creativity and computational thinking. They allow to exploit in real time what has been learned, in order to develop new ideas, new theories and new knowledge [9]. Problem solving is an important aspect of mathematics teaching and learning, it occurs in all mathematical curricula [14]. In recent years the use of digital technologies in problem solving activities has allowed a precious variety of representation and exploration of mathematical tasks [18] extending the ways of thinking about the strategies involved in problem solving [12]. One of the technologies used for problem solving activities is Maple, which allows numerical and symbolic calculations, static and animated graphical representations in 2 and 3 dimensions, writing procedures in simple language, programming and connecting all these different registers representation in a single worksheet using verbal language, too [2]. A very important aspect of Maple for problem solving is the design and programming of animated graphs and interactive components. The user can explore through the variation of a parameter and modify inputs.

The first objective of this research is to analyze the computational thinking at the base of the creation of animated graphs for the resolution of a contextualized problem with Maple. To this end, some examples of resolutions of a contextualized problem carried out by fourth-grade students of upper secondary schools will be analyzed. The second objective is to discuss some examples, arising from training activities with secondary school teachers about the creation of animated graphics, to show how animations can support the teaching of STEM, in particular of Mathematics.

# 2  Theoretical Framework

The expression "computational thinking" was popularized by an article by Wing [19] in which she supports the importance of teaching the fundamental concepts of computer science in school, possibly from the first classes. Even today the teaching of Computer Science, although frequently described as the systematic study of computational processes that describe and transform information, is reduced in most cases to the use of computers (consumer Computer Science) [6]. Initially, Wing [19] did not give a precise definition of computational thinking but outlined its main features:

- a way in which human beings solve a problem;
- a base for conceptualizing, not programming, on multiple levels of abstraction;
- a base for ideas, not artifacts;
- an integration between mathematical and engineering thinking;
- a process open to everyone, all over the world.

The search for a precise and uniquely shared operational definition of the expression, which still does not exist, can create confusion on the topic, leading firstly to mistakenly consider computational thinking as a new subject of teaching, conceptually distinct from Computer Science, Lodi et al. [15]. The authors believe that it is more important to use the expression "computational thinking" as a short way to refer to a well-structured concept, the founding nucleus of Informatics. Therefore in [15] they describe computational thinking as a mental process (or more generally a way of thinking) to solve problems (problem solving) and define its constitutive elements: mental strategies, methods, practices and transversal competences. Regarding the mental strategies useful for solving problems, the authors describe the following mental processes:

- algorithmic thinking: approach to the design of an ordered sequence of steps (instructions) to solve a problem;
- logical thinking: reasoning to establish and control facts;
- decomposition of problems: dividing and modularizing a complex problem into simple sub-problems, which can be solved more easily;
- abstraction: getting rid of useless details to focus on relevant ideas;
- pattern recognition: identifying regularities and recurrent patterns in data and problems;
- generalization: use the recognized patterns to foresee or to solve more general problems.

These mental strategies recall in many aspects the phases of a problem solving activity in teaching, for example, of Mathematics: understanding the problem, designing the mathematical model, model resolution and interpretation of the results obtained [17]. What distinguishes computational thinking from problem solving is the conceptual paradigm shift constituted by the transition from solving problems to making problem solving [15]. In fact, the first one does not concern a specific resolution of problems: the formulation of the problem and of the solution must be expressed, mainly through an algorithm in an appropriate language, so that an "information processing agent" (human being or machine) can understand, interpret and execute the instructions provided. In our opinion, this difference thins out when problem solving activities are proposed through technologies and in particular Maple. In this case, starting from mental thinking, the student must choose how to set the solution procedure using the available ways (words, graphs, numerical or symbolic calculations, procedures, cycles, etc.) and at the same time write an algorithm in an appropriate language, so that the software processes the information and returns an output. In this research we focus on the creation of animated graphs that involve the generalization of a static graph by choosing the parameter to be varied and its range of variation.

According to Malara [16], the term "generalization process" includes a series of acts of thought that lead a subject to recognize, by examining individual cases, the occurrence of common characteristic elements; to shift attention from individual cases to the totality of possible cases and to extend the identified common features to this totality. The main actions of this process are pattern recognition, identification and connection of similarities. They lead the student to consider all possibilities instead of a single case, and to extend and adapt the identified model. This definition is not so far from the mental processes of the computational recognition and generalization we mentioned above. About the process of generalization, the author focuses on a reflection by Dörfler [11] that considers the representation of the process to be crucial through the use of perceptible objects, such as written signs, characteristic elements, steps and results of actions. In this way a procedure is generated that allows a cognitive reconstruction and conceptualization of the process itself. The interaction of students with visual representations, in the form of static or animated images can greatly enhance the learning of scientific concepts otherwise expressed only in verbal or mathematical form. As Landriscina explains [13], when people are engaged in tasks that require understanding and reasoning, they also create "mental models", i.e. representations of a dynamic nature close to a certain situation in the external world that serve to make predictions or simulations. In producing or evaluating a "didactic image" it is therefore important to consider not only to what extent it faithfully represents its object, but also which mental models the student will form by looking at the image and which factors will influence this representation. The didactic potential of images and animations is still largely unexplored, as evidenced, for example, by the purely ornamental images that are still printed in many textbooks, and by the limited use of images generated by students in teaching scientific disciplines and learning assessment [13]. It is therefore important to train teachers to know the processes of computational thinking and to understand how to use graphic visualizations and animations in teaching.

## 3   Animated Graphs Created with Maple

The creation of an animated graph with Maple2018[1] can be done through the use of the "animate" command, which creates the animation of a graph in 2 or 3 dimensions when a parameter changes. The use of the command can introduce processes that are activated in the generalization of a static graph by choosing the parameter to be varied and its range of variation. The command uses the following syntax:

$$\text{animate(plotcommand,plotargs,t=a..b,options)} \qquad (1)$$

---

[1] https://www.maplesoft.com/.

where:

- `plotcommand`: a Maple command or a procedure that returns a 2D or 3D graph;
- `plotargs`: list of arguments of `plotcommand`;
- `t`: name of the animation parameter;
- `a, b`: extremes of the interval in which the parameter varies.

Suppose, for example, that the user wants to create an animation to visualize how the concavity of a parabola with a vertex varies in the origin in the Cartesian plane. First, we create the static graph of a particular parabola, for example $y = 3x^2$, using the following command:

```
plot(3*x^2,x=-15..15,color=blue)
```

By pressing Enter the user obtains the graph illustrated in Fig. 1.

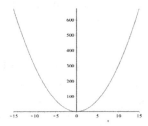

**Fig. 1.** Static graph of the parabola $y = 3x^2$.

We can do the same process to plot the parabola $y = -3x^2$, drawing a parabola with the concavity facing down. Now suppose we want to visualize through an animation how the concavity of the parabola between the first and the second situation varies. We therefore use the command animate, where, as in (1), the `plotcommand` is `plot` and all the rest into brackets is `plotargs`. In the `animate` command, `plotargs` is not the same as the static graph, but it must be generalized recognizing common patterns between the graphs of the first and second situation that we want to merge into a single command. In particular, in this case we choose the coefficient of $x^2$ as the parameter to be changed, and the interval in which it varies is represented by the real numbers between 3 and −3. We then get the following command:

```
animate(plot,[a*x^2,x=15..15,color=blue],a=3..-3)
```

In this way we have created the desired animation (Fig. 2) having an immediate feed-back of the result of the generalization we have carried out.

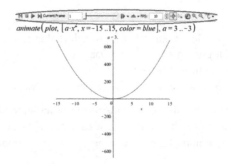

**Fig. 2.** Animated graph of the parabola $y = a \cdot x^2$, with a varying between 3 and $-3$.

A further feature of the use of Maple consists in the possibility to define new commands, called "procedures", through a specific programming language. Inside the animate command it is possible to generalize any procedure that outputs a graph in two or three dimensions. It is also possible to export the animated graphic obtained in GIF format to be able to use it as teaching material on its own, to publish it in a Virtual Learning Environment or to reproduce it on any web page [3, 5].

In light of the studied theoretical framework, we believe that the creation of the animation of a graph using Maple is a process of computational thinking and the immediate visualization as output of the result can strongly help the students in the development of this competence. We also believe that the use of animations can greatly support the teaching of STEM disciplines. In the following paragraphs we will analyze some examples to demonstrate both aspects.

## 4 Methodology for the Analysis of Processes of Computational Thinking in the Creation of Animated Graphs

To analyze the processes of computational thinking behind the creation of animated graphs, we analyzed the resolutions, using Maple, of a contextualized problem performed by grade 12th students (corresponding to the fourth-grade of upper secondary school in Italy) of different Italian regions in the frame of the Digital Math Training project [1]. The problem, entitled "Ladybug", deals with a ladybug resting on the rear wheel of a bicycle, on the part of the rim closest to the ground. Thinking of the wheel of the bike as a circumference centered in the origin, the position of the ladybug corresponds to the point of tangency of the wheel with the ground, that is the angle of amplitude $\frac{3}{2}\pi$. The total diameter of the wheels (including the inner tube) is 60 cm, while the inner tube is 4 cm thick. The request of the problem we focused on is the ladybug trajectory in a kilometer, supposing that it has never moved. This trajectory is given by the combination of two motions: the circular one of the wheels and the straight one of the bike. It is therefore represented by a cycloid, whose graph can certainly be considered a correct answer to the request of the problem. However, the static graph does not show how that result was achieved nor the process carried out to arrive at the solution. Instead, through an animated graph it is possible to visualize the motion of the wheel and the path traveled by the ladybug during the motion itself. Moreover, it is possible to keep track of the frames of the animation and in this way the trajectory is directly traced. For our research we have shown 80 solutions proposed by as many students, among which we have selected and analyzed the 16 in which an animation was used to visualize the trajectory.

## 5 Results

In all the 16 analyzed solutions in which an animation was used, we came across the use of algorithmic thinking to construct the animated graph as a result of an ordered sequence of commands. However, different processes of computational thinking have emerged, some of them successful, reflecting different resolutive strategies. We have classified the analyzed solutions into five categories:

1. animation as creation of a curve point by point (2 solutions);
2. animation as verification of the results (3 solutions);
3. animation as a union of animations (5 solutions);
4. animation of a procedure (1 solution);
5. animations with incorrect procedures (5 solutions).

Although the solutions classified in the same category are not identical, they show analogous processes of computational thinking (as depicted in the theoretical framework). We are going to present an example of a paradigmatic resolution for each category mentioned above. In the solutions of the first category, the graph of the cycloid was obtained through the command animatecurve which creates the animation of the trajectory of a curve in 2 dimensions, with the parametric equations of the curve as input. The syntax used in an example of this category was the following:

```
animatecurve([30*sin(t)+30*t,30+30*cos(t),t=0..10*Pi],
    color=black,frames=100,thickness=2)
```

We obtained as output a graph drawing the curve point by point (Fig. 3):

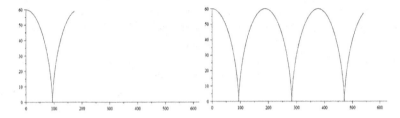

**Fig. 3.** Example of animation as creation of a curve point by point.

In order to derive the parametric equations of the curve, the student who presented this resolution studied the motions of translation and rotation of the wheel, using the mental process of the decomposition of problems. Then the student used the process of generalization and abstraction to combine the two motions and obtain, with the animate command, the trajectory of the ladybug by calculating its punctual position corresponding to time t, the parameter of the animation. The generalization and use of the command in this case was carried out correctly, even though the solution is not correct because of the wrong starting point.

In the second category we included the solutions in which the student drew the static graph of the cycloid and then animated a point moving on it. In this case, the graph of the cycloid was obtained using the parametric form of the curve with the following command:

```
graf:=plot([0.3(t-sin(t)),0.3(1-cos(t)),t=0..8*Pi],
    color=black)
```

The command used to create the animation is the following:

```
animate(pointplot,[[0.3(t-sin(t)),0.3(1-cos(t))],
    color=red, symbol=solidcircle],t=0..8*Pi,
    background=graf)
```

Figure 4 illustrates two frames of the animation.

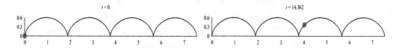

**Fig. 4.** Example of animation as verification of the results.

This procedure was used to verify the static graph obtained by directly observing how the ladybug moves. In this case the process of generalization concerns exclusively the graph of the point and consequently its coordinates. The graph of the cycloid was used as the background of the animation. The abstraction process took place in two phases: in the first phase a generalization was carried out to create a static command (the parametric curve), in the second phase the generalization was used to create the animation. At the base of the creation of this animation there is the logical process to check the correctness of the previously obtained results.

The third category concerns the solutions in which the animated graph contained: the two circumferences to represent the wheel, the center of the wheel, the ladybug and the radius of the bicycle that joins the ladybug to the center of the wheel. Five different animated graphs were created and then displayed together using the display command. In the following example, the student defined three procedures coccinel, R, center to statically represent the ladybug, the radius and the center of the wheel respectively. Then the student individually animated the graphs of these three elements and of the two circumferences to represent the wheels and finally all the animations were combined in a single graph through the following command:

```
display({
    animate(R,[t],t=0..8*Pi,scaling=constrained,
        frames=80),
    animate(centro,[3*t,3],t=0..8*Pi,frames=80),
    animate(implicitplot,
        [x^2+y^2+9*t^2-6*t*x-6*y+2.24=0,x=0..80,
        y=0..8, color=black], t=0..8*Pi, frames=80),
    animate(implicitplot, [9*t^2-6*t*x+x^2+y^2-6*y=0,
        x=0..80, y=0..10, color=black], t=0..8*Pi,
        frames=80),
    animate(coccinel,[3*t-2.6*sin(t),3-2.6*cos(t)],
        t=0..8*Pi,frames=80,trace=90000)},
labels=["x (dm)","y (dm)"])
```

In this command, the trace option allows you to select a set of frames to print while the animation is being performed (Fig. 5). In this resolutive strategy, used by most of the students, the mental process of decomposing problems was used to create animation: a complex animation was divided into different ones, solved in a simpler way. The combination of the animations required the use of the same parameter with the same range of variation, using the process of pattern recognition and generalization.

**Fig. 5.** Example of animation as a union of animations.

The second-last solution analyzed, even if it contains minor accounting errors, is very interesting from the point of view of computational thinking in order to implement it. The ladybug is represented on the left of the graph the ladybug while moving along the wheel and on the right the trajectory is drawn point by point with a segment (Fig. 7). Unlike the previous example, the creation of the animation did not require different animations, but different graphical commands with the same parameter were combined in a procedure, which was then animated. The procedure is the following:

```
F:=proc(t)
    plots[display](
        plottools[line]([-2,0],[sin(t)-2,-cos(t)],
            color=blue),
        plottools[line]([sin(t)-2,-cos(t)],
            [t,abs(2*sin((1/2)*t))],color=blue),
        plot(abs(2*sin((1/2)*x)),x=0..t,color="Green"))
end proc
```

This procedure, which requires as input the value of the parameter *t*, returns the static graph of the radius of the wheel, the graph of the curve from 0 to *t* and the segment that joins the position of the ladybug on the wheel to that on the trajectory. For example, the value *t = 3.14* generates the graph shown in Fig. 6.

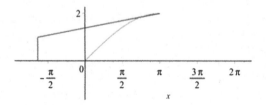

**Fig. 6.** Static graph for the animation of a procedure.

Then the procedure was animated as the input parameter was changed, adding as a background the static graph of the circumference representing the wheel (Fig. 7). The resolutive strategy is different from the previous one because, instead of joining different animations in a single one, it appears as the creation of a sequence of different snapshots of the same motion. From the point of view of computational thinking this requires a more advanced abstraction process since the generalization through pattern recognition is performed twice: one to define the procedure and one to animate it.

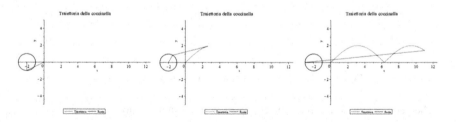

**Fig. 7.** Example of animation of a procedure.

The remaining five solutions analyzed contain incorrect animations for different aspects. The errors regard the use of the command (incorrect generalization process), the scale chosen to display the graph (incorrect logical process) or the contextualization of the graph (displayed in the negative semi-axis of the ordinates, arising from a wrong process of abstraction). In the following example (Fig. 8) we can see how the student correctly plotted the graph to illustrate the solution of the problem, to represent the position of the ladybug on the wheel and on the trajectory at the same time. However, already in the static graph, the trajectory of the ladybug on the wheel turns out to be in the negative semi-axis of the ordinates, thus showing an incorrect logical process in the reasoning. In fact, the contextualization of the problem was not considered since the wheel proceeds tangent to the ground level (abscissa axis). The process of generalization was carried out correctly in the creation of the animated graph, but the abstraction process was lacking, it did not check if the solution reflected the context and was explanatory as a proposed solution. This confirms that aspects of computational thinking, as well as of problem solving, are all equally important.

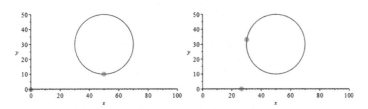

**Fig. 8.** Example of incorrect animation.

# 6 Animated Graphs for Didactics

Since the creation of animated graphs activate all the processes behind the mental strategies of computational thinking, we believe that it can be very important to train the teachers about this, and we add several reasons to support our belief. First, to understand the processes that students will activate when creating animated graphs, for example for problem solving (as shown in the previous paragraphs), but also to illustrate a theoretical concept or build a simulation. Secondly, to enrich the theoretical or practical explanations with animated representations. Thirdly, to enhance the learning of scientific concepts otherwise expressed only in verbal or mathematical form. In the National Project "PP&S - Problem Posing and Solving" [4, 7, 8, 10] during the online training course "Didactics of STEM with Maple" the creation of animated graphs was discussed in various online training meetings. In the case of more elaborate graphs, in which more than one command is used, we explained to teachers how to construct a procedure with a single input parameter, and a graph as output, and then how to animate it (as in the last example of the previous paragraph). During the explanations, the tutor focused the attention of the teachers on the computational thinking processes activated during the creation of animated graphics. In particular, the tutor highlighted the process of pattern recognition and problem decomposition to define a procedure that combines different graphic commands to vary the same parameter and the process of generalization and abstraction to create animation. Below the reader can find some examples divided into two categories: animated problem solving charts, in which the focus is on training teachers to teach their students how to use Maple for problem solving, and animated graphics for illustration of theoretical concepts.

## 6.1 Animated Graphs for Problem Solving

The following example concerns a problem on geometric transformations in the Cartesian plane that asks how a craftsman can draw a tile formed by rotated concentric squares (Fig. 9), starting from the outermost square.

**Fig. 9.** Tile decoration to be reproduced in the problem.

The solution to the problem is obtained by rotating and expanding the outer square. The static graph of the two transformations is a solution to the problem but it is possible to use animations to illustrate the resolutive procedure step by step. The creation of animated graphs was introduced gradually to the teachers, to underline the process of decomposition of problems, showing how to obtain, in order: the static graph of the dilated and subsequently rotated square; the animation of rotation and expansion; the combination of the two animations to visualize the homothety. With Maple2018, the extension and rotation of a planar figure are obtained respectively with the dilatation and rotation commands of the geometry package. Both commands need three input arguments: the name that the transformed figure must have, the name of the figure to be transformed and the transformation coefficient. To create the animation of the extension and rotation of the square as the expansion coefficient or the angle of rotation varies, the two commands have been generalized defining them as procedures according to the transformation coefficient, indicated with k. Below are the commands to define the two procedures (respectively f and g), where: quad2 is the starting square, quadD is the dilated square and quadDR is the rotated dilated square. The commands to display the two animations are also shown:

```
g:=proc (k)
    options operator, arrow;
    plots:-display(draw([quad2(color=blue),
        (dilatation(quadD, quad2,k))(color=green)],
            axes=none))
end proc;
plots:-animate(g,['k'],k=1..(1/2)*sqrt(2));
f:=proc (k)
    options operator, arrow;
        plots:-display(draw([quad2(color=blue),
            (rotation(quadDR, quadD, k,
                'counterclockwise'))(color=green)]))
end proc;
plots:-animate(f,['k'],k=0..-(1/4)*Pi);
```

Multiple strategies can be used to merge the two graphs into one. Since the animations extend over different intervals, teachers proposed to define a new procedure that uses a function that for some times returns the expansion and in another the rotation. In the following commands, the rotation interval was shifted to avoid the pause between the two transformations:

```
gf:=proc (k)
    options operator, arrow;
        piecewise(k<(1/2)*sqrt(2),
            plots:-display(draw([quad2(color=blue),
                (rotation(quadDR2, quadD2,
                    k-(1/2)*sqrt(2),
                    'counterclockwise'))
                    (color=green)],axes=none),
            (1/2)*sqrt(2)<=k,
            plots:-display(draw([quad2(color=blue),
                (dilatation(quadD2,quad2,k))
                (color=green)], axes=none))))
end proc;
plots:-animate(gf,['k'],k=1..(1/4)*Pi+(1/2)*sqrt(2))
```

Figure 10 shows the output of last command.

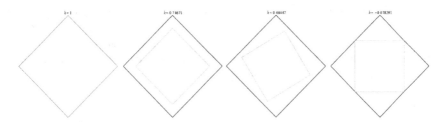

**Fig. 10.** Animated graph to illustrate the combination of the two transformations.

In this example the process of generalization for the creation of the two animations and the process of abstraction for their union take on importance. This problem gives the chance of multiple insights, addressed together with the teachers, always characterized by the use of animated graphics: development of multiple strategies for the creation of the drawing, iteration of the procedure to create the entire tile and creation of the same type of drawing with other types of regular polygons. In this case, the animations can be used by the teachers to illustrate the solution of the problem but also to introduce or further study the theory on geometric transformations in the Cartesian plane. The animations can be exported in GIF format and loaded into a Virtual Learning Environment where the animation is maintained. In the example in Fig. 11, the GIF of the in-depth problem with polygons was attached in a message from the forum of the teacher training course. This type of teaching material can also be used to involve students, to make them curious and passionate about mathematics.

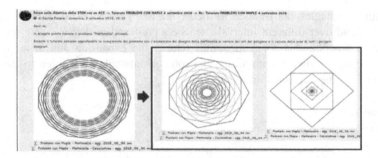

**Fig. 11.** Example of animation loaded as GIF in a Virtual Learning Environment.

## 6.2  Animated Graphics for Illustrating Theoretical Concepts

A second important aspect in the creation of animated graphs is the use of animations as stand-alone teaching materials to illustrate or enrich the explanation of theoretical concepts or to present simulations, in order to support the construction of mental models by students. The following example concerns the generation of solids of revolution. In addition to a theoretical explanation on the generation of solids of revolution, an example describes how the solid is obtained by rotating a rectangular trapezoid around its major base, minor base, height or oblique side. Graphic visualization is indeed very important for learning because it enriches the theoretical one. In this case, the commands of the `plottools` package are of help, in particular the `polygon` command to draw the trapezoid in the space and the `rotate` command to rotate it by a given angle with respect to a given segment in space. Together with the teachers, a procedure was designed, defined and implemented in several training sessions. Taking as input a figure in space and any segment in space, the procedure returns the animation of the rotation of 360° around the given segment. Following is the procedure:

```
rotaz:=proc (fig,l::list)
    local gr,rot,fr;
     gr:=plots:-display(fig,axes=none);
    fr:=proc (k)
       options operator, arrow;
       plottools:-rotate(gr, k, l)
end proc;
return plots:-animate(fr,['k'],'k'=0..2*Pi,trace=100,
    scaling=constrained, style=pointline,
        orientation=[-96, 86, -10])
end proc;
```

As an example, Fig. 12 shows the animation of the trapezoid's rotation around the major side that generates a solid composed of a cylinder surmounted by a cone. Space animations are even more effective because three-dimensional graphs are interactive and have a nice manipulability. It is possible to rotate the figure in each angle, move it and, as in two-dimensional graphics, modify it to study the solid of revolution.

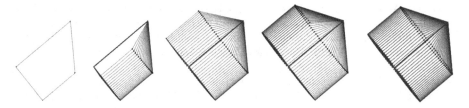

**Fig. 12.** Three-dimensional animated graph of the generation of a solid of revolution.

In this example, we used the process of decomposition of problems and the process of generalization several times: to move from the representation of a trapezoid in the plane to that in space, to create the trapezoid animation around one of its sides, to generalize the animation to any segment in space and to generalize the animation to any plane figure. The procedure is versatile and allows to create solids of revolution (concave and convex) starting from any shape and segment in the space. The animation in output, exportable in GIF format, is an effective educational tool to illustrate the creation of rotational solids and to support students in understanding their generation, which sometimes can be difficult to imagine. The animations can be a valid aid in the construction of mental models, too, more than a simple static representation of the final result of the rotation or a simply theoretical explanation.

## 7   Conclusions

The results show some examples of processes of computational thinking in the creation of animated graphs for solving a contextualized problem using Maple. By analyzing the animations created to derive the trajectory of the ladybug, four different processes of computational thinking emerged, reflecting different resolution strategies and generalization processes. From the results, it emerged that the creation of the animated graphs activates all the processes behind the mental strategies of computational thinking, useful for solving problems. It is therefore desirable, because of their additional playful component, to use these graphs in ordinary mathematics education. In the examples we discussed about animations created during training activities with secondary school teachers, we have shown some examples of how animations can support the learning of scientific concepts otherwise only expressed in verbal or mathematical form. We therefore believe that it is very important to train the teachers, both because they need to raise awareness of the processes that the students activate during the creation of animated graphs and to enrich the theoretical or practical explanations with animated representations.

## References

1. Barana, A., Marchisio, M.: Dall'esperienza di digital mate training all'attività di Alternanza Scuola Lavoro. Mondo Digitale **15**(64), 63–82 (2016)
2. Barana, A., Fioravera, M., Marchisio, M.: Developing problem solving competences through the resolution of contextualized problems with an advanced computing environment. In: Proceedings of the 3rd International Conference on Higher Education Advances. Universitat Politècnica València, pp. 1015–1023 (2017)

3. Barana, A., Fioravera, M., Marchisio, M., Rabellino, S.: Adaptive teaching supported by ICTs to reduce the school failure in the project "Scuola Dei Compiti". In: Proceedings of 2017 IEEE 41st Annual Computer Software and Applications Conference (COMPSAC). Presented at the 2017 IEEE 41st Annual Computer Software and Applications Conference (COMPSAC), pp. 432–437. IEEE (2017)
4. Barana, A., Conte, A., Fioravera, M., Marchisio, M., Rabellino, S.: A model of formative automatic assessment and interactive feedback for STEM. In: Proceedings of 2018 IEEE 42nd Annual Computer Software and Applications Conference (COMPSAC), pp. 1016–1025. IEEE, Tokyo, Japan (2018)
5. Barana, A., Marchisio, M., Miori, R.: MATE-BOOSTER: design of an e-Learning course to boost mathematical competence. In: Proceedings of the 11th International Conference on Computer Supported Education (CSEDU 2019), pp. 280–291 (2019)
6. Bizzarri, G., Forlizzi, L., Proietti, G.: Informatica: didattica possibile e pensiero computazionale 10 (2011)
7. Brancaccio, A., Marchisio, M., Palumbo, C., Pardini, C., Patrucco, A., Zich, R.: Problem posing and solving: strategic italian key action to enhance teaching and learning mathematics and informatics in the high school. In: Proceedings of 2015 39th Annual Computer Software and Applications Conference, pp. 845–850. IEEE, Taichung, Taiwan (2015)
8. Brancaccio, A., Esposito, M., Marchisio, M., Pardini, C.: L'efficacia dell'apprendimento in rete degli immigrati digitali. L'esperienza SMART per le discipline scientifiche. Mondo Digitale 15(64), 803–821 (2016)
9. Council Recommendation of 22 May 2018 on key competences for lifelong learning. Off. J. Eur. Union (2018)
10. Demartini, C.G., et al.: Problem posing (& solving) in the second grade higher secondary school. Mondo Digitale 14, 418–422 (2015)
11. Dorfler, W.: Forms and means of generalization in mathematics. In: Mathematical Knowledge: Its Growth Through Teaching. pp. 63–95. Kluver (1991)
12. Kuzniak, A., Parzysz, B., Vivier, L.: Trajectory of a problem: a study in teacher training. Math. Enthus. 10, 407–440 (2013)
13. Landriscina, F.: Didattica delle immagini: dall'informazione ai modelli mentali 12, 27–34 (2013)
14. Liljedahl, P., Santos-Trigo, M., Malaspina, U., Bruder, R.: Problem Solving in Mathematics Education. Springer, Heidelberg (2016). https://doi.org/10.1007/978-3-319-40730-2
15. Lodi, M., Martini, S., Nardelli, E.: Abbiamo davvero bisogno del pensiero computazionale? 15 (2017)
16. Malara, N.A.: Processi di generalizzazione nell'insegnamento/apprendimento dell'algebra. Annali online formazione docente 4, 13–35 (2013)
17. Samo, D.D., Darhim, D., Kartasasmita, B.: Culture-based contextual learning to increase problem-solving ability of first year university student. J. Math. Educ. 9 (2017). https://doi.org/10.22342/jme.9.1.4125.81-94
18. Santos-Trigo, M., Moreno-Armella, L., Camacho-Machín, M.: Problem solving and the use of digital technologies within the mathematical working space framework. ZDM 48, 827–842 (2016)
19. Wing, J.: Computational thinking. Presented at the Communications of the ACM (2006)

# Effective Problem Solving Using SAT Solvers

Curtis Bright[1,2(✉)], Jürgen Gerhard[2], Ilias Kotsireas[3], and Vijay Ganesh[1]

[1] University of Waterloo, Waterloo, Canada
cbright@uwaterloo.ca
[2] Maplesoft, Waterloo, Canada
[3] Wilfrid Laurier University, Waterloo, Canada

**Abstract.** In this article we demonstrate how to solve a variety of problems and puzzles using the built-in SAT solver of the computer algebra system Maple. Once the problems have been encoded into Boolean logic, solutions can be found (or shown to not exist) automatically, without the need to implement any search algorithm. In particular, we describe how to solve the $n$-queens problem, how to generate and solve Sudoku puzzles, how to solve logic puzzles like the Einstein riddle, how to solve the 15-puzzle, how to solve the maximum clique problem, and finding Graeco-Latin squares.

**Keywords:** SAT solving · Maple · $n$-queens problem · Sudoku · Logic puzzles · 15-puzzle · Maximum clique problem · Graeco-Latin squares

## 1 Introduction

"... it is a constant source of annoyance when you come up with a clever special algorithm which then gets beaten by translation to SAT."

—Chris Jefferson

The satisfiability (SAT) problem is to determine if a given Boolean expression can be satisfied—is there some way of assigning true and false to its variables that makes the whole formula true? Despite at first seeming disconnected from most of the kinds of problems that mathematicians care about we argue in this paper that it is in the interests of mathematicians to have a familiarity with SAT solving and encoding problems in SAT. An immense amount of effort over the past several decades has produced SAT solvers that are not only practical for many problems but are actually the fastest known way of solving an impressive variety of problems such as software and hardware verification problems [3]. They have also recently been used to resolve long-standing mathematical conjectures [12] and construct large combinatorial designs [7].

Since 2018, the computer algebra system Maple has included the award-winning SAT solver MapleSAT [15] as its built-in SAT solver. This solver can be used through the `Satisfy` command of the `Logic` package. `Satisfy` returns a

© Springer Nature Switzerland AG 2020
J. Gerhard and I. Kotsireas (Eds.): MC 2019, CCIS 1125, pp. 205–219, 2020.
https://doi.org/10.1007/978-3-030-41258-6_15

satisfying assignment of a given Boolean expression (if one exists) or NULL if no satisfying assignment exists. In this paper we demonstrate through a number of detailed examples how `Satisfy` can be an effective and efficient way of solving a variety of problems and puzzles.

Very little prerequisites are necessary to understand this paper; the main necessary background is a familiarity with Boolean logic which we outline in Sect. 2. We then present effective solutions to the $n$-queens problem (Sect. 3), logic puzzles like the Einstein riddle (Sect. 4), Sudoku puzzles (Sect. 5), Euler's Graeco-Latin square problem (Sect. 6), the maximum clique problem (Sect. 7), and the 15-puzzle (Sect. 8). In each case we require no knowledge of any of the special-purpose search algorithms that have been proposed to solve these problems; once the problems have been encoded into Boolean logic they are automatically solved using Maple's `Satisfy`.

All of the examples discussed in this paper were implemented and run in Maple 2018 and Maple 2019. Due to space constraints we have not included our code in this paper, but Maple worksheets containing complete implementations have been made available online through the Maple Application Center [6].

## 2   Background

A basic understanding of Boolean logic is the only prerequisite necessary to understand the solutions described in this paper. One of the main advantages of Boolean logic (but also one of its main disadvantages) is its simplicity: each variable can assume only one of two values denoted by true and false. Boolean expressions consist of variables joined by Boolean operators. The most common Boolean operators (and the ones available in the `Logic` package of Maple) are summarized in Table 1.

The $\vee$ (or), $\wedge$ (and), and $\neg$ (not) operators have meanings based on their everyday English meanings: $x_1 \vee \cdots \vee x_n$ is true exactly when at least one $x_i$ is true, $x_1 \wedge \cdots \wedge x_n$ is true exactly when all $x_i$ are true, and $\neg x$ is true exactly when $x$ is false. More generally, $x \Leftrightarrow y$ is true exactly when $x$ and $y$ have the

Table 1. The Boolean logical operators available in Maple.

| Name | Symbol | Arity | Maple syntax |
| --- | --- | --- | --- |
| Negation | $\neg$ | 1 | `&not` |
| Conjunction | $\wedge$ | $n$-ary | `&and` |
| Disjunction | $\vee$ | $n$-ary | `&or` |
| Implication | $\Rightarrow$ | 2 | `&implies` |
| Biconditional | $\Leftrightarrow$ | 2 | `&iff` |
| Alternative denial | $\uparrow$ | $n$-ary | `&nand` |
| Joint denial | $\downarrow$ | $n$-ary | `&nor` |
| Exclusive disjunction | $\underline{\vee}$ | $n$-ary | `&xor` |

same truth values, $x \Rightarrow y$ is false exactly when $y$ is true and $x$ is false, $x_1 \veebar \cdots \veebar x_n$ is true exactly when an odd number of $x_i$ are true, $x_1 \uparrow \cdots \uparrow x_n$ is true exactly when at least one $x_i$ is false, and $x_1 \downarrow \cdots \downarrow x_n$ is true exactly when all $x_i$ are false.

A *literal* is an expression of the form $x$ or $\neg x$ where $x$ is a Boolean variable. A *clause* is an expression of the form $l_1 \vee \cdots \vee l_n$ where all $l_i$ are literals. A *conjunctive normal form* (CNF) expression is of the form $c_1 \wedge \cdots \wedge c_n$ where all $c_i$ are clauses. A standard theorem of Boolean logic is that any expression can be converted into an equivalent expression in conjunctive normal form where two expressions are said to be *equivalent* if they assume the same truth values under all variable assignments.

The current algorithms used in state-of-the-art SAT solvers require that the input formula be given in conjunctive normal form. While this is convenient for the purposes of designing efficient solvers it is not convenient for the mathematician who wants to express their problem in Boolean logic—not all expressions are *naturally* expressed in conjunctive normal form. An advantage that Maple has over most current state-of-the-art SAT solvers is that Maple does not require the input to be given in conjunctive normal form. Since the algorithms used by MapleSAT require CNF to work properly, Maple internally converts expressions into CNF automatically. This is done by using a number of equivalence transformations, e.g., the expression $x \Rightarrow y$ is rewritten as the clause $\neg x \vee y$.

Care has been taken to make the necessary conversion to CNF efficient. This is important because conversions that use the most straightforward equivalence rules generally require exponential time to complete. For example, the Maple command `Normalize` from the `Logic` package can be used to convert an expression into CNF. But watch out—many expressions explode in size following this conversion. For example, the expression $x_1 \veebar \cdots \veebar x_n$ when converted into CNF contains $2^{n-1}$ clauses. The main trick used to make the conversion into CNF efficient is the *Tseitin transformation* [23]. This transformation avoids the exponential blowup of the straightforward transformations by using additional variables to derive a new formula that is satisfiable if and only if the original formula is satisfiable. For example, the expression $x_1 \veebar \cdots \veebar x_n$ is rewritten as $(t \veebar x_3 \veebar \cdots \veebar x_n) \wedge C$ where $t$ is a new variable and $C$ is a CNF encoding of the formula $t \Leftrightarrow (x_1 \veebar x_2)$, namely,

$$(\neg x_1 \vee x_2 \vee t) \wedge (x_1 \vee \neg x_2 \vee t) \wedge (x_1 \vee x_2 \vee \neg t) \wedge (\neg x_1 \vee \neg x_2 \vee \neg t).$$

The transformation is then recursively applied to $t \veebar x_3 \veebar \cdots \veebar x_n$ (the part of the formula not in CNF) until the entire formula is in CNF. The Maple command `Tseitin` of the `Logic` package can be applied to convert an arbitrary formula into CNF using this translation. Thus, Maple offers us the convenience of not requiring encodings to be in CNF while avoiding the inefficiencies associated with a totally unrestricted encoding.

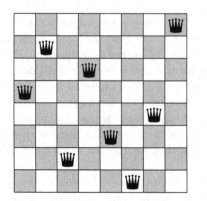

$$Q_{1,5} = \text{true}$$
$$Q_{2,7} = \text{true}$$
$$Q_{3,2} = \text{true}$$
$$Q_{4,6} = \text{true}$$
$$Q_{5,3} = \text{true}$$
$$Q_{6,1} = \text{true}$$
$$Q_{7,4} = \text{true}$$
$$Q_{8,8} = \text{true}$$

**Fig. 1.** A visual representation of a solution for the 8-queens problem (left) and the variables assigned to true in this solution using our SAT encoding. Following traditional chess convention, columns are indexed left to right and rows are indexed bottom to top.

## 3    The $n$-queens Problem

The $n$-queens problem is to place $n$ chess queens on an $n \times n$ chessboard such that no two queens are mutually attacking (i.e., in the same row, column, or diagonal). The problem was first proposed for $n = 8$ by Bezzel in 1848 and the first solution for general $n$ was given by Pauls in 1874 [2]. The problem is solvable for all $n \geq 4$; a solution for $n = 8$ (found in 0.015 s using `Satisfy`) is shown in Fig. 1.

The $n$-queens problem is a standard example of a constraint satisfaction problem [18]. The encoding that we use for this problem uses the $n^2$ Boolean variables $Q_{x,y}$ with $1 \leq x, y \leq n$ to denote if there is a queen on square $(x, y)$. There are two kinds of constraints necessary for this problem: *positive* constraints that say that there are $n$ queens on the board and *negative* constraints that say that queens do not attack each other. A satisfying assignment of these constraints exists exactly when the $n$-queens problem is solvable.

Since there are $n$ rows and each row must contain a queen the positive constraints are of the form $Q_{1,j} \vee \cdots \vee Q_{n,j}$ for $1 \leq j \leq n$. Similarly, each column must contain a queen; these constraints are of the form $Q_{i,1} \vee \cdots \vee Q_{i,n}$ for $1 \leq i \leq n$. The negative constraints say that if $(x, y)$ contains a queen then all squares attacked by $(x, y)$ do not contain a queen. These constraints are represented in Boolean logic by $Q_{x,y} \Rightarrow \neg(A_1 \vee \cdots \vee A_k)$ where $\{A_1, \ldots, A_k\}$ are the variables "attacked" by a queen on $(x, y)$. In general this encoding uses $\Theta(n^2)$ constraints in order $n$. Typically `Satisfy` is able to solve each order using slightly more time than the previous order and the last order it can solve in under a second is $n = 32$.

# 4    The Einstein Riddle

The Einstein riddle is a logic puzzle apocryphally attributed to Albert Einstein and is often stated with the remark that it is only solvable by 2% of the world's population. The true source of the puzzle is unknown, but a version of it appeared in the magazine Life International in 1962. In the puzzle there are five houses in a row with each house a different colour and each house owned by a man of a different nationality. Additionally, each of the owners have a different pet, prefer a different kind of drink, and smoke a different brand of cigarette. Furthermore, the following information is given:

1. The Brit lives in the red house.
2. The Swede keeps dogs as pets.
3. The Dane drinks tea.
4. The green house is next to the white house, on the left.
5. The owner of the green house drinks coffee.
6. The person who smokes Pall Mall rears birds.
7. The owner of the yellow house smokes Dunhill.
8. The man living in the centre house drinks milk.
9. The Norwegian lives in the first house.
10. The man who smokes Blends lives next to the one who keeps cats.
11. The man who keeps horses lives next to the man who smokes Dunhill.
12. The man who smokes Blue Master drinks beer.
13. The German smokes Prince.
14. The Norwegian lives next to the blue house.
15. The man who smokes Blends has a neighbour who drinks water.

The puzzle is: Who owns the fish?

To solve this riddle using Maple, we label the houses 1 to 5 and use the variables $S_{i,a}$ where $1 \leq i \leq 5$ and $a$ is an attribute (one of the colours, nationalities, pets, drinks, or cigarette brands). For example, if $a$ is a colour then $a$ is in the set $C := \{\text{red}, \text{green}, \text{white}, \text{yellow}, \text{blue}\}$ and similarly for the other attribute types; there are five distinct possible attributes for each type of attribute. In total there are $5^2 = 25$ possible values for $a$ and $5^3 = 125$ variables $S_{i,a}$.

We know that each attribute is not shared among the five houses or their owners. Since there are exactly five houses, each attribute must appear exactly once among the five houses. The knowledge that each attribute appears at least once can be encoded as the clauses $\bigvee_{i=1}^{5} S_{i,a}$ for each attribute $a$ and the knowledge that each attribute is not shared can be encoded as $S_{i,a} \Rightarrow \neg S_{j,a}$ where $j$ is a house index not equal to $i$ and $a$ is an attribute. Additionally, the fact that each house has some colour is encoded as $\bigvee_{a \in C} S_{i,a}$ for each house index $i$ and the knowledge that each house cannot have two colours can be encoded as $S_{i,c} \Rightarrow \neg S_{i,d}$ where $i$ is a house index and $c$ and $d$ are two distinct colours (and similarly for the other kinds of attributes).

The known facts can be encoded into logic fairly straightforwardly; for example, the first fact can be encoded into logic as $S_{i,\text{Brit}} \Rightarrow S_{i,\text{red}}$ for house indices $i$

|  |  |  |  |  |  |  |  |  |
|---|---|---|---|---|---|---|---|---|
| 8 |  |  |  |  |  |  |  |  |
|  |  | 3 | 6 |  |  |  |  |  |
|  | 7 |  |  | 9 |  | 2 |  |  |
|  | 5 |  |  |  | 7 |  |  |  |
|  |  |  | 4 | 5 | 7 |  |  |  |
|  |  |  | 1 |  |  |  | 3 |  |
|  | 1 |  |  |  |  | 6 | 8 |  |
|  | 8 | 5 |  |  |  |  | 1 |  |
|  | 9 |  |  |  | 4 |  |  |  |

$S_{1,1,8}$ $S_{2,3,3}$ $S_{2,4,6}$
$S_{3,2,7}$ $S_{3,5,9}$ $S_{3,7,2}$
$S_{4,2,5}$ $S_{4,6,7}$ $S_{5,5,4}$
$S_{5,6,5}$ $S_{5,7,7}$ $S_{6,4,1}$
$S_{6,8,3}$ $S_{7,3,1}$ $S_{7,8,6}$
$S_{7,9,8}$ $S_{8,3,8}$ $S_{8,4,5}$
$S_{8,8,1}$ $S_{9,2,9}$ $S_{9,7,4}$

**Fig. 2.** A Sudoku puzzle estimated to have a difficulty rating of 11 stars, whereas the most challenging Sudoku puzzles that are usually published are given 5 stars. On the right are the starting constraints (as unit clauses) of our encoding for this puzzle.

and the last fact can be encoded as $S_{i,\text{Blends}} \Rightarrow (S_{i-1,\text{water}} \lor S_{i+1,\text{water}})$ for $1 < i < 5$ and $(S_{1,\text{Blends}} \Rightarrow S_{2,\text{water}}) \land (S_{5,\text{Blends}} \Rightarrow S_{4,\text{water}})$. Using `Satisfy` on these constraints produces the unique satisfying solution (that includes the equations $S_{4,\text{German}} = \text{true}$ and $S_{4,\text{fish}} = \text{true}$, thereby solving the puzzle) in under 0.01 s.

## 5   Sudoku Puzzles

Sudoku is a popular puzzle that appears in many puzzle books and newspapers. Given a 9 by 9 grid whose squares are either blank or contain a number between 1 and 9, the objective is to fill in the blank squares in such a way that each row and column contains exactly one digit between 1 and 9. Additionally, each of the nine 3 by 3 subgrids which compose the grid (called blocks) must also contain exactly one digit between 1 and 9. Figure 2 contains a Sudoku puzzle designed by mathematician Arto Inkala and claimed to be the world's hardest Sudoku [9].

It is known that Sudoku can be modelled as a SAT problem [16] or a constraint satisfaction problem [20]. A straightforward encoding uses $9^3 = 729$ variables $S_{i,j,k}$ with $1 \leq i, j, k \leq 9$ where $S_{i,j,k}$ is true exactly when the square $(i, j)$ contains the digit $k$. The rules of Sudoku state that each square must be filled with a digit between 1 and 9 and that the same digit cannot appear twice in the same row, column, or block. The first constraint has the form $S_{i,j,1} \lor \cdots \lor S_{i,j,9}$ for all $1 \leq i, j \leq 9$ and the second constraint has the form $S_{i,j,k} \Rightarrow \neg S_{i',j',k}$ for all $1 \leq i, j, i', j', k \leq 9$ where $(i, j)$ does not equal $(i', j')$ but is in the same row, column, or block as $(i, j)$.

One can also include the constraints $S_{i,j,k} \Rightarrow \neg S_{i,j,k'}$ for all $1 \leq i, j, k, k' \leq 9$ with $k \neq k'$ that say that each square can contain at most one digit. However, these constraints are unnecessary since they are logically implied by the first two constraints. In our tests, including these additional constraints slightly decreased the performance of `Satisfy`. Without the additional constraints the "world's hardest Sudoku" was solved in 0.25 s and with them it was solved in 0.33 s.

Additionally, we developed a method of generating Sudoku puzzles with a unique solution using `Satisfy`. This allowed us to write an interactive Sudoku game where random puzzles can automatically be generated on command. To begin, `Satisfy` is used to find a solution to the above Sudoku constraints with an empty grid (no starting clues) and the produced satisfying solution generates a completed Sudoku grid $R$. A random seed is passed to the SAT solver so that a different solution $R$ is generated each time; this is done by passing the following `solveroptions` in the call to `Satisfy`:

```
[rnd_init_act=true, random_seed=floor(1000*time[real]())]
```

Let $R_{i,j}$ denote the $(i,j)$ the entry in the solution $R$ where $1 \leq i, j \leq 9$. The $R_{i,j}$ are randomly ordered and the first 50 entries $R_{i,j}$ are selected as the potential starting configuration of a Sudoku puzzle. This puzzle has the solution $R$ by construction, though other solutions may also exist.

To verify that the generated solution is unique, we re-run `Satisfy` with the additional 50 unit clauses corresponding to the starting configuration along with the constraint $\bigvee_{R_{i,j}=k} \neg S_{i,j,k}$ which blocks the solution $R$. If `Satisfy` returns another solution then we start over and find a new $R$ to try. Otherwise the starting configuration forms a legal Sudoku puzzle.

Additionally, it may be the case that we can use fewer than 50 entries and still obtain a Sudoku puzzle with a unique solution. To estimate how many entries need to be assigned using only a few extra calls to the SAT solver we use a variant of binary search, letting $l := 20$ and $h := 50$ be lower and upper bounds on how many entries we will define in the puzzle. Next, we let $m := \text{round}((l+h)/2)$ and repeat the first step except using only the first $m$ entries $R_{i,j}$. If the resulting SAT instance is satisfiable then we need to use strictly more than $m$ entries to ensure that a unique solution exists and if the resulting SAT instance is unsatisfiable then we can perhaps use strictly fewer than $m$ entries. Either way, we improve the bounds on how many entries to assign (in the former case we can update $l$ to $m$ and in the latter case we can update $h$ to $m$) and this step can be repeated a few times to find more precise bounds on how many entries need to be assigned to ensure a unique solution exists.

# 6 Euler's Graeco-Latin Square Problem

A *Latin square* of order $n$ is an $n \times n$ matrix containing integer entries between 1 and $n$ such that every row and every column contains each entry exactly once. Two Latin squares are *orthogonal* if the superposition of one over the other produces all $n^2$ distinct pairs of integers between 1 and $n$. A pair of orthogonal Latin squares was called a *Graeco-Latin square* by the mathematician Leonhard Euler who in 1782 used Latin characters to represent the entries of the first square and Greek characters to represent the entries of the second square [11]. Figure 3 contains a visual representation of a Graeco-Latin square.

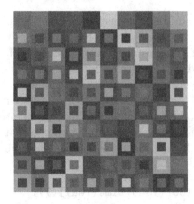

| | |
|---|---|
| $A_{1,1,1} = \text{true}$ | $B_{1,1,1} = \text{true}$ |
| $A_{2,1,2} = \text{true}$ | $B_{2,1,10} = \text{true}$ |
| $A_{3,1,3} = \text{true}$ | $B_{3,1,5} = \text{true}$ |
| $A_{4,1,4} = \text{true}$ | $B_{4,1,2} = \text{true}$ |
| $A_{5,1,5} = \text{true}$ | $B_{5,1,6} = \text{true}$ |
| $A_{6,1,6} = \text{true}$ | $B_{6,1,9} = \text{true}$ |
| $A_{7,1,7} = \text{true}$ | $B_{7,1,4} = \text{true}$ |
| $A_{8,1,8} = \text{true}$ | $B_{8,1,7} = \text{true}$ |
| $A_{9,1,9} = \text{true}$ | $B_{9,1,3} = \text{true}$ |
| $A_{10,1,10} = \text{true}$ | $B_{10,1,8} = \text{true}$ |

**Fig. 3.** On the left is a visual representation of a Graeco-Latin square of order 10 with each colour denoting a separate integer. The entries of the second Latin square (represented by the small squares in the image) are superimposed onto the entries of the first Latin square. On the right are the variables corresponding to the first column of this Graeco-Latin square that are assigned to true using our encoding.

Euler studied the orders $n$ for which Graeco-Latin squares exist and found methods for constructing them when $n$ was odd or a multiple of 4. Since such squares do not exist for $n = 2$ and he was unable to find a solution for $n = 6$ he conjectured that Graeco-Latin squares do not exist when $n \equiv 2 \pmod 4$. Euler's conjecture became famous as he was not able to resolve it in his lifetime.

The first progress on the conjecture did not come until over a hundred years later when in 1900 Tarry showed that Graeco-Latin squares of order 6 do not exist [22]. This gave credence to Euler's conjecture and many mathematicians thought the conjecture was true—in fact, three independent proofs of the conjecture were published in the early 20th century [17,19,24]. In 1959–1960 Bose, Shrikhande, and Parker [4,5] made explosive news (even appearing on the front page of the New York Times) by showing that these proofs were invalid by giving explicit constructions for Graeco-Latin squares in all orders except two and six. As it turns out, a lot of time could have been saved if Euler had a copy of Maple—we now show that Euler's conjecture can be automatically disproven in Maple. With Satisfy we are able to construct small Graeco-Latin squares without any knowledge of search algorithms or construction methods.

Our encoding for the Graeco-Latin square problem of order $n$ uses the $2n^3$ variables $A_{i,j,k}$ and $B_{i,j,k}$ with $1 \le i, j, k \le n$. The variables $A_{i,j,k}$ will be true exactly when the $(i, j)$th entry of the Latin square $A$ is $k$ and $B_{i,j,k}$ will be true exactly when the $(i, j)$th entry of the Graeco square $B$ is $k$.

There are three kinds of constraints that specify that $(A, B)$ is a Graeco-Latin square: Those that specify that every entry of $A$ and $B$ is an integer between 1 and $n$, those that specify that the rows and columns of $A$ and $B$ contain no duplicate entries, and those that specify that $A$ and $B$ are orthogonal. Additionally, there are constraints that are not logically necessary but help cut down the search space. Some work has previously been done using SAT solvers to search for special kinds of Graeco-Latin squares [25]. The encoding we use is similar but takes advantage of the fact that Maple does not require constraints to be specified in conjunctive normal form.

First, we specify that the entries of $A$ are well-defined, i.e., consist of a single integer between 1 and $n$. The constraints that say that each entry of $A$ contains at least one integer are of the form $A_{i,j,1} \vee \cdots \vee A_{i,j,n}$ for each index pair $(i,j)$ and the constraints that say that each entry of $A$ contains at most one integer are of the form $A_{i,j,k} \Rightarrow \neg A_{i,j,l}$ for each index pair $(i,j)$ and integer $k \neq l$. Similar constraints are also used to specify that the entries of $B$ are well-defined.

Second, we specify that $A$ is a Latin square, i.e., all columns and rows contain distinct entries. These have the form $A_{i,j,k} \Rightarrow \neg A_{i',j',k}$ where $1 \leq k \leq n$ and $(i,j) \neq (i',j')$ but $(i,j)$ is in the same column or row as $(i',j')$. Similarly, we also specify that $B$ is a Latin square.

Third, we specify that $A$ and $B$ are orthogonal, i.e., for every pair $(k,l)$ there exists some pair $(i,j)$ such that $A_{i,j,k} \wedge B_{i,j,l}$ holds. These constraints are of the form $\bigvee_{i,j=1}^{n}(A_{i,j,k} \wedge B_{i,j,l})$ for each pair $(k,l)$.

Lastly, we include some "symmetry breaking" constraints. These constraints are not strictly necessary but they shrink the search space and thereby make the search more efficient. In general, when a search space splits into symmetric subspaces it is beneficial to add constraints that remove or "break" the symmetry. Graeco-Latin squares $(A,B)$ have a number of symmetries, in particular, a row or column permutation simultaneously applied to $A$ and $B$ produces another Graeco-Latin square. Also, any permutation of $\{1,\ldots,n\}$ may be applied to the entries of either $A$ or $B$.

The result of these symmetries is that any Graeco-Latin square can be transformed into one where the first row and column of $A$ has entries in ascending order (by permuting rows/columns) and the first row of $B$ has entries in ascending order (by renaming the entries of $B$). Thus, we can assume the constraint $\bigwedge_{i=1}^{n}(A_{1,i,i} \wedge B_{1,i,i} \wedge A_{i,1,i})$. Altogether this encoding uses $\Theta(n^4)$ constraints.

Using this encoding the orders up to eight can be solved in 25 total seconds (including 14 s to show that no Graeco-Latin squares exist in order six), a Graeco-Latin square of order nine can be found in about 45 min, and a Graeco-Latin square of order ten can be found in about 23 h, thereby disproving Euler's Graeco-Latin square conjecture.

## 7   The Maximum Clique Problem

The maximum clique problem is to find a clique of maximum size in a given graph. A *clique* of a graph is a subset of its vertices that are all mutually connected (see Fig. 4). The decision version of this problem (does a graph contain a clique of size $k$?) is in NP, meaning that it is easy to verify the correctness of a solution if one can be found. By the Cook–Levin theorem [10] the problem can be encoded into a SAT instance in polynomial time. However, the reduction involves simulating the computation of a machine that solves the maximum clique problem and is therefore not very convenient to use in practice. Thus, we provide a simpler encoding into Boolean logic.

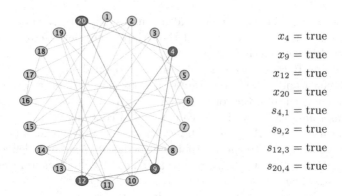

$x_4 = \text{true}$

$x_9 = \text{true}$

$x_{12} = \text{true}$

$x_{20} = \text{true}$

$s_{4,1} = \text{true}$

$s_{9,2} = \text{true}$

$s_{12,3} = \text{true}$

$s_{20,4} = \text{true}$

**Fig. 4.** On the left is a visual representation of a graph with 20 vertices and a high-lighted clique of size 4 that was found in 0.025 s. On the right are the important variables assigned to true in the assignment returned by Satisfy for our encoding of the maximum clique problem for this graph.

Suppose the given graph $G$ has vertices labelled $1, \ldots, n$ and we want to find a clique of size $k$ in $G$. Our encoding uses the variables $x_1, \ldots, x_n$ where $x_i$ represents that the vertex $i$ appears in the clique we are attempting to find. We need to enforce a constraint that says that if $x_i$ and $x_j$ are true (for any distinct vertices $1 \leq i, j \leq n$) then the edge $\{i, j\}$ exists in the graph $G$. Equivalently, if the edge $\{i, j\}$ does not exist in the graph $G$ then the variables $x_i$ and $x_j$ cannot both be true (for any vertices $i, j$). In other words, for every edge $\{i, j\}$ in the complement of $G$ we use the clause $\neg x_i \vee \neg x_j$.

Additionally, we need a way to enforce that the found clique is of size $k$. The most naive way to encode this is as a disjunction over all $\binom{n}{k}$ conjunctions of length $k$ on the variables $x_1, \ldots, x_n$. However, this encoding is very inefficient in practice. A cleverer encoding uses Boolean counter variables $s_{i,j}$ (where $0 \leq i \leq n$ and $0 \leq j \leq k$) that represent that at least $j$ of the variables $x_1, \ldots, x_i$ are assigned to true. We know that $s_{0,j}$ will be false for $1 \leq j \leq k$ and that $s_{i,0}$ will be true for $0 \leq i \leq n$. Additionally, we know that $s_{i,j}$ is true exactly when $s_{i-1,j}$ is true or $x_i$ is true and $s_{i-1,j-1}$ is true. This is represented by the formulas

$$s_{i,j} \Leftrightarrow (s_{i-1,j} \vee (x_i \wedge s_{i-1,j-1})) \qquad \text{for } 1 \leq i \leq n \text{ and } 1 \leq j \leq k$$

or in conjunctive normal form by the clauses $\neg s_{i-1,j} \vee s_{i,j}$, $\neg x_i \vee \neg s_{i-1,j-1} \vee s_{i,j}$, $\neg s_{i,j} \vee s_{i-1,j} \vee x_i$, and $\neg s_{i,j} \vee s_{i-1,j} \vee s_{i-1,j-1}$. To enforce that the found clique contains at least $k$ vertices we also assign $s_{n,k}$ to true.

To solve the maximum clique problem for a given graph $G$ we initialize $k$ to 3 (assuming the graph has at least one edge, otherwise the problem is trivial) and use the above encoding to search for a clique of size $k$. If such a clique exists we increase $k$ by 1 and repeat the search in this manner until a clique of size $k$ does not exist. The last satisfying assignment found then provides a maximum clique of $G$ using an encoding with $\Theta(nk)$ variables and $\Theta(n^2)$ clauses.

**Table 2.** A comparison of the SAT method and the `MaximumClique` function of Maple 2018 on a collection of maximum clique benchmarks with a timeout of an hour.

| Benchmark | SAT time (sec) | Maple 2018 (sec) | Vertices | Edges | Clique size |
|---|---|---|---|---|---|
| brock200_2 | 23.52 | **22.19** | 200 | 9876 | 12 |
| c-fat200-1 | 0.71 | **0.03** | 200 | 1534 | 12 |
| c-fat200-2 | 2.32 | **0.13** | 200 | 3235 | 24 |
| c-fat200-5 | **9.07** | 20.63 | 200 | 8473 | 58 |
| c-fat500-1 | 4.78 | **0.13** | 500 | 4459 | 14 |
| c-fat500-2 | 11.12 | **0.82** | 500 | 9139 | 26 |
| c-fat500-5 | **42.46** | 132.84 | 500 | 23191 | 64 |
| c-fat500-10 | **134.42** | Timeout | 500 | 46627 | 126 |
| hamming6-2 | **0.64** | 51.10 | 64 | 1824 | 32 |
| hamming6-4 | 0.04 | **0.02** | 64 | 704 | 4 |
| hamming8-2 | **58.51** | Timeout | 256 | 31616 | 128 |
| hamming8-4 | **7.96** | 3393.26 | 256 | 20864 | 16 |
| johnson8-2-4 | **0.01** | **0.01** | 28 | 210 | 4 |
| johnson8-4-4 | **0.23** | 7.80 | 70 | 1855 | 14 |
| johnson16-2-4 | **5.62** | 642.24 | 120 | 5460 | 8 |
| keller4 | **7.76** | 414.80 | 171 | 9435 | 11 |
| MANN_a9 | **0.13** | 226.18 | 45 | 918 | 16 |
| p_hat300-1 | 11.81 | **3.30** | 300 | 10933 | 8 |
| p_hat500-1 | 308.87 | **34.60** | 500 | 31569 | 9 |
| p_hat700-1 | 1281.62 | **169.68** | 700 | 60999 | 11 |

This implementation was tested on the maximum clique problems from the second DIMACS implementation challenge [13]. Additionally, it was compared with Maple's `MaximumClique` function from the `GraphTheory` package that uses a branch-and-bound backtracking algorithm [14]. Of the 80 benchmarks, the SAT method solved 18 in under 3 min and the branch-and-bound method solved 15. The SAT approach was faster in over half of the solved benchmarks and in one case solved a benchmark in 8 s that `MaximumClique` required 57 min to solve (see Table 2).

The SAT method has been made available in Maple 2019 by using the `method=sat` option of `MaximumClique`. By default the `MaximumClique` function in Maple 2019 will run the previous method used by Maple and the SAT method in parallel and return the answer of whichever method finishes first. This hybrid approach is the method of choice especially when more than a single core is available.

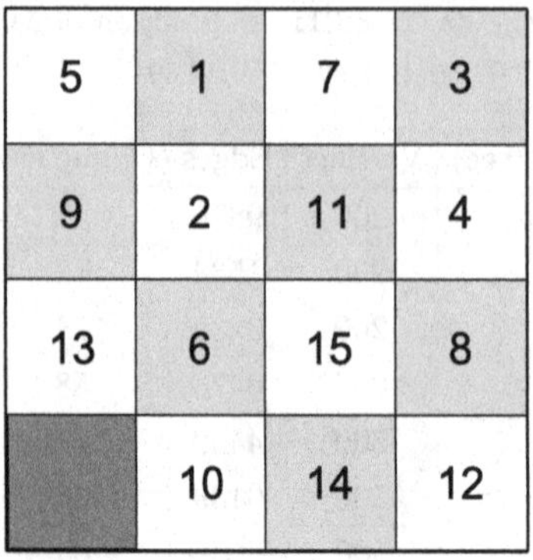

$$S_{1,1,5,0} \quad S_{1,2,1,0} \quad S_{1,3,7,0} \quad S_{1,4,3,0}$$
$$S_{2,1,9,0} \quad S_{2,2,2,0} \quad S_{2,3,11,0} \quad S_{2,4,4,0}$$
$$S_{3,1,13,0} \quad S_{3,2,6,0} \quad S_{3,3,15,0} \quad S_{3,4,8,0}$$
$$S_{4,1,16,0} \quad S_{4,2,10,0} \quad S_{4,3,14,0} \quad S_{4,4,12,0}$$

**Fig. 5.** On the left is a visual representation of one starting configuration of the 15-puzzle and on the right are the starting constraints (as unit clauses) for this starting configuration in our encoding.

## 8    The 15-Puzzle

The 15-puzzle is a classic "sliding tile" puzzle that was first designed in 1880 and became very popular in the 1880s [21]. It consists of a $4 \times 4$ grid containing tiles numbered 1 through 15 along with one missing tile (see Fig. 5). The objective of the puzzle is to arrange the tiles so that they are in ascending order when read from left to right and top to bottom and to end with the blank tile in the lower right. The only moves allowed are those that slide a tile adjacent to the blank space into the blank space. Half of the possible starting positions are solvable [1] and the hardest legal starting positions require eighty moves to complete [8].

Our encoding of the 15-puzzle is more complicated because unlike the other problems we've considered, a solution to the 15-puzzle is not static. In other words, our encoding must be able to deal with the state of the puzzle changing over time. To do this we use the variables $S_{i,j,n,t}$ to denote that the entry at $(i,j)$ contains tile $n$ at timestep $t$. Here $1 \leq i, j \leq 4$, $1 \leq n \leq 16$ (we let 16 denote the blank tile), and $0 \leq t \leq 80$ since each instance requires at most 80 moves to complete.

The board does not permit two tiles to occupy the same location at the same time; these constraints are of the form $S_{i,j,n,t} \Rightarrow \neg S_{i,j,m,t}$ for each tile numbers $n \neq m$, valid indices $i$ and $j$, and valid timesteps $t$. Next we need to generate constraints that tell the SAT solver how the state of the board can change from time $t$ to time $t + 1$. There are two cases to consider, depending on if a square (or an adjacent square) contains the blank tile.

The easier case is when a square $(i,j)$ and none of the squares adjacent to that square are blank. In that case, the rules of the puzzle imply that the tile in square $(i,j)$ does not change. We define the function doesNotChange$(i,j,t)$ to be the constraint that says that the tile in square $(i,j)$ does not change at time $t$; these constraints are of the form $\bigwedge_{n=1}^{16}(S_{i,j,n,t} \Leftrightarrow S_{i,j,n,t+1})$. We also use the function adj$(i,j)$ to denote the squares adjacent to $(i,j)$ and define notEqualOrAdjacentToBlank$(i, j, t)$ to be $\neg S_{i,j,16,t} \wedge \bigwedge_{(k,l) \in \mathrm{adj}(i,j)} \neg S_{k,l,16,t}$. The static transition constraints are of the form

$$\text{notEqualOrAdjacentToBlank}(i, j, t) \Rightarrow \text{doesNotChange}(i, j, t)$$

for all valid squares $(i, j)$ and timesteps $t$.

The harder transition case is when a square $(i, j)$ contains the blank tile. In this case we need to encode the fact that the blank tile will switch positions with exactly one of the squares adjacent to square $(i, j)$. If the tile on square $(i, j)$ switches positions with the square $(k, l)$ at time $t$ this can be encoded as the constraint $\bigwedge_{n=1}^{16}(S_{i,j,n,t} \Leftrightarrow S_{k,l,n,t+1})$. We also need to enforce that all squares adjacent to $(i, j)$ other than $(k, l)$ do not change; these constraints are of the form $\bigwedge_{(x,y)} \text{doesNotChange}(x, y, t)$ where $(x, y)$ is adjacent to $(i, j)$ but not equal to $(k, l)$. Let oneTileMoved$(i, j, k, l, t)$ denote the conjunction of the above two constraints. Then the slide transition constraints are of the form

$$S_{i,j,16,t} \Rightarrow \bigvee_{(k,l) \in \mathrm{adj}(i,j)} \text{oneTileMoved}(i, j, k, l, t)$$

for all valid squares $(i, j)$ and timesteps $t$.

The constraint boardSolved$(t)$ that says the board is solved at timestep $t$ can be encoded as $\bigwedge_{i,j=1}^{4} S_{i,j,4i+j-4,t}$. For efficiency reasons we only start looking for solutions with at most 5 moves; if no solution is found then we look for solutions using at most 10 moves and continue in this manner until a solution is found. In other words, we call Satisfy with the given starting constraints, the constraints of the puzzle as described above, and the constraint $\bigvee_{t=m-4}^{m}$ boardSolved$(t)$ where $m$ is initialized to 5 and then increased by 5 every time no solution is found.

This method was applied to the starting configuration from Fig. 5. It found that no solutions with at most 5 moves exist in 1.5 s, no solutions with at most 10 moves exist in 2.8 s, and found a solution with 15 moves in 6.3 s. It was also able to solve puzzles requiring up to 40 moves in 20 min. While this is not competitive with dedicated solvers for the 15-puzzle, it requires no knowledge beyond the rules of the game and makes an interesting example of how to push SAT solvers to their limits.

## 9    Conclusion

In this paper we've demonstrated how to solve a variety of problems and puzzles using the computer algebra system Maple and its SAT solver MapleSAT [15]. We discussed a number of encodings and ways for improving those encodings, e.g., by using symmetry breaking (as in Sect. 6) or by using auxiliary variables (as in Sect. 7). We also took advantage of Maple's ability to solve SAT problems not encoded in conjunctive normal form in Sects. 3, 6, and 8. Maple code for all the examples covered in this paper (including code to read the output of Satisfy and generate the figures included in this paper) are available for download from the Maple Application Center [6].

The implementations presented in this paper can be considered examples of *declarative programming* where the programmer focuses on describing the problem but not the solution—the computer automatically decides the best way to solve the problem. This is in contrast to *imperative programming* where a programmer needs to describe precisely *how* the computation is to take place. An advantage of declarative programming is that the programmer does not need to worry about specifying a potentially complicated search algorithm. However, a disadvantage of declarative programming is that it lacks the kind of detailed control over the method of solution that can be required for optimally efficient solutions. Furthermore, not all problems are naturally expressed in a declarative way.

As we saw in the maximum clique problem, sometimes declarative solutions can outperform imperative solutions. This also occurs in the *graph colouring* (or *chromatic number*) problem of colouring the vertices of a graph using the fewest number of colours subject to the constraint that adjacent vertices are coloured differently. For example, prior to Maple 2018 the `ChromaticNumber` function required several hours to find a minimal colouring of the $8 \times 8$ queens graph but a SAT encoding can solve this problem in under 10 s [6]. The SAT approach is available in Maple 2019 using the `method=sat` option of `ChromaticNumber`.

This somewhat unconventional manner of using Maple is not applicable to all problems but we hope the examples in this paper have convinced the reader that SAT solvers are more useful and powerful than they might at first appear. With its extensive logic functionality and convenient method of expressing logical constraints, Maple is an ideal tool for experimenting with SAT solvers and logical programming.

# References

1. Archer, A.F.: A modern treatment of the 15 puzzle. Am. Math. Mon. **106**(9), 793–799 (1999)
2. Bell, J., Stevens, B.: A survey of known results and research areas for $n$-queens. Discret. Math. **309**(1), 1–31 (2009)
3. Biere, A., Heule, M.J.H., van Maaren, H., Walsh, T.: Handbook of Satisfiability, Frontiers in Artificial Intelligence and Applications, vol. 185. IOS Press, Amsterdam (2009)
4. Bose, R.C., Shrikhande, S.S., Parker, E.T.: Further results on the construction of mutually orthogonal Latin squares and the falsity of Euler's conjecture. Can. J. Math. **12**, 189–203 (1960)
5. Bose, R.C., Shrikhande, S.S.: On the falsity of Euler's conjecture about the non-existence of two orthogonal Latin squares of order $4t + 2$. Proc. Natl. Acad. Sci. U.S.A. **45**(5), 734–737 (1959)
6. Bright, C.: Maple applications by Curtis Bright. https://www.maplesoft.com/applications/Author.aspx?mid=345070

7. Bright, C., Kotsireas, I., Ganesh, V.: A SAT+CAS method for enumerating Williamson matrices of even order. In: McIlraith, S., Weinberger, K. (eds.) Thirty-Second AAAI Conference on Artificial Intelligence, pp. 6573–6580. AAAI Press (2018)

8. Brüngger, A., Marzetta, A., Fukuda, K., Nievergelt, J.: The parallel search bench ZRAM and its applications. Ann. Oper. Res. **90**, 45–63 (1999)

9. Collins, N.: World's hardest sudoku: can you crack it? (2012). https://www.telegraph.co.uk/news/science/science-news/9359579/Worlds-hardest-sudoku-can-you-crack-it.html

10. Cook, S.A.: The complexity of theorem-proving procedures. In: Proceedings of the Third Annual ACM Symposium on Theory of Computing, pp. 151–158. ACM (1971)

11. Euler, L.: Recherches sur un nouvelle espéce de quarrés magiques. Verhandelingen uitgegeven door het zeeuwsch Genootschap der Wetenschappen te Vlissingen, pp. 85–239 (1782)

12. Heule, M.J.H., Kullmann, O., Marek, V.W.: Solving very hard problems: cube-and-conquer, a hybrid SAT solving method. In: Sierra, C. (ed.) Proceedings of the Twenty-Sixth International Joint Conference on Artificial Intelligence, IJCAI-17, pp. 4864–4868. IJCAI (2017)

13. Johnson, D.S., Trick, M.A.: Cliques, coloring, and satisfiability: second DIMACS implementation challenge, 11–13 October 1993, vol. 26. American Mathematical Society (1996)

14. Kreher, D., Stinson, D.: Combinatorial Algorithms: Generation, Enumeration, and Search. Discrete Mathematics and Its Applications. Taylor & Francis, Routledge (1998)

15. Liang, J.H., Govind V.K., H., Poupart, P., Czarnecki, K., Ganesh, V.: An empirical study of branching heuristics through the lens of global learning rate. In: Gaspers, S., Walsh, T. (eds.) SAT 2017. LNCS, vol. 10491, pp. 119–135. Springer, Cham (2017). https://doi.org/10.1007/978-3-319-66263-3_8. https://ece.uwaterloo.ca/maplesat/

16. Lynce, I., Ouaknine, J.: Sudoku as a SAT problem. In: 9th International Symposium on Artificial Intelligence and Mathematics (2006)

17. MacNeish, H.F.: Euler squares. Ann. Math. **23**(2), 221–227 (1922)

18. Nadel, B.A.: Representation selection for constraint satisfaction: a case study using $n$-queens. IEEE Intell. Syst. **5**(3), 16–23 (1990)

19. Peterson, J.: Les 36 officieurs. Annuaire des Mathématiciens, pp. 413–427 (1902)

20. Russell, S.J., Norvig, P.: Artificial Intelligence: A Modern Approach. Pearson Education, London (2010)

21. Slocum, J., Sonneveld, D.: The 15 Puzzle: How It Drove the World Crazy. Slocum Puzzle Foundation, Beverly Hills (2006)

22. Tarry, G.: Le problème des 36 officiers. Association Française pour l'Avancement des Sciences: Compte Rendu de la 29me session en Paris 1900, pp. 170–203 (1901)

23. Tseitin, G.S.: On the complexity of derivation in propositional calculus. In: Slisenko, A.O. (ed.) Studies in Constructive Mathematics and Mathematical Logic, pp. 115–125 (1970)

24. Wernicke, P.: Das Problem der 36 Offiziere. Jahresbericht der Deutschen Mathematiker-Vereinigung **19**, 264–267 (1910)

25. Zaikin, O., Kochemazov, S.: The search for systems of diagonal Latin squares using the SAT@home project. Int. J. Open Inf. Technol. **3**(11), 4–9 (2015)

# Using Maple to Make Manageable Matrices

Ana C. Camargos Couto and David J. Jeffrey[✉]

Department of Applied Mathematics and ORCCA, The University
of Western Ontario, London, ON, Canada
{acamarg,djeffrey}@uwo.ca

**Abstract.** This paper describes an application of Maple in the teaching of linear algebra. The topic is the construction of an orthogonal basis for a set of vectors or a matrix using Householder transformations. We present a method for generating matrices which, when subject to using Householder transformations, require only rational computations and give rational results. The pedagogical problem addressed is that numerical examples in this topic will usually contain unsimplified square roots, which add an extra layer of difficulty for students working examples.

## 1 Introduction

Maple is well known for its uses in education, where it is often used as an aid in helping students understand important concepts through visualization and computation. Here we consider a different role for Maple: the construction of educational material, in particular exercises and examination questions. Our focus is on constructing problems which avoid unnecessary arithmetic complications. One linear algebra topic where this has already been considered is in the generation of problems requiring the calculation of eigenvalues and eigenvectors, see, for example, [5,7]. Here we consider a different topic, a staple of first courses in linear algebra, namely the construction of an orthonormal basis from a given matrix or from a set of vectors. One of the most important methods for constructing such bases uses Householder transformations [2, p. 185], [3, p. 119].

Let $A$ be a full-rank $n \times n$ matrix. The standard Householder method finds orthonormal matrices $Q_k$ such that

$$Q_{n-1}Q_{n-2}\cdots Q_2Q_1A = R \,, \tag{1}$$

with $R$ being an upper triangular matrix. Then, since $Q_k^{-1} = Q_k^T$, we easily transform this to $A = QR$. The $Q_k$ are built from Householder matrices $H$, which are defined to operate on a vector $z$ and reduce to zero all elements below the first. Thus

$$Hz = \|z\|e_1 \,, \tag{2}$$

© Springer Nature Switzerland AG 2020
J. Gerhard and I. Kotsireas (Eds.): MC 2019, CCIS 1125, pp. 220–229, 2020.
https://doi.org/10.1007/978-3-030-41258-6_16

where $e_1^T = [1, 0, \ldots, 0]$, and the norm is the 2-norm. $H$ is computed as follows: introduce a vector $v$ as

$$v = \|z\|e_1 - z \ . \tag{3}$$

Then, with $I$ as the identity matrix, we have

$$H = I - \frac{2vv^T}{v^T v} \ . \tag{4}$$

Note that if $z = e_1$, then $v = 0$, meaning there is nothing to do; in this case, (4) is replaced by $H = I$.

We illustrate the procedure, and the pedagogical difficulty, with an example. Courses in linear algebra mostly use integer matrices for their examples and examination questions, but even with this restriction, it is usually the case that calculations quickly become contaminated by awkward square-root calculations. Even with a system such as Maple, such calculations are not straightforward. For example, consider the matrix

$$A = \begin{bmatrix} 3 & 1 & 1 \\ 1 & 1 & 1 \\ 1 & 2 & 1 \end{bmatrix} \ . \tag{5}$$

The first target vector is the first column, thus $z = [3, 1, 1]^T$ and then from (3)

$$v = \begin{bmatrix} \sqrt{11} - 3 \\ -1 \\ -1 \end{bmatrix} \ . \tag{6}$$

Substituting this into (2), we obtain

$$H_1 = \begin{bmatrix} 1 - \frac{2(\sqrt{11}-3)^2}{2+(\sqrt{11}-3)^2} & -\frac{2(-\sqrt{11}+3)}{2+(\sqrt{11}-3)^2} & -\frac{2(-\sqrt{11}+3)}{2+(\sqrt{11}-3)^2} \\ -\frac{2(-\sqrt{11}+3)}{2+(\sqrt{11}-3)^2} & 1 - \frac{2}{2+(\sqrt{11}-3)^2} & -\frac{2}{2+(\sqrt{11}-3)^2} \\ -\frac{2(-\sqrt{11}+3)}{2+(\sqrt{11}-3)^2} & -\frac{2}{2+(\sqrt{11}-3)^2} & 1 - \frac{2}{2+(\sqrt{11}-3)^2} \end{bmatrix} \ . \tag{7}$$

Taking the $(1, 1)$ element as typical, we must simplify

$$1 - \frac{2(\sqrt{11} - 3)^2}{2 + (\sqrt{11} - 3)^2} \ . \tag{8}$$

Either working by hand, or using the Maple simplify command, a student would obtain

$$\frac{9 - 3\sqrt{11}}{-11 + 3\sqrt{11}} \ .$$

We can obtain a simpler form by rationalizing with $11 + 3\sqrt{11}$.

$$\frac{(9 - 3\sqrt{11})(11 + 3\sqrt{11})}{(-11 + 3\sqrt{11})(11 + 3\sqrt{11})} = \frac{99 - 99 - 6\sqrt{11}}{-121 + 99} = \frac{3\sqrt{11}}{11} \ ,$$

which is achieved in Maple by using the command `rationalize`. (The command `rationalize` does not invoke `simplify`, which is needed to obtain the above result.) This demonstrates that, working manually, considerable computational challenges arise immediately. We obtain

$$H_1 A = \begin{bmatrix} \frac{3\sqrt{11}}{11} & \frac{\sqrt{11}}{11} & \frac{\sqrt{11}}{11} \\ \frac{\sqrt{11}}{11} & \frac{1}{2} - \frac{3\sqrt{11}}{22} & -\frac{1}{2} - \frac{3\sqrt{11}}{22} \\ \frac{\sqrt{11}}{11} & -\frac{1}{2} - \frac{3\sqrt{11}}{22} & \frac{1}{2} - \frac{3\sqrt{11}}{22} \end{bmatrix} \quad A = \begin{bmatrix} \sqrt{11} & \frac{6\sqrt{11}}{11} & \frac{5\sqrt{11}}{11} \\ 0 & -\frac{1}{2} - \frac{7\sqrt{11}}{22} & -\frac{2\sqrt{11}}{11} \\ 0 & \frac{1}{2} - \frac{7\sqrt{11}}{22} & -\frac{2\sqrt{11}}{11} \end{bmatrix} . \quad (9)$$

The second Householder reflection uses the $2 \times 2$ submatrix:

$$\begin{bmatrix} -\frac{1}{2} - \frac{7\sqrt{11}}{22} & -\frac{2\sqrt{11}}{11} \\ \frac{1}{2} - \frac{7\sqrt{11}}{22} & -\frac{2\sqrt{11}}{11} \end{bmatrix} \quad (10)$$

the first column of this submatrix becoming the next vector $z$. After a second Householder matrix multiplication we arrive at the unsightly final answer.

$$A = QR = \begin{bmatrix} \frac{3\sqrt{11}}{11} & -\frac{7\sqrt{330}}{330} & -\frac{\sqrt{30}}{30} \\ \frac{\sqrt{11}}{11} & \frac{\sqrt{330}}{66} & \frac{\sqrt{30}}{6} \\ \frac{\sqrt{11}}{11} & \frac{8\sqrt{330}}{165} & -\frac{\sqrt{30}}{15} \end{bmatrix} \begin{bmatrix} \sqrt{11} & \frac{6\sqrt{11}}{11} & \frac{5\sqrt{11}}{11} \\ 0 & \frac{\sqrt{330}}{11} & \frac{7\sqrt{330}}{165} \\ 0 & 0 & \frac{\sqrt{30}}{15} \end{bmatrix} .$$

If the instructor of Linear Algebra wishes to set non-computer exercises or exams on this topic, the burden of simplifying complicated expressions of square roots will almost certainly cause arithmetic errors which are peripheral to the understanding of the procedure. Hence, it would be very useful if matrices could be created easily that would not force students to work with square roots.

## 2   The Method

Rational orthonormal matrices have been of interest since Cayley [1,6], but here we are interested not in the *end result*, namely, the orthonormal matrix $Q$ in $A = QR$, but in *the process* of obtaining it from a given matrix $A$ by Householder transformations. Our aim is to make all intermediate quantities encountered be rational. To ensure this, we need that every $z$ we choose has a rational 2-norm, so then $v$ has rational components and so does $H$. Then if $A$ has rational components, so will $HA$ and eventually $Q$ and $R$.

The obvious way to achieve this is to arrange for the target vectors $z$ to contain elements taken from Pythagorean $n$-tuples. A Pythagorean n-tuple is defined [4] as a set of $n$ positive integers, $x_k \in \mathbb{Z}^+$, with the property

$$x_1^2 + x_2^2 + \ldots + x_{n-1}^2 = x_n^2 . \quad (11)$$

It is then obvious that a vector with Pythagorean elements $[\pm x_1, \ldots, \pm x_{n-1}]$ has 2-norm $x_n$. We allow vectors to contain positive and negative Pythagorean

$n$-tuples, zero elements, and rational numbers, provided the 2-norm is integral. For example, $<3,4>$, $<-3,0,4>$ and $<3,-4,12>$ are all Pythagorean vectors (we are using Maple's notation for vectors).

We can illustrate the general case by considering the following $3 \times 3$ matrix:

$$A = \begin{bmatrix} 1 & a_{12} & a_{13} \\ 2 & a_{22} & a_{23} \\ 2 & a_{32} & a_{33} \end{bmatrix} . \tag{12}$$

The first column of A is a Pythagorean quadruple: $||<1,2,2>|| = 3$, and this is also the first target vector $z_1$. The first $v$ vector is $v_1 = <2,-2,-2>$. From this we get the $H_1$ matrix as

$$H_1 = \begin{bmatrix} 1/3 & 2/3 & 2/3 \\ 2/3 & 1/3 & -2/3 \\ 2/3 & -2/3 & 1/3 \end{bmatrix} .$$

The first transformation then results in

$$H_1 A = \frac{1}{3} \begin{bmatrix} 1 & a_{12} + 2a_{22} + 2a_{32} & a_{13} + 2a_{23} + 2a_{33} \\ 0 & 2a_{12} + a_{22} - 2a_{32} & 2a_{13} + a_{23} - 2a_{33} \\ 0 & 2a_{12} - 2a_{22} + a_{32} & 2a_{13} - 2a_{23} + a_{33} \end{bmatrix} .$$

The target vector for the next step of the process will use the target vector

$$z_2 = \frac{1}{3} \begin{bmatrix} 2a_{12} + a_{22} - 2a_{32} \\ 2a_{12} - 2a_{22} + a_{32} \end{bmatrix} . \tag{13}$$

To ensure that the next stage of the process is also rational, we must look for a set of values $\{a_{12}, a_{22}, a_{32}\}$ that give $z_2$ an integer norm. Usually there will be multiple sets of values satisfying the conditions. To find such sets of values, we implement a straightforward search.

## 2.1   Implementation

The search proceeds by building a matrix $A$ column by column. As shown above, the first column must be a Pythagorean vector; this is chosen by the user and becomes the seed vector for the search. Next, a Maple procedure SearchAddCol selects a random integer vector and adjoins it to the seed vector. Then a second Maple procedure HouseholderTransform is called to construct a Householder matrix based on the first column and applied to the two columns. The second column is then tested to see whether it also contains a Pythagorean vector. If the search is successful, then the process is repeated by adjoining a third column to the two existing ones. This is repeated until the matrix $A$ has the desired dimensions.

We show first the procedure for constructing and applying a Householder transform.

```
HouseholderTransform := proc(A::Matrix,TargetCol)
    local n,TargetLength,zv,H,Q;
    uses LinearAlgebra;
    option 'Copyright A.C. Camargos Couto and D.J. Jeffrey';
# Transform calculates the Householder matrix for column
# TargetCol of Matrix A. The intended use is to start with
# a matrix A and successively transform the columns. That is
# (Q1,R1):=HouseholderTransform(A,1);
# (Q2,R2):=HouseholderTransform(R1,2), etc
    # n=rows  of A
    # We should add error checking of arguments here.
    n := RowDimension(A):
    # We want a Householder matrix for column TargetCol,
    # using rows TargetCol to end. Note we are looking down
    # TargetCol, so we index rows.
    if TargetCol=n then
        return(IdentityMatrix(n),A); # There is nothing to do
    else
        # In the books, vector z and vector v are treated as
        # separate, but here we use one variable to hold both.
        # Also, the vector v here is the negative of the usual
        #  textbook one, but it gives the same H.
        zv := A[TargetCol .. -1,TargetCol]:
        TargetLength:=n+1-TargetCol;
        zv[1] := zv[1]-norm(zv,2); # Same as vector subtraction
        if norm(zv,2)=0 then
          H:=IdentityMatrix(TargetLength);
        else
          H:=IdentityMatrix(TargetLength)
              -2*zv.Transpose(zv)/(Transpose(zv).zv):
        end if;
        Q := IdentityMatrix(n,compact=false): # allow overwrite
        Q[TargetCol .. -1, TargetCol .. -1] := H ;
        # Note Q is n-by-n and H is a submatrix
    end if;
    return (Q,Q.A) ;
end proc;
```

The main procedure is SearchAddCol. It takes a partially calculated matrix
$A$ and adds a symbolic column to it; then it successively applies Householder
transforms calculated for all the non-symbolic columns. Next, it calculates the
norm of the added (presently symbolic) target vector and begins a random search
for values that will give a rational norm. The number of attempts is determined
by the second argument to the procedure. The successful columns are collected
and presented to the user as a list. The user can choose which one to use for the
appended column. In this way, the desired matrix is built column by column.

```
SearchAddCol := proc(A::Matrix,SizeOfSearch)
   local m,n,alpha,r,rsubs,target_norm,i,j,R,col,test_norm,
      success,OutList;
   uses LinearAlgebra;
   n,m := Dimension(A):
   col:=m+1;
   # the symbolic vector NewCol becomes an extra column
   R := <A|Vector(n,symbol=a)>;
   # Performing Householder transformations until step m.
   for i from 1 to m do
     R := HouseholderTransform(R,i)[2]:
   end do:
   # We need the target norm to be rational
   target_norm := norm(R(col..-1,col),2):
   success:=0;
   OutList:=[];
   j:=0;
   while j<= SizeOfSearch and success<10 do
     j:=j+1;
     alpha := RandomVector(n,generator=rand(-3..3)):
     r := RandomVector(n,generator=rand(-9..9)):
     rsubs := alpha*~r:
     test_norm := eval(target_norm,[seq(a[i]=rsubs(i),i=1..n)]);
     test_norm := simplify( test_norm);
     # If test_norm is rational, the vector is added to
     # the output list.
     if type(test_norm,rational) and test_norm > 0 then
        success:=success+1;
   # Many successes are multiples of previous successes
        rsubs:=rsubs/igcd(op(convert(rsubs,list)));
        OutList:=[op(OutList),rsubs];
     end if:
   end do:
   return(OutList):
end proc:
```

## 3  Usage

A Maple procedure BuildMatrix has been written as an interface to the two main routines. The user selects a seed vector, which must be Pythagorean, i.e. built from a Pythagorean $n$-tuple. This is because the seed vector becomes the first $z$ vector in (3). The length of the vector determines the size of the matrix. The interface then asks the user interactively to choose subsequent columns from a list. The code is given in the appendix.

As an example, we take as a seed $z = <3,4,12>$. THen BuildMatrix returns (the output depends upon the state of the random number generator):

$$\left[\begin{bmatrix} -5 \\ -4 \\ -18 \end{bmatrix}, \begin{bmatrix} -3 \\ 0 \\ 1 \end{bmatrix}, \begin{bmatrix} 4 \\ -1 \\ 0 \end{bmatrix}, \begin{bmatrix} 0 \\ 0 \\ -1 \end{bmatrix}, \begin{bmatrix} 1 \\ 1 \\ 0 \end{bmatrix}, \begin{bmatrix} 0 \\ 0 \\ 1 \end{bmatrix}, \begin{bmatrix} 1 \\ -6 \\ -12 \end{bmatrix}, \begin{bmatrix} 0 \\ 0 \\ 1 \end{bmatrix}, \begin{bmatrix} -2 \\ 1 \\ 0 \end{bmatrix}\right].$$

Notice that there is no code to remove repetitions and equivalences. Suppose the user chooses column 2, making the new matrix <<3,4,12>|<-3,0,1>>. The procedure continues and offers choices for the third column. In this way, one possible matrix could be

$$A := \begin{bmatrix} 3 & -3 & 0 \\ 4 & 0 & 1 \\ 12 & 1 & 0 \end{bmatrix}.$$

Another procedure, `RationalHouseholder`, can now be used to check and summarize the exercise that has been created.

"Step: ", 1

$$\text{"H= ", } \begin{bmatrix} 3/13 & 4/13 & 12/13 \\ 4/13 & 57/65 & -24/65 \\ 12/13 & -24/65 & -7/65 \end{bmatrix}, \text{"R= ", } \begin{bmatrix} 13 & 3/13 & 4/13 \\ 0 & -84/65 & 57/65 \\ 0 & -187/65 & -24/65 \end{bmatrix}$$

"Step: ", 2

$$\text{"H= ", } \begin{bmatrix} 1 & 0 & 0 \\ 0 & -84/205 & -187/205 \\ 0 & -187/205 & 84/205 \end{bmatrix}, \text{"R= ", } \begin{bmatrix} 13 & 3/13 & 4/13 \\ 0 & 41/13 & -12/533 \\ 0 & 0 & -39/41 \end{bmatrix}$$

$$\text{"Q= ", } \begin{bmatrix} 3/13 & -516/533 & 4/41 \\ 4/13 & -12/533 & -39/41 \\ 12/13 & 133/533 & 12/41 \end{bmatrix}$$

These results can be compared with results from Maple's `LinearAlgebra` package, which has a `QRDecomposition` function. There are occasional minor differences in the distribution of negative signs, owing to different algorithms being used.

## 4  Observations and Examples

The above example contains fractions that grow relatively large. Thus, from a teaching point of view, we have simplified the arithmetic by removing square roots, but still have difficult arithmetic owing to fractions with large numerators or denominators. Some of the examples below show that this is not always the case. At present, however, the best way to avoid large fractions is by trying a number of seeds and different choices for the subsequent columns. Choosing sparse matrices also helps keep the sizes of entries down.

Another observation is that the search method can fail to find a continuation from a given seed and choices. Since the search choices that we have programmed are restricted to trial vectors having small components (mostly single digit entries) this is not a proof that a rational orthonormal matrix cannot be found, but suggests that any successful search might be uninteresting as requiring too large entries. For example, for the 5-dimensional seed $<1,1,1,2,3>$, produces the following search results for continuation:

$$\left[\begin{bmatrix} -6 \\ 0 \\ 0 \\ -5 \\ -12 \end{bmatrix}, \begin{bmatrix} -4 \\ -3 \\ 3 \\ 4 \\ 0 \end{bmatrix}, \begin{bmatrix} 0 \\ 0 \\ -14 \\ -15 \\ 16 \end{bmatrix}, \begin{bmatrix} 5 \\ -3 \\ 3 \\ 3 \\ 3 \end{bmatrix}, \begin{bmatrix} 0 \\ 0 \\ -1 \\ 0 \\ -1 \end{bmatrix}, \begin{bmatrix} -10 \\ -9 \\ 0 \\ -6 \\ 1 \end{bmatrix}, \begin{bmatrix} 0 \\ 0 \\ 0 \\ 1 \\ 2 \end{bmatrix}\right].$$

If we choose the third vector for the second column, then the search for a third column fails. A successful search is shown below.

We show some $3 \times 3$ examples.

$$A_1 = \begin{bmatrix} 3 & 5 & 1 \\ 0 & 0 & 1 \\ 4 & 5 & 1 \end{bmatrix} = QR = \begin{bmatrix} \frac{3}{5} & \frac{4}{5} & 0 \\ 0 & 0 & 1 \\ \frac{4}{5} & -\frac{3}{5} & 0 \end{bmatrix} \begin{bmatrix} 5 & 7 & \frac{7}{5} \\ 0 & 1 & \frac{1}{5} \\ 0 & 0 & 1 \end{bmatrix}. \tag{14}$$

$$A_2 = \begin{bmatrix} 1 & 3 & 1 \\ 4 & 0 & 1 \\ 8 & 3 & 1 \end{bmatrix} = QR = \begin{bmatrix} \frac{1}{9} & \frac{8}{9} & \frac{4}{9} \\ \frac{4}{9} & \frac{4}{9} & \frac{7}{9} \\ \frac{8}{9} & \frac{1}{9} & -\frac{4}{9} \end{bmatrix} \begin{bmatrix} 9 & 3 & \frac{13}{9} \\ 0 & 3 & \frac{5}{9} \\ 0 & 0 & \frac{7}{9} \end{bmatrix}. \tag{15}$$

A $4 \times 4$ example. Note the use of sparsity to keep the entries (fairly) small.

$$A_3 = \begin{bmatrix} 3 & 0 & 0 & 1 \\ 4 & 1 & 4 & 1 \\ 0 & 0 & 4 & 1 \\ 12 & 4 & 3 & 1 \end{bmatrix} = QR = \begin{bmatrix} \frac{3}{13} & -\frac{12}{13} & -\frac{12}{65} & -\frac{16}{65} \\ \frac{4}{13} & -\frac{3}{13} & \frac{36}{65} & \frac{48}{65} \\ 0 & 0 & \frac{4}{5} & -\frac{3}{5} \\ \frac{12}{13} & \frac{4}{13} & -\frac{9}{65} & -\frac{12}{65} \end{bmatrix} \begin{bmatrix} 13 & 4 & 4 & \frac{19}{13} \\ 0 & 1 & 0 & -\frac{11}{13} \\ 0 & 0 & 5 & \frac{67}{65} \\ 0 & 0 & 0 & -\frac{19}{65} \end{bmatrix} \tag{16}$$

It should also be noted that it is a common observation that when performing linear algebra with exact rational arithmetic, the size of entries is known to grow through the computation. This is one reason for the complexity of standard linear algebra operations being greater for exact computation than for floating-point computation. Thus one way to reduce the arithmetic load is not to require the QR decomposition of a full $4 \times 4$ matrix , but to be content with a $4 \times 2$ matrix, which can test the approach, while finishing before the component growth has become prohibitive.

Finally, to respond to a challenge from a referee, we show a $5 \times 5$ example. In this case, we started with the seed $<1,1,1,2,3>$ as above, but chose the seventh possibility from the list above.

$$A_4 = \begin{bmatrix} 1 & 0 & 3 & -2 & -6 \\ 1 & 0 & 1 & 0 & 0 \\ 1 & 0 & 1 & 7 & 3 \\ 2 & 1 & 1 & 5 & 9 \\ 3 & 2 & 3 & 3 & 8 \end{bmatrix} = \begin{bmatrix} 1/4 & -1/2 & 3/4 & -1/20 & \frac{7}{20} \\ 1/4 & -1/2 & -1/4 & -\frac{13}{20} & -\frac{9}{20} \\ 1/4 & -1/2 & -1/4 & 3/4 & -1/4 \\ 1/2 & 0 & -1/2 & -1/10 & \frac{7}{10} \\ 3/4 & 1/2 & 1/4 & 1/20 & -\frac{7}{20} \end{bmatrix} \begin{bmatrix} 4 & 2 & 4 & 6 & \frac{39}{4} \\ 0 & 1 & -1 & -1 & 11/2 \\ 0 & 0 & 2 & -5 & -\frac{31}{4} \\ 0 & 0 & 0 & 5 & \frac{41}{20} \\ 0 & 0 & 0 & 0 & \frac{13}{20} \end{bmatrix}$$

# A    Appendix: Supplementary Code

```
BuildMatrix:=proc(FirstCol) local i,k,n,B,C;
        uses LinearAlgebra;
        if not type(FirstCol,Vector) then
          error("Please give initial column as vector");
        end if;
        n:=Dimension(FirstCol);
        B:=< FirstCol>;
        for i from 2 to n do
             C:= SearchAddCol(B, 200*n);
             print(C);
             k:=choice(nops(C));
             B:=<B | C[k] >;
        end do;
    end proc:


choice:=proc(n) local m;
    print("Which candidate out of ",n," to continue?");
    m:=parse(readline(default));
    if m<=n then
      m ;
    else
      error("not a valid choice");
    end if;
end proc:


RationalHouseholder := proc(A::Matrix)
    local R,Q,n,m,i,H,Ht;
    uses LinearAlgebra;
    n,m := Dimension(A);
    R := Copy(A):
    H := IdentityMatrix(n,compact=false):
    for i from 1 to n-1 do
      (Ht,R) := HouseholderTransform(R,i);
      H := Ht.H;
      print("Step: ",i);
      print("H=",Ht,"R = ",R);
    end do:
    Q := Transpose(H):
    print("Q = ",Q);
end proc:
```

# References

1. Cayley, A.: Sur quelques proprié tes des determinantes gauches. J. Reine Angew. Math. **32**, 119–123 (1846). Reprinted in the Collected Mathematical Papers of Cayley, Cambridge University Press 1889–1898, vol. 1, pp. 332–336
2. Corless, R.M., Fillion, N.: A Graduate Introduction to Numerical Methods. From the Viewpoint of Backward Error Analysis. Springer, New York (2013). https://doi.org/10.1007/978-1-4614-8453-0
3. Demmel, J.W.: Applied Numerical Linear Algebra. SIAM press, Philadelphia (1997)
4. Frisch, S., Vaserstein, L.: Polynomial parametrization of Pythagorean quadruples, quintuples and sextuples. J. Pure Appl. Algebra **216**, 184–191 (2012). https://doi.org/10.1016/j.jpaa.2011.06.002
5. Gilbert, R.C.: Companion matrices with integer entries and integer eigenvalues and eigenvectors. Am. Math. Mon. **95**(10), 947–950 (1988)
6. Khattak, N., Jeffrey, D.J.: Rational Orthonormal Matrices. In: 2017 IEEE SYNASC, p. 71 (2017). CPS, ISBN-13: 978-1-5386-2626-9
7. Renaud, J.-C.: Matrices with integer entries and integer eigenvalues. Am. Math. Mon. **90**, 202–203 (1983)

# Use of Maple and Möbius in an Undergraduate Course on Cryptography

Bruce Char[(✉)] and Jeremy R. Johnson

Drexel University, Philadelphia, PA 19104, USA
{charbw,johnsojr}@drexel.edu
http://www.cs.drexel.edu/~bchar
http://www.cs.drexel.edu/~jjohnson

**Abstract.** A senior undergraduate course on cryptography for computer science majors that combines the use of conventional materials with Maple worksheets and Möbius modules is in development. The design intent and impact of the Maple-based materials on course conduct is discussed. Pedagogical and practical considerations are discussed along and initial impressions given regarding benefits and difficulties in using the tools.

**Keywords:** Computer science education · Cryptography · Symbolic computation

## 1 Overview

CS 303 is a course for senior-level computer science undergraduates on the use of algorithmic number theory in cryptography (www.cs.drexel.edu/~jjohnson/2017-18/cs303.html). It has been run yearly as a conventional, face-to-face course during a ten-week term. This paper discusses the objectives of the course, and the design intent in meeting those objectives through combined use of conventional materials (a textbook, homework, slide presentations, and quizzes) with Maple worksheets and Möbius presentations. We discuss how the instructor-developed materials complement, replace, or extend the use of conventional materials. The paper concludes with a discussion of pedagogical considerations along with a summary of how the tools help to support these considerations.

## 2 Introduction

A primary course objective is for students to gain a working understanding the roles that number theory, abstract algebra, and computational complexity play in modern cryptography. A hands-on approach is taken where students gain experience in doing high-level programming of basic cryptographic computations, and

Supported by Department of Computer Science, College of Computing and Informatics, Drexel University.

J. Gerhard and I. Kotsireas (Eds.): MC 2019, CCIS 1125, pp. 230–244, 2020.
https://doi.org/10.1007/978-3-030-41258-6_17

then are asked to reflect on and interpret their results and mathematical derivations. A secondary objective is to acquaint students with symbolic computation in Maple so that they can do such programming expediently. Since our major requires only four math courses and a CS-oriented course on logic and proof, the course also needs to develop student capabilities in discovery and growth of understanding through mathematical reasoning, and to better communicate such discoveries in mathematical writing.

The course textbook is *An Introduction to Mathematical Cryptography* [7]. Initial offerings of the course supplemented this with a combination of presentation slides (Powerpoint), and Maple worksheets for labs and homework. More recently we converted most of the presentation to Möbius, and added additional paper-and-pencil in-class exercises and quizzes to reflect the non-computational aspects of the learning of this course. Figure 1 lists the topics covered.

1. Introduction to Maple
2. Integer and polynomial arithmetic – review of properties
3. Establishing properties of the Euclidean algorithm
4. Modular arithmetic,
5. The Little Fermat theorem
6. Chinese remainder theorem
7. RSA public-key cryptosystem
8. Discrete logarithms and the El Gamal cryptosystem
9. Attacks on cryptosystems
10. Primality testing (probabilistic and deterministic) and integer factorization
11. Legendre symbols
12. Blum zero-knowledge coin-flipping
13. Micali-Goldwasser cryptography
14. Elliptic curve cryptography

**Fig. 1.** List of topics covered in the most recently offering of the cryptography course

The course has three hourly meetings weekly. A typical class meeting begins with a short lecture based on assigned readings from the textbook or from Möbius modules developed for the course. The ideas in the lecture are further explored in a more active way during class by small-groups of students working on lab problems, or through the administration of quizzes done individually. Periodic homework assignments are given during the term to provide more lengthy work with the course topics. There is a take-home programming project in lieu of a written final exam.

In the remainder of this paper, we discuss our objectives in using, and techniques for developing various components.

## 3 Learning from Reading and Lecturing

A good textbook provides a wealth of resources that is often hard for a busy instructor to produce themselves: domain expertise, pedagogically astute selection and sequencing of topics, a rich collection of examples, and content formatted both for initial learning and for subsequent review and archival recall (tables of content, indexing, search). A book also provides a place for discussion of finer details that are important to treat, even if time constraints prevent detailed presentation in class.

A detailed treatment can trigger "TL:DR" [9] (too long; didn't read) behavior in students. The reasoning goes, "if it was important (examinable), the instructor would have included it in what they said." This can lead to lecture notes where a subset of the content is presented in a form more suitable for "real-time" learning. We have found that some students think they should use the lecture notes as a substitute for reading the textbook depends on what information the course activities and assessments depend on.

In Fall 2018 we used lecture notes in Möbius instead of the Powerpoint slides we developed for previous versions of the course. Figure 2 lists the daily sequence of modules given to the students to work through. Like Powerpoint slides, they have a digest of important points being made in the textbook. They also include a new feature: simple interactive automatically graded exercises (example in Fig. 4). Since we have the textbook to organize information in archival, retrospective form, we used the Möbius modules in a "just in time order", for example scattering the definition of rings, groups, and fields throughout the course as they were needed for the cryptography discussion.

We believe that an advantage of giving lecture notes in Möbius form is that slide-sized chunks of information can be interleaved with exercises where the student must recall or think through basic points of comprehension and understanding before proceeding with the presentation. This is a kind of active learning, well-known to have beneficial educational effects if the activities cause psychological processing that supports learning [5,11].

## 4   Active Learning in the Classroom: Labs and Quizzes

One could always run the in-class portion of the course in the traditional lecture-oriented way, hoping that the students are paying attention and learning all that the instructor is trying to tell them. We prefer to inject a more data or observation based approach – asking the students to work on problems individually or in groups in the classroom, and studying their answers. This approach can lead to *peer instruction* [10], or other small-group activities, where students work on problems individually and then convene in small groups to agree (or agree to disagree) on results and how to get them. Maple or paper-and-pencil worksheets (as in-class "labs"), or Möbius exercises can be a way of causing engagement in a kind of guided discovery process. An advantage of on-line Möbius autograded materials is that simple analytics such as successful completion rates or the frequency of wrong choices in selection problems are immediately available to the instructor, if they are prepared to react on-the-fly.

In our post-course evaluations, we often receive comments to the effect that the students believe that they learn more when there's the pressure of a quiz to study for. We have developed a number of quizzes on course topics, listed in Fig. 5. Most of these are paper-and-pencil rather than Möbius because they

1. Prime numbers
2. The Euclidean algorithm for finding greatest common divisors
3. The extended Euclidean algorithm
4. Modular inverses
5. Primes, roots of unity, totient-$\phi$
6. Modular arithmetic – basics
7. Modular arithmetic, the ring of integers mod m
8. One time pads
9. What is a ring?
10. Divisibility, divisors, and greatest common divisors
11. Maple evaluation and quotation
12. The "litte Fermat" theorem, modular multiplications, the $\phi$ function
13. Euler's theorem
14. RSA Encryption (public key encryption)
15. A demo of RSA
16. Understandings and relationships about RSA
17. Completing the story of why the Euclidean algorithm works
18. Understanding RSA
19. Introduction to "Chinese remaindering"
20. Fields and finite fields
21. Finding primes for RSA
22. Digital signatures in RSA
23. Groups, and the group $\mathbb{Z}_p$
24. Quadratic residues and quadratic reciprocity
25. Micali-Goldwasser encryption
26. The positive and symmetric representations for modular numbers
27. Blum coin-tossing over the telephone
28. The Chinese remainder theorem – the details
29. Attacks on RSA
30. Digital logarithms – introduction.
31. Diffie-Hellman key exchange, El Gamal public key encryption
32. Solving DLPs in $\mathbb{Z}_N$, deterministically and probabilistically
33. B-smoothness and index calculus
34. Elliptic curves : elliptic curve geometry and elliptic curve computation (see Figures 3 and 4)

**Fig. 2.** List of Möbius modules (slideshows + simple exercises)

require higher-level processing and feedback generation than is easily implemented through Möbius autograding. Done in class, in-person feedback can be informal. The instructor can make comments about solutions without the need for formally scoring or recording grades for them, or use the quiz question to spur small-group discussion.

We believe that students care enough about the results of quizzes to be attentive of results even if feedback is not immediate but within a few days. The needs of rapid manual grading leads to a "brief and simple" quiz style. Another ploy to spur interest in delayed feedback is to make students aware that similar questions will be asked in later quizzes or exams, generating a kind of spaced practice [6].

Figure 6 lists a number of lab activities to be done in class after assigned reading (textbook or Möbius modules) and a brief lecture-presentation by the instructor.

## What are elliptic curves -- overview

The (X,Y) points that satisfy an equation $Y^2 = X^3 + A \cdot X + B$, for particular real values of A and B, defines a curve in two dimensional space. For example, with A=-1, B=0, we have $Y^2 = -X^3 + X$  In Maple, it's easy to plot this:

algcurves:-plot(real_curve Y^2 - X^3 + X, X, Y) produces the curve (in blue)

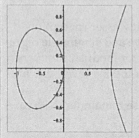

This graph does not look at all like an ellipse, which are described by a quadratic (not cubic) formula. The connection to ellipses comes because the parameterization of the cubic formulas can be done through use of *elliptic functions*, which are related to *elliptic integrals*, which are a generalization of a way to calculate the *arc length of* portions of *ellipses* (what we're familiar with). Elliptic functions and elliptic integrals are advanced mathematical topics, but we speak of this only to explain why mathematicians see a connection between ellipses and these un-ellipse like curves.  See for example, https://medium.com/@youssef.housni21/why-elliptic-curves-are-called-elliptic-a8327d94e3d1. or https://prateekvjoshi.com/2015/02/07/why-are-they-called-elliptic-curves/.

**Fig. 3.** Part of a presentation on elliptic curves (Color figure online)

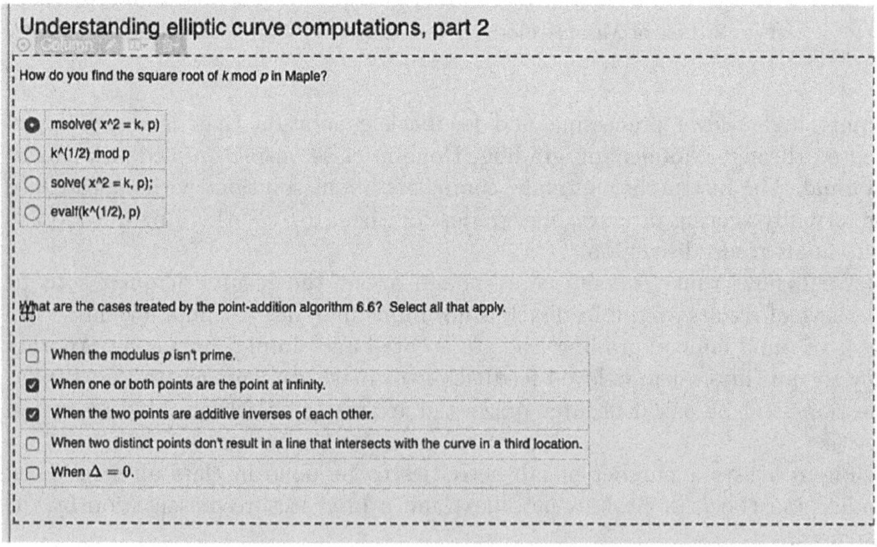

## Understanding elliptic curve computations, part 2

How do you find the square root of *k* mod *p* in Maple?

- ● msolve( x^2 = k, p)
- ○ k^(1/2) mod p
- ○ solve( x^2 = k, p);
- ○ evalf(k^(1/2), p)

What are the cases treated by the point-addition algorithm 6.6?  Select all that apply.

- ☐ When the modulus *p* isn't prime.
- ☑ When one or both points are the point at infinity.
- ☑ When the two points are additive inverses of each other.
- ☐ When two distinct points don't result in a line that intersects with the curve in a third location.
- ☐ When $\Delta = 0$.

**Fig. 4.** Simple Möbius exercises for elliptic curve computations

1. Recursive implementation of the Euclidean algorithm
2. Properties of the relationship between the Euclidean Algorithm and solutions to Diophantine equations
3. Quadratic residues of $\mathbb{Z}_{11}$.
4. Chinese remainder computation from Blum coin-flipping
5. Man-in-the-middle attacks, modular arithmetic (abstract algebra properties, symmetric representation of integers) Goldwasser-Micali encryption.
6. Discrete logarithms

**Fig. 5.** List of quiz topics

1. Substitution cyphers
2. Maple list processing, modular arithmetic, and affine cyphers
3. More on modular arithmetic and an introduction to RSA
4. The Chinese remainder theorem and fast powering
5. Quadratic residues, pseudo primes, Strong pseudoprime, and primality testing
6. Primitive roots and digital logarithms
7. Elliptic curves

**Fig. 6.** List of lab activities

1. String manipulation in Maple; the index of coincidence; implementing Vignére cyphers; cryptanalysis of Vignére Cyphers
2. Modular matrices (review of linear algebra operations); implementing RSA; breaking RSA; Hill cyphers.
3. Square roots mod $p$ and the Chinese remainder theorem (CRT)
4. Implement Blum coin-flipping protocol; devise and simulate use stories for it (see Figures 8, 9).
5. Implement El Gamal encryption.
6. Implement the Pohlig-Hellman algorithm that computes discrete logarithms mod $p^m$; a probabilistic algorithm solving digital logarithm problems; index calculus (Legendre symbol) calculations for discrete logarithms.
7. Elliptic curve encryption: implement and demonstrate Diffie-Hellman key exchange; El Gamal-style encryption from elliptic curve addition, EC cryptographic signatures.

**Fig. 7.** List of homework assignments

## 5   Homework Assignments

Computer science students are used to assignments that ask them to write programs. Maple's math libraries and built-in operations make it easy to require fairly ambitious projects without too many lines of code, enhancing what can be covered with hands-on experience. Figure 7 lists the assignments we used in

▼ **Part 1:  Design and implement the pieces you need to implement a simulation of Blum Coin-flipping**

Here are some observations about some of the pieces and what counts as a good submission.

1. Reading Blum's description of the protocol in his original paper (posted in the class website) reveals that it uses an RSA public system system. You may reuse the code you developed in a previous lab for this, without further commentary about it.
2. The protocol also needs a way to generate the coin-flipping $n$ (different from the RSA public keys used)..
3. The example in the Mobius description of the Blum protocol has/will have things worked out for one coin flip.
4. The protocol needs a source of random integers. Maple's *rand* function or RandomTools package has a source of pseudorandom integers. For "milspec" randomness these may not be random enough, but you may use them in the simulation as "randomness good enough for our purposes".
5. There should be a procedure for each step of the protocol, including those that used to generate and process the messages going between Alice and Bob in the protocol, suitably parameterized.
6. The paper also talks about steps a judge might take if the protocol steps are ever contested. You should have procedures designed and implemented that handle them, too.
7. The paper mentions that its a feature of the protocol that each step can be checked immediately for cheating. You should design and implement procedures for the checking. Some of the steps (5) could be designed to invoke procedures developed here, if you want the checking to be automatic so that the user knows that it's safe to proceed to the next step of the protocol.
8. There should be unit tests for each procedure, along with execution that indicates that the procedures pass their tests.
9. Procedures and tests should be documented, making it easy for the reader to understand what each procedure is doing. If a procedure ouotputs something like "Alice cheated in this site", the documentation should explain justify why this output should be believed. Assume that your audience understands the steps of the protocol, but doesn't really yet understand why it works. So you can't expect that "because Blum Protocol" will work as an adequate explanation.

What you should submit in this part are:

a list of procedures that handle the various steps and generate the information needed, appropriately documented. The documentation should include an explanation if, of rexample, one of the checking procedures outputs "Alice cheated", why that output is correct. If the procedure implements a step that outputs an RSA message, there should be a unit test that ensures that the output is as intended.

Scoring: +10 Full coverage of the protocol by presented procedures. Adequate tests establishing correctness. Documentation fit to be considered highly beneficial by users provided for all procedures. (As a user of documentation yourself, you realize that users usually prefer easy-to-understand concisement over verbosity or incomprehensible cryptic statements.)

> **#include your code and documentation here. We prefer you to use text for documentation** -- like this – over vast amounts of fixed-width text masquerading as code.   Code can be entered as a combination of Code Edit Regions and Execution groups.

**Fig. 8.** A portion of an exercise on Blum-coin flipping [4], part 1

▼ **Part 2:Devise  and simulate stories of the use of the Blum coin-flipping protocol**

You can tell a number of stories here, which are intended to demonstrate that you understand the various cases of the operation of the protocol.for $k \geq 2$ coins concurrently  You should include at least one story where Alice and Bob are flipping multiple coins at once (e.g. $k > 1$ coin flips) but things are working out well, to the end where the protocol is concluded and both parties agree on what has happened.  In addition you should tell a story about something going wrong during a  particular step .  For example, the first step is about the production by Bob of the coin-flipping $n$., which is supposed to be the product of two primes $p$ and $q$ with certain additional constraints.   How could Bob cheat if  he told Alice an $n$ that didn't meet (all of) the constraints specified for it  by the protocol?  How is Alice going to detect that he's cheating?  One of the points that Blum's paper makes is that if someone tries to cheat in a particular step, then in many cases the other person can detect it *immediately;*  they don't have to wait tto see what happens at the end in order to have the evidence to call out a cheater.

Scoring + 10

A story with accompanying demonstration and explanation for at least five distinctly different scenarios. One of them must be the the operation of the protocol successfully from start to finish, where no one is trying to cheat.

Extra credit: +2. Creative story telling. Awarded at grader's disgression based on the educational or entertainment factor involved.

**Fig. 9.** A portion of an exercise on Blum-coin flipping, part 2.

the last offering of the course. Further comments about the uses and abuses of learning through such assignments are made in Sect. 6.

## 6    Pedagogical Considerations

In this section, we discuss the difficulties of achieving "good enough" programming proficiency with Maple, pitfalls to avoid with computation-based learning, and cultivating patterns of mathematical thought, work proficiency, and exploration.

## 6.1   Getting Started with Maple

Few of our students coming into the class have prior experience with Maple. We find a week of orientation and practice needs to be spent on getting even our programming-savvy CS majors up to speed conveying the following ideas that appear novel to our majors:

1. Symbolic computation – operations on formulas that produce formulas instead of "ints" or "floats". Quotation and part-extraction need discussion.
2. Many of our majors lack experience with languages such as Matlab or R practice using *map, zip, reduce,* or vector/list/matrix operations is needed to get them to use such in lieu of using indexed iteration. While one could downplay this because it is not impediment to correctness, a curriculum-wide agenda is to cultivate the habit of avoiding egregious inefficiency (e.g. the "quadratic-space" problem when iterating over immutable lists) and learning the preferred notational idioms and shortcuts when learning a new language. Getting
3. Learning that the linear order of results listed in the Maple worksheet does not reflect the current internal state of variables in the Maple session – using all the information in the worksheet GUI to see the current session state.
4. Using the Maple notebooks for exposition as well as computation. This is more complicated than it would seem. The worksheet interface has dozens of word processing-related features to learn. Furthermore, CS majors are often not aware of the need for being a producer as well as a consumer of good mathematical communication. The point is to make mathematical writing an activity that can be summoned by habit rather than by external orders. This latter point is discussed further in Sect. 6.3.

As usual, a combination of examples and demos building over several weeks help to convey the ideas and direction far better than treating it in a single lecture or exercise.

## 6.2   Going Beyond "Understanding Through Programming and Problem-Solving"

Students that are engaged and active in and out of the classroom do not necessarily extrapolate or properly extend their understanding so that it transfers to other situations. We are experimenting with using reflective and explanatory activities (with feedback) to have students engage in the higher level activities of the Bloom or SOLO taxonomies: analysis, synthesis, judgment, awareness of relations between ideas, and the ability to work with deeper abstraction [2,3]. For example, the assignment about implementing Blum coin-flipping (see Fig. 9) also asked for a discussion of scenarios of use (crucial for understanding the inclusion of certain features of the protocol) not fully fleshed out in the paper.

Practice in only one context can also hinder transfer. Research has found that some discussion and practice with the terminology and programming concepts that transfer to other kinds of problems produces better results than just

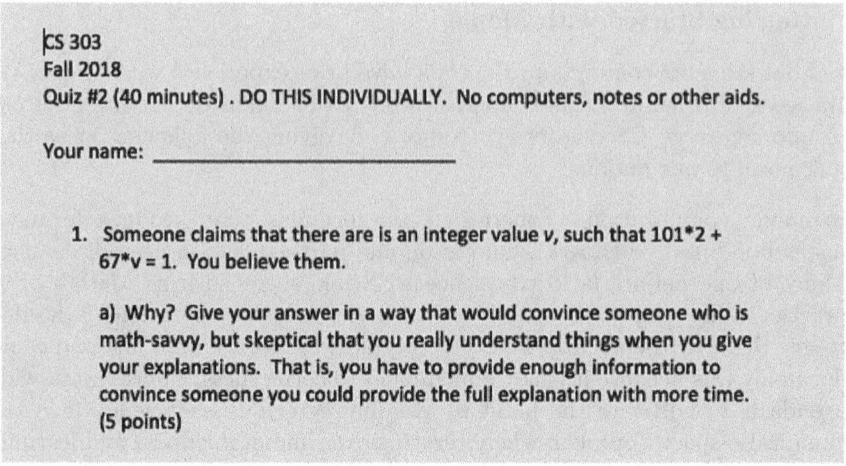
CS 303
Fall 2018
Quiz #2 (40 minutes) . DO THIS INDIVIDUALLY. No computers, notes or other aids.

Your name: ______________________

1. Someone claims that there are is an integer value *v*, such that 101*2 + 67*v = 1. You believe them.

   a) Why? Give your answer in a way that would convince someone who is math-savvy, but skeptical that you really understand things when you give your explanations. That is, you have to provide enough information to convince someone you could provide the full explanation with more time. (5 points)

**Fig. 10.** An in-class quiz question about the extended Euclidean algorithm

exposure to various situations that require that concept [8]. CS students should experience how knowledge of abstract algebra and number theory provides the conceptual reasoning framework and abstraction that allow knowledge transfer between cryptographic situations.

## 6.3 Cultivating Habits of Mathematical Thought

Our CS major requires calculus, linear algebra, and discrete math, but does not include abstract algebra. We "cultivate" in the following ways:

1. We ask students to find ways of exploiting consequences of mathematical definitions. Students often can find useful information if given an explicit order about what to explore. The goal is to make it a habit not needing directions on where to look. For example, the statements implied by the declaration of "$a$ is divisible by $b$" that can be used to establish understanding of the Euclidean algorithm for greatest common divisors.
2. Thinking and working at a high level of abstraction, such as when trying to understand and use concepts of ring, group and field.
3. Gaining knowledge through explicit reasoning (proofs and counter-examples), use of deeper abstraction (groups, rings, and fields). Making CS majors feel comfortable with working math is in part getting used to the culture, and in part getting accustomed to new expectations about what to do and what to produce.
4. Being able to articulate and communicate understandings, in particular understandings of proof. This is a matter of content creation, not mathematical word-processing. CS majors do not necessarily have the background to be strong at writing informal proofs despite some experience in a prerequisite course. We view this as a creative writing problem to be solved once

the student understands the objective and the starting point. This may lead to less conventional requests for "proof", such as in Fig. 10.

All things need to be habitual because the understanding happens when these kinds of things happen spontaneously without the need for continuous micromanagement. We try to design in opportunities for worked examples (with discussion of "how it was done" as well as the technical justification), reflection and for return encounters.

### 6.4   Learning the Experimental Paradigm

We have found that even CS majors, who do a lot of computation in their course-work, do not often have the habits that allow them to be given an assignment where they are told to use computation for discovery.

1. We find that our students sometimes have an assignment mentality when working: "here are the specs, write a program to meet them". Often there are test cases given which are meant to aid the development of the various conditions that must be built into the program to meet various specifications. In using results to aid discovery, however, there are three additional things to learn: how to use a computational result to confirm or disprove conjectures, how to design the conjectures in the first place, how to design the experiments to support the discovery process. An example of where computation is used for "finding out" rather than as a derivational mechanism is in the use of the index of coincidence to determine likely candidates for a substitution cypher, as shown in an assignment worksheet in Fig. 11.
2. We find that some students sometimes behave as if they believe that the end-point of a presentation of experimental results is a just a listing of particular artifacts: – "show a graph of $x$ vs. $y$", "answer yes or no: are the results increasing or decreasing?", etc. Care must be taken to have the student realize that they are responsible for finding and then describing a satisfactory (complete and correct, with justification provided) interpretation of the result and discussion of significance. This can be supported by providing similar in instructor presentations. We have also found it useful to remind students that they should not stop once they write down "the answer", but to write enough to convince the readers (graders, for example) that they understand what is going on.

## 7   What Can a Busy Instructor Afford to Do?

If we were trying to establish the superiority of our ideas, conventional practice would expect us to give a review of relevant literature, carefully evaluate outcomes, and perhaps conduct student surveys and ethnographic observations. We might also be expected to provide a cost-benefit analysis, etc. Of course, most instructors cannot complete such a task list when they also have their instructional and other professional responsibilities to perform. While design efforts

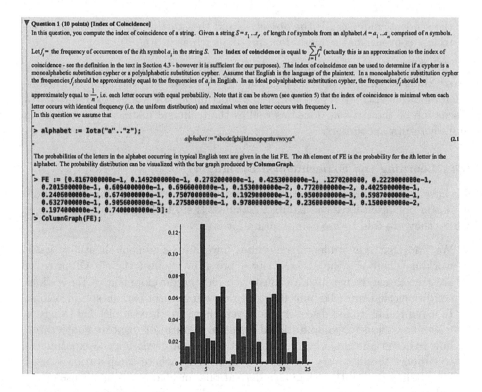

**Fig. 11.** Guided learning: an assignment question on the index of coincidence

should be evaluated carefully and objectively, course design experimentation should occur concurrently with development. The first step is to realize design ideas – what we are doing here. This is part of proceeding through "design science" rather than "controlled experiment" educational research. By discussing design possibilities (with usage reports establishing that the design is at least approximately feasible in a classroom), it is possible to contribute to progress in instructional practice without waiting until accumulating a complete picture.

Our experience suggests a few principles:

1. Use as much as you can that already exists. For example, high quality published textbooks, existing Powerpoint slides, assignments, and exams.
2. Conversion of Powerpoint slides into Möbius is straightforward, if tedious. (Compare Figs. 12 and 13.)
3. Use manual grading for handling situations that are hard to address with autograded questions. Given the current state of technology this includes many programming problems (since most platforms including Möbius have difficulties testing student Maple submissions), as well as higher level questions that typically require essay-style responses. In lieu of complicated autogenerated problems and autograding algorithms, start with Möbius autogenerated ("algorithmic") simple multiple choice, true/false, and constructed

answer problems. Once you learn which ones have significant instructional value, consider scaling them up to more elaborate versions that take more Möbius software development.

4. Be prepared to mix usage of Möbius work with Maple worksheet work or paper-and-pencil exercises. This substitutes human intelligence for automation where there is substantial educational benefit. We observe: (a) Keyboarding proofs or abstract algebra into Möbius or Maple worksheets requires more effort than handwriting, (b) questions requiring proofs, derivations, explanations, or judgment can be handled more adeptly through human intelligence in small classes. Keep in mind that Möbius automation can be used for feedback immediately after an initial encounter, harder to pull off in the other modes.

5. Use the in-class presence to cover any deficiencies in Möbius presentation that you don't have time to fix. For example, if you realize that more worked examples are needed, you can create them (with the help of Maple worksheets) and talk about them, live, rather needing a polished on-line presentation for them. Since the worksheets can be posted on-line, you are counting on the students to study some of the details that you didn't have time to talk about in their quest for "worked examples". While this is not ideal, you can take the stance that the students are better off than not getting the examples at all.

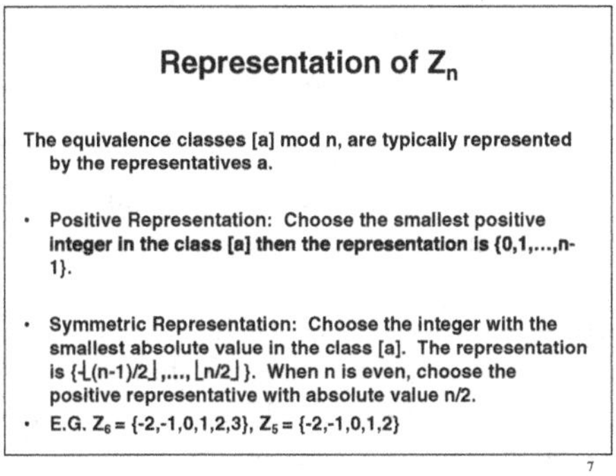

**Representation of $Z_n$**

The equivalence classes [a] mod n, are typically represented by the representatives a.

- **Positive Representation:** Choose the smallest positive integer in the class [a] then the representation is {0,1,...,n-1}.

- **Symmetric Representation:** Choose the integer with the smallest absolute value in the class [a]. The representation is $\{-\lfloor (n-1)/2 \rfloor ,..., \lfloor n/2 \rfloor \}$. When n is even, choose the positive representative with absolute value n/2.
- E.G. $Z_6$ = {-2,-1,0,1,2,3}, $Z_5$ = {-2,-1,0,1,2}

**Fig. 12.** Part of a presentation on modular numbers in Powerpoint

## 8    Iterative Refinement, Not "Waterfall" Development

The limited amount of time available for development means that courses should be viewed as "releases" as part of an on-going development. In other words, the development work should be viewed more as a design science activity [1] than an implementation of a finalized design. One can start with Powerpoint and a

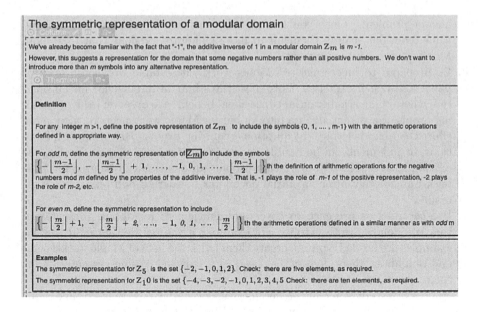

**The symmetric representation of a modular domain**

We've already become familiar with the fact that "-1", the additive inverse of 1 in a modular domain $Z_m$ is $m$-1.

However, this suggests a representation for the domain that some negative numbers rather than all positive numbers. We don't want to introduce more than $m$ symbols into any alternative representation.

**Definition**

For any integer m >1, define the positive representation of $Z_m$ to include the symbols {0, 1, ...., m-1} with the arithmetic operations defined in an appropriate way.

For *odd m*, define the symmetric representation of $Z_m$ to include the symbols
$\left\{ -\left\lfloor \frac{m-1}{2} \right\rfloor, \ -\left\lfloor \frac{m-1}{2} \right\rfloor + 1, \ ...., \ -1, \ 0, \ 1, \ .... \ \left\lfloor \frac{m-1}{2} \right\rfloor \right\}$ th the definition of arithmetic operations for the negative numbers mod m defined by the properties of the additive inverse. That is, -1 plays the role of *m-1* of the positive representation, -2 plays the role of *m-2*, etc.

For *even m*, define the symmetric representation to include
$\left\{ -\left\lfloor \frac{m}{2} \right\rfloor + 1, \ -\left\lfloor \frac{m}{2} \right\rfloor + 2, \ ...., \ -1, \ 0, \ 1, \ .... \ \left\lfloor \frac{m}{2} \right\rfloor \right\}$ th the arithmetic operations defined in a similar manner as with *odd* m

**Examples**

The symmetric representation for $Z_5$ is the set $\{-2,-1,0,1,2\}$. Check: there are five elements, as required.

The symmetric representation for $Z_{10}$ is the set $\{-4,-3,-2,-1,0,1,2,3,4,5\}$ Check: there are ten elements, as required.

**Fig. 13.** Part of a presentation on modular numbers in Möbius

textbook, and gradually introduce or revise labs, quizzes, and Möbius modules as one discovers what benefits from a more active involvement by the student, or a campaign for reducing common misconceptions.

Such practices lead to the following maxims for content-creation tool designers:

- Support introductory use through allowing minimal change and the analytical tools for understanding if they work. Then support expansion for those things that are worthwhile.
- Default or easily produced idioms should be easy to implement and to reuse and the system must make it easy to handle changes.
- The work of implementing changes should be automated with minimal effort needed to propagate changes consistently.
- Busy instructors benefit from access to evaluation and assessment techniques that are easy to carry out in normal operations, such as exam and quiz scores, regularly scheduling pre- and post- course surveys, and personal observations during class. Having a low-cost ways of building in assessment of Möbius activities would enhance its value for course development.

## 9    Mashups and Integration

There is a cognitive and labor cost at making a course out of several different kinds of things, both for the instructors (content creators), and for the students. Instructors must know enough to use the four tools, possibly picking up new expertise incrementally. This "knowing" includes not only how to use each tool for a task, but sequencing their use in a way consistent with the learning mission and resource budget. Students in our course needed to become used to interaction with four tools: paper, Maple worksheets, on-line file viewing over the Internet or through course management system, and Möbius. It is a situation where instructors must consider the prospects of cognitive overload and of classroom tactics to avoid it. Sometimes we have found that students need to be explicitly asked to do the thinking that integrates results across the tools.

Instructors have to contend with pulling and pushing content from and to multiple sources and destinations. While making up lecture notes freshly in Möbius may be no more difficult than doing it in Powerpoint or SLTEX, sometimes the work involves importing work originally in Powerpoint into Möbius or Maple slideshows. This common would-be workflow would benefit from "import wizards" or "integration centers" which would help to automate the work in moving content around.

Content management is another integration need. Should a question show up in Möbius, in a Maple worksheet, or in an on-line or paper-and-pencil quiz? What about the instructor's side notes and commentary for the problem? What if they decide in a later offering to change which tool is used to give the question? What about recording the schedule for spaced practice with the question? The repository for question content can exist independently of in what medium the students actually see it. Autograding favors delivery platforms which can run the custom programming to support it, but the alternative of human grading would also benefit from a centralized repository for usage plans, design and implementation notes, explicit answer keys and scoring rubrics.

Informal assessment of the delivery of course content is important in any design-science based work. It helps illuminate what is working well and what needs improvement. Assessment feedback has similar multi-mode integration needs. After a student's submission for a problem is evaluated, where does the point award and feedback content live? An integrated way of handling it as well as reasonable export methods could be significantly labor-saving.

Worked example libraries also benefit from multi-mode integration. Maple-based "native" representation of worked examples are favored because of the interweaving of presentation/explanation and computation. Möbius modules can do similar interweaving, although the computation would happen in an algorithm and so not be transparent to or mutable by to the student.

# 10  Conclusion

The intent of our course is to have CS students learn about essential ingredients of modern cryptographic work: symbolic computation, number theory, complexity theory, and abstract algebra. The course is an opportunity to develop student understanding of the nature of mathematical results, how to perform scientific/mathematical computation in a high-level language, and how to communicate their own mathematical findings to an audience. Course learning is designed to be through construction and synthesis, through guided discovery and experimentation, through practice at problem-solving and communication, and through activities that aim to develop relational and reflective understanding of the subject.

We have found it feasible to adopt a style of incremental refinement in course development that includes student use of Maple worksheets and Möbius course notes as well as use of small group lab work, and quizzes. The use of Maple and Möbius allows for convenient construction and technical communication, avoiding having to explain or work through some low-level details. Students can benefit from multiple modes of active learning that include rapid feedback for constructed or selected responses, and computational experimentation. However, trying to reap the benefits from the use of multiple tech tools and a diverse set of learning activities comes at a price. For students we see non-trivial additional complexity in tool learning. For instructors, there is additional complexity of developing and administering content across multiple platforms. We look forward to further development and evaluation to better understand and refine this approach. We also look forward to further support from the technology to facilitate iterative course design.

# References

1. Barab, S., Squire, K.: Design-based research: putting a stake in the ground. J. Learn. Sci. **13**(1), 1–14 (2004)
2. Biggs, J.B., Collis, K.F.: Evaluating the Quality of Learning: The SOLO Taxonomy (Structure of the Observed Learning Outcome). Academic Press, Cambridge (2014)
3. Bloom, B.S., et al.: Taxonomy of Educational Objectives. Vol. 1: Cognitive Domain, pp. 20–24 . McKay, New York (1956)
4. Blum, M.: Coin flipping by telephone a protocol for solving impossible problems. ACM SIGACT News **15**(1), 23–27 (1983)
5. Clark, R.C., Mayer, R.E.: E-learning and the Science of Instruction: Proven Guidelines for Consumers and Designers of Multimedia Learning. Wiley, Hoboken (2016)
6. Grote, M.G.: The effect of massed versus spaced practice on retention and problem-solving in high school physics. Ohio J. Sci. **95**(3), 243–247 (1995). https://pdfs.semanticscholar.org/7289/4d1499b9f2089739808bfcf2b01c00f1e928.pdf
7. Hoffstein, J., Pipher, J., Silverman, J.H.: An Introduction to Mathematical Cryptography. UTM. Springer, New York (2014). https://doi.org/10.1007/978-1-4939-1711-2
8. Klahr, D., Carver, S.M.: Cognitive objectives in a LOGO debugging curriculum: instruction, learning, and transfer. Cogn. Psychol. **20**(3), 362–404 (1988)
9. Kumar, D.: REFLECTIONS: TL; DR -: "best practices" of student learning (and how to bust them). ACM Inroads **8**(4), 21–22 (2017)
10. Lee, C.B., Garcia, S., Porter, L.: Can peer instruction be effective in upper-division computer science courses? Trans. Comput. Educ. **13**(3), 12:1–12:22 (2013)
11. Prince, M.: Does active learning work? A review of the research. J. Eng. Educ. **93**(3), 223 (2004)

# Enhance Faculty Experience and Skills Using Maple in the 21st Century Classroom

Lancelot Arthur Gooden[⊠] [ID]

Johnston Community College, Smithfield, NC 27577, USA
l_gooden@johstoncc.edu

**Abstract.** What role does faculty confidence and skills play in the use of a Computer Algebra System (CAS) play in increasing student success in college calculus courses? Studies show that improving students' spatial abilities in math is a key indicator to improving their success in calculus Sorby et al. (2013). This proposal discusses challenges faced while attempting to implement the use of Maple within the classroom in the Calculus sequences, Linear Algebra and Differential Equations at Johnston Community College as well as across North Carolina Community Colleges.

**Keywords:** Maple classroom · Instruction · Faculty

## 1 Introduction

A 15 to 20-minute tutorial using Maple 9 at the beginning of the fall 2010 semester, was all it took to transform my teaching as a Community College math instructor. "Please continue to teach calculus using Maple, my students loved it!", was what my predecessor said to me after the brief tutorial. Maple 9 was the only version purchased and installed in a single lab at our institution at the time. I knew I needed to devote time working with Maple independently during open lab times. My goal at the time was to introduce Maple labs in my calculus I class during the fall 2010 semester. I managed to compute first derivatives and indefinite integrals during my first attempts. I felt accomplished but exhausted as it took almost an hour to figure it out. While I was able to successfully implement a few labs in my calculus I class during the fall 2010 semester, students were only tasked with evaluating limits, derivatives and integrals. In some instances, students were required to plot 2-D functions and their derivatives. As I increased the use of Maple in my calculus classes the preparation time doubled and often quadrupled. Maple was only being used during scheduled in class labs to verify problems students had already worked out by hand. Something was still missing. There had to be more efficient and dynamic ways of teaching calculus using Maple.

Using Maple for computational and graphical exercises during scheduled labs only had no noticeable positive impact on student performance. Even though I grew more confident and excited using Maple, students grew increasingly frustrated and uninterested due to the limited scheduled lab times for practice. Furthermore, the activities

© Springer Nature Switzerland AG 2020
J. Gerhard and I. Kotsireas (Eds.): MC 2019, CCIS 1125, pp. 245–253, 2020.
https://doi.org/10.1007/978-3-030-41258-6_18

became repetitive to students. I needed to find ways to implement the use of Maple in my weekly instruction as well as explore best practices to teach calculus using the dynamic graphical, symbolic and numeric features of Maple. Philosophically, this was challenging. Why allow students to use Maple to explore concepts in limits, derivatives and anti-derivative. It makes perfect sense! Why shouldn't they? Calculus is dynamic in nature and often can be represented through three forms of registers of representations; [1]Graphically, [2]Numerically and [3]Analytically. The latter approach is used most often in many traditional calculus courses.

One of the most dynamic commands that I have found useful when teaching Calculus I using Maple was the Explore command. In exploratory laboratory activities, students drag sliders **a** and **b** shown below in Fig. 1 to investigate the absolute and relative extrema on a closed interval. In Fig. 2 students drag sliders **a** and **b** to investigate the Mean Value Theorem. As simple as this may seem, anyone who has encountered the challenges calculus students typically face when learning these concepts would understand the impact such a tool can have on their conceptual understanding.

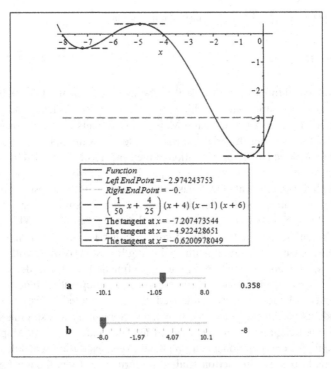

**Fig. 1.** Figure showing an interactive graph developed using Maple when studying absolute and relative extrema in a Calculus I course in a laboratory setting.

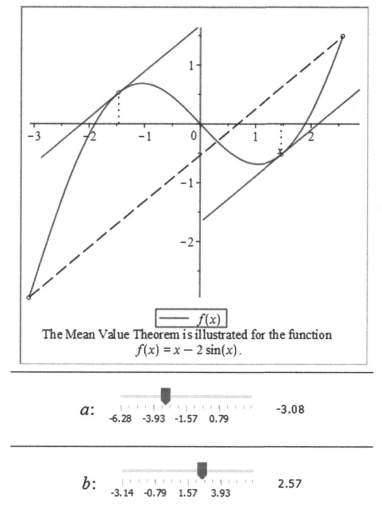

The Mean Value Theorem is illustrated for the function
$$f(x) = x - 2\sin(x).$$

$a$:    -6.28  -3.93  -1.57  0.79    -3.08

$b$:    -3.14  -0.79  1.57  3.93    2.57

**Fig. 2.** Figure showing an interactive graph developed using Maple when studying the Mean Value Theorem in a Calculus I course in a laboratory setting.

Success rates have significantly increased in my calculus courses and subsequent math courses since full implementation of Maple in my daily instruction. I wanted my colleagues to experience similar successes, energy and excitement as I have since the full integration of Maple my Calculus courses. I believed that if I conducted workshops demonstrating uses of Maple in my calculus courses would be enough to convince other math faculty to use Maple as an integral part of their instruction in calculus, just as I have. Assuming the faculty have had some introduction to Maple during under-graduate studies, the beginning sessions demonstrated various plotting commands for

2D graphs frequently studied in calculus I, specifically piecewise functions produced by using the line commands below and displayed in Fig. 3.

$$with(plots):\tag{1}$$

$$f(x) := piecewice((x \geq -3\,and\,x < -2, 2, x \geq -2\,and\,x < -1, 1, x \geq -1\,and\,x < 0, 0, x \geq 0\,and\,x < 1, -1))\tag{2}$$

$$p1 := pointplot([[-2, 2], [-1, 1], [0, 0], [1, -1]], symbolsize = 30, symbol = circle):\tag{3}$$

$$p2 := plot([f(x)], x = -3..0.99, discont = [showremovable], symbolsize = 30,\\ symbol = solidcircle):\tag{4}$$

$$display(\{p1, p2\})\tag{5}$$

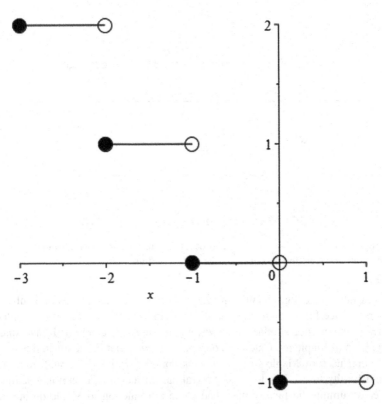

**Fig. 3.** Figure showing the graph of a piecewise function generated in Maple.

The sessions increased faculty awareness of the various Maple Math Applications and Tutors available for use in all the math courses we offer at the college. A total of five workshops were held, which much to my surprise seemed overwhelming even to a group of passionate and experienced calculus instructors. Even though some progress was made it was not as impactful as I expected it to be. While faculty use of Maple within the calculus courses increased, their use was not what I envisioned it to be. Having had various meetings and discussions with individual faculty, I came to the realization that they had similar philosophical beliefs about the use of Maple during instruction. In general, faculty felt the increased use of Maple within the classroom would compromise the learning outcomes of their students. According to Kilicman et al. (2010), CAS's such as Maple can be used as a powerful assistant to perform the symbol manipulations and computations in algebra as well as calculus. It has been suggested that these systems will benefit undergraduates and postgraduates in mathematics, engineering and physics by keeping track of the details in complicated manipulations.

How does one effectively convey to college math faculty how critical it is to engage students in the use of a CAS in calculus courses? First, faculty need to feel comfortable and confident using a CAS. (Zehavi 2004) discussed the theoretical views on "symbol sense," and regarding the notion of "instrumentation" developed among the CAS in education community in a paper consisting of a trilogy of high-school teachers solving an algebraic problem while learning to use a CAS. It is difficult to define symbol sense because it interacts with other senses like number sense, function sense, and graphical sense in problem-solving situations (Zehavi 2004). Zehavi gave reference to (Arcavi 1994, p. 1) who characterizes symbol sense through a rich variety of examples and illustrations of mathematical behaviors. Of the four characteristics mentioned in Zehavi's paper, two mention by (Arcavi 1994, p. 2), were associated with advanced symbol sense.

- "The awareness that one can successfully engineer symbolic relationships which express the verbal or graphical information needed to make progress in a problem, and the ability to engineer those expressions."
- "The ability to select a possible symbolic representation of the problem, and, if necessary, to have the courage, first to recognize, and heed one's dissatisfaction with that choice, and second, to be resourceful in searching for a better one as a replacement."

When teachers are first exposed to CAS, they are impressed by its "magic" in taking over the execution of routine techniques. At a later stage, however, they begin to notice that there are instances when the software works in a different way than the way they would ordinarily perform the same techniques by hand (Zehavi 2004). The following example illustrates the typical experience of faculty using a CAS, that may lead to their dissatisfaction with the software. Figure 4 shows the stepped-out solutions using the Limit Tutor approach or Show Solution Steps to analytically compute $\lim_{x \to 16} \frac{4-\sqrt{x}}{x-16}$ in a calculus I course. One issue faculty may find with this particular example is the choice to apply L'Hôpital's Rule as a technique in computing the limit when students at this point in the course would not have been taught derivatives. Figure 5 shows that a faculty who is resourceful in searching for a better option would seek the Rule Definitions to apply the preferred factoring technique first, which dictates consequent steps more consistent with the typical techniques taught to calculus I students when learning the analytical approach to calculating limits (Fig. 6).

$$\lim_{x \to 16} \frac{4 - \sqrt{x}}{x - 16}$$

$$= \lim_{x \to 16} -\frac{1}{2\sqrt{x}} \qquad \left[lhopital, 4 - \sqrt{x}\right]$$

$$= -\frac{\left(\lim\limits_{x \to 16} \frac{1}{\sqrt{x}}\right)}{2} \qquad [constantmultiple]$$

$$= -\frac{1}{2\sqrt{\lim\limits_{x \to 16} x}} \qquad [power]$$

$$= -\frac{\sqrt{16}}{32} \qquad [identity]$$

$$\lim_{x \to 16} \frac{4 - \sqrt{x}}{x - 16} = -\frac{\sqrt{16}}{32}$$

**Fig. 4.** Figure showing the stepped-out solutions to a limit exercise using Tools-Tutors-Calculus Single Variable-Limit Methods

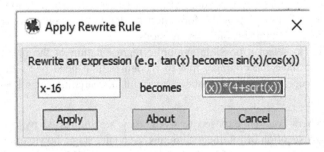

**Fig. 5.** Figure showing a factoring technique using the Rewrite Rule to the denominator of the limit expression $\lim\limits_{x \to 16} \frac{4 - \sqrt{x}}{x - 16}$.

$$\lim_{x\to 16}\frac{4-\sqrt{x}}{x-16}$$

$$\lim_{x\to 16}\frac{4-\sqrt{x}}{x-16}$$

$$=\lim_{x\to 16}\frac{4-\sqrt{x}}{\left(-4+\sqrt{x}\right)\left(\sqrt{x}+4\right)}\qquad\left[\textit{rewrite, }\frac{4-\sqrt{x}}{x-16}=\frac{4-\sqrt{x}}{\left(-4+\sqrt{x}\right)\left(\sqrt{x}+4\right)}\right]$$

$$=\frac{\displaystyle\lim_{x\to 16}\frac{4-\sqrt{x}}{-4+\sqrt{x}}}{\displaystyle\lim_{x\to 16}\left(\sqrt{x}+4\right)}\qquad\qquad[\textit{quotient}]$$

$$=\frac{\displaystyle\lim_{x\to 16}\frac{4-\sqrt{x}}{-4+\sqrt{x}}}{\left(\displaystyle\lim_{x\to 16}\sqrt{x}\right)+\left(\displaystyle\lim_{x\to 16}4\right)}\qquad[\textit{sum}]$$

$$=\frac{\displaystyle\lim_{x\to 16}\frac{4-\sqrt{x}}{-4+\sqrt{x}}}{\left(\displaystyle\lim_{x\to 16}\sqrt{x}\right)+4}\qquad\qquad[\textit{constant}]$$

$$=\frac{\displaystyle\lim_{x\to 16}\frac{4-\sqrt{x}}{-4+\sqrt{x}}}{\sqrt{\displaystyle\lim_{x\to 16}x}+4}\qquad\qquad[\textit{power}]$$

$$=\frac{\displaystyle\lim_{x\to 16}\frac{4-\sqrt{x}}{-4+\sqrt{x}}}{\sqrt{16}+4}\qquad\qquad[\textit{identity}]$$

$$=\frac{\displaystyle\lim_{x\to 16}(-1)}{\sqrt{16}+4}\qquad\qquad\left[\textit{lhopital, }4-\sqrt{x}\right]$$

$$=-\frac{1}{\sqrt{16}+4}\qquad\qquad[\textit{constant}]$$

$$\lim_{x\to 16}\frac{4-\sqrt{x}}{x-16}=-\frac{1}{\sqrt{16}+4}$$

**Fig. 6.** Figure showing the stepped-out solution to the $\lim_{x\to 16}\frac{4-\sqrt{x}}{x-16}$ using step 1 shown in Fig. 5.

Also noticeable are the differences in the representations of the equivalent solutions displayed using both approaches. $-\frac{\sqrt{16}}{32}$ and $-\frac{1}{\sqrt{16}+4}$ can be easily reduced to $-\frac{1}{8}$. While some faculty may argue the inconsistencies in the solutions may confuse students, I believe the inconsistencies enhances their ability to recognize symbolic relationships.

With the support of the Maple Product Management team, I guided to rethink my approach to enhancing the individual faculty skills and confidence using the software as well as increase support in and outside the classroom. The faculty at Johnston

Community College along with participants from surrounding community colleges in North Carolina engaged in a series of training sessions coordinated by the Maple Product Management Team. The training sessions were conducted using zoom on Friday afternoons from December 2018, with Johnston Community College Faculty only, and then on four consecutive Fridays during the Months of January and February 2019. I invited all full-time and part-Mathematics faculty as well as Physics and Engineering faculty at Johnston Community College to participate in the sessions. The experience was a valuable professional development opportunity. The response has been phenomenal among faculty. Faculty were in a computer lab with access to Maple during each session so that they would be actively engaged in the activities. Session one gave an overview for faculty demonstrating a variety of commands and applications ranging from calculus I through to differential equations as well as some engineering applications. Faculty were also made aware of the wealth of resources available in the Maple community. This was an eye-opening experience! The sessions were recorded, produced and shared among the math faculty. The video recordings and maple files are now stored in a math repository for faculty reference. Since the beginning of this FA2019 semester, faculty have shown much more excitement and receptiveness to using Maple in their weekly instruction. The faculty at Johnston Community College schedule Bootcamps with the Maple support staff, who remote into their courses via zoom to do various Maple demonstrations per their request. My hope is to continue this effort and to assess the long-term impact on student learning and success in calculus at participating community colleges.

Community Colleges in North Carolina received approval from the State Board of Community Colleges to offer an Associates Degree in Engineering. Engineering Pathways, which is the organization that launched this initiative, is a joint project of the North Carolina Community College System and the University of North Carolina engineering programs focused on developing the pathways for students to begin engineering studies at a community college and then transfer as seamlessly as possible to one of the University of North Carolina engineering programs. Over two-thirds of North Carolina community colleges now offer the degree program, which lists Calculus I as its first math. Success rates in Calculus I are vital to the sustainability of engineering programs across our state; hence, fueling our economy with the next generation of engineers. Figure 7 shows our current success rates in Calculus I over a nine-year period. The success rates appear stable and above average in general from FA2012-FA2018, indicating what I believe are successful gains in student performance.

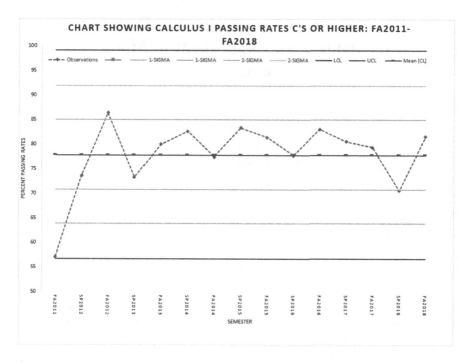

**Fig. 7.** Figure showing the passing rates, C's and better, in Calculus I at Johnston Community College from FA2011 to FA2018

# References

Arcavi, A.: Symbol sense: informal sense-making in formal mathematics. Learn. Math. **14**(3), 24 (1994). https://search.proquest.com/docview/1309139402

Kilicman, A., Hassan, M.A., Husain, S.K.S.: Teaching and learning using mathematics software "The new challenge". Proc. - Soc. Behav. Sci. **8**, 613–619 (2010). https://doi.org/10.1016/j.sbspro.2010.12.085

Sorby, S., Casey, B., Veurink, N., Dulaney, A.: The role of spatial training in improving spatial and calculus performance in engineering students (2013). https://www.lib.ncsu.edu/, https://doi.org/10.1016/j.lindif.2013.03.010

Zehavi, N.: Symbol sense with a symbolic-graphical system: a story in three rounds (2004). https://www.lib.ncsu.edu/, https://doi.org/10.1016/j.jmathb.2004.03.003

# Undergraduate Upper Division Quantum Mechanics: An Experiment in Maple® Immersion

Scot A. C. Gould[✉]

W.M. Keck Science Department, Claremont McKenna, Pitzer, Scripps,
Claremont, CA 91711, USA
sgould@kecksci.claremont.edu

**Abstract.** Dirac-notation based upper division undergraduate quantum mechanics was taught in the Spring semester of 2019 using Maple to present the mathematics of the course and to solve all mathematical and computational problems. In addition to step-by-step presentations on using Maple, students were provided with numerous examples of solving quantum mechanical problems using Maple. Students were required to submit all homework and "take-home" exam solutions as PDF documents, primarily generated from a Maple worksheet. However, students were not required to solve all problems using Maple. Through external evaluation and student survey, it was determined that by the end of the semester, all students used Maple for solving over half the problems; nearly three-quarters of the students developed sophisticated Maple skill sets; and a third of the students used Maple to solve every type of problem – completing assignments in a single worksheet. Maple was most frequently used to solve problems involving single variable continuous functions, vectors and matrices. Maple was least frequently used to solve problems involving Dirac-notation based algebra. Maple was nearly universally appreciated by the students.

**Keywords:** Maple · Quantum mechanics · Education

## 1 Motivation and Objectives: Why Include Maple [6]?

Quantum mechanics at the undergraduate level [10–12] is a combination of philosophy [1], science and mathematics [2]. Students are expected to suspend their inherent notion of deterministic outcomes with absolute precision. Rather, they must adopt a philosophy where particles are essentially everywhere until a quantum measurement is performed. The theory is derived from interpreting experimental outcomes and building a model for which a future outcome can be predicted as a probability – either as a set of discrete values, or as a continuous function. In addition, the measurement of one property of the system might affect the information about another property of the system. With upper division undergraduate physics, the notation of Dirac is the primary mathematical notation used to model the physical systems represented by an abstract Hilbert space [2]. It is a notation for which most of the students have no experience with, even though most have completed an undergraduate course in linear algebra.

© Springer Nature Switzerland AG 2020
J. Gerhard and I. Kotsireas (Eds.): MC 2019, CCIS 1125, pp. 254–262, 2020.
https://doi.org/10.1007/978-3-030-41258-6_19

In teaching the material, the instructor is faced with a dilemma. If the instructor wants the students to learn how to apply the postulates of quantum mechanics to physical systems without overwhelming the students with mathematical derivations or calculations, the instructor is usually forced to select simplistic, synthetic systems. Physics principle comprehension is the goal of the course. However, to make the course more relevant, it is best if the instructor selects problems from physically realistic systems, thus increasing the number of lines of algebra or numerical calculations. To accomplish both goals, the author chose to introduce a computer algebra system (CAS) into the course. This choice helps achieve a balance and thereby create a quantum mechanics course with more relevant content for which physical principles can be covered, but not a course where the students are overwhelmed with mind-numbing algebra or calculations.

In addition, by including a CAS in this standard upper division undergraduate physics course, the hope was that the students would spend less time doing mathematics and spend more time contemplating the underlying physical principles, generating physically viable mathematical models of physical systems, and interpreting both analytic and numerical results calculated by the software platform. While both the author and the students have been exposed to number of mathematical solving systems such as Mathematica® [11] and MATLAB® [7], Maple was chosen because it is the system with which the instructor has the most familiarity and because all the students had completed a lower division physics course, Modern Physics, in which they received a limited exposure to Maple. In that course, Maple was used primarily for basic graphing and solving sets of equations with multiple knowns. While the author notes the use of a CAS system for solving quantum mechanical problems is not novel, he is unaware of other upper division quantum mechanics courses for which so much of the material was presented and analyzed using Maple [3, 5, 9].

In an undergraduate quantum mechanics course, the students work with three main mathematical representations of a physical system: (1) Dirac notation based linear algebra, (2) vector and matrix based linear algebra, and (3) one involving differential equations and/or partial differential equations. For nearly every problem solved in the course, the goal is usually to generate an analytical or numerical prediction for an outcome either in the form of a single value or a function. These solutions are often best understood by the students when the solutions are presented in some graphical format.

The objective of this experiment was that, by the end of the course, all students would have a strong understanding of quantum mechanics postulates, be able to apply those postulates to physical systems and should be capable of presenting their answers and mathematical derivations to all problems in a Maple worksheet. At no time should the students need to generate handwritten work.

## 2 Implementation and Instructions to Class

Students were informed from the beginning of the course that the course would include quantum theory using Dirac-notation and the mathematics of differential equations and linear algebra. In addition, the mathematics for the physics and virtually every model or problem would be presented to them using Maple. The style would be in the form of

"2D Input" - "2D Output" on a Maple worksheet. This format was chosen after polling the students about which form of Maple input/output they found more understandable and useful. Since most students have experience with both Python and Mathematica, the majority find the "prompt and response" system easier to read, to compose and to understand, when compared to the arrow-based system of Maple's Document format. In addition, the vast majority prefer the "near traditional representation of the mathematics" style of the 2D-input/output to a 1-D courier font-based representation of the input code.

The class met twice per week for 75 min each period. The first class period consisted of a traditional lecture, or a set of class-time activities as needed. The second class period consisted of a problem-solving session during which students were required to present their initial attempts at solving the homework problems. Their final solutions were then submitted a day after the second class period. In solving the homework problems, students were encouraged first, to try solving the problems on their own, and then work together. Homework solutions had to be self-generated. Simply copying the work of another student is considered academic dishonesty at the colleges.

To help the students become more familiar with the capabilities of Maple and to encourage them to allow Maple to derive the messy algebra, the course included the following:

- Line-by-line coding exercises by the students on their laptops during class-time. Topics included, how to enter mathematical commands in execution groups and text in text groups, how to include images in a Maple document, how to enter and calculate using the Dirac notation in the Physics library, how to create matrices and calculate results with them.
- "How to" documents, listing mathematical techniques that could be performed with Maple code, were handed out. The philosophy of every "how-to" instruction document was to employ the smallest number of Maple commands to solve the largest number of problems.
- Top-down and bottom up examples of solving problem through Maple code were provided.
- All solutions by the instructor were written in Maple in the form of a well-organized Maple document that used headers and sub-headers. This included using the "title" format and grouping solutions to problems by sections. Answers to all problems were presented in a text-box. When complete line-by-line derivations were required, a text-box with both "text mode" and "math mode" based text was used.
- All Maple worksheets and documents were available to the students only in a non-editable, non-executable, PDF format. Consequently, if a student wanted to use a handout as the basis for working on a problem, they had to type the code themselves.
- Students had to attempt two open-book, open-note, open-internet, Maple allowed exams. (In addition to these exams, there were two closed-note, closed-device exams, for which Maple was not permitted.) Most of the calculations or derivations in the Maple-permitted exams would be either extremely challenging to solve either

algebraically or numerically or would contain many lines of algebra. Hence, students would view Maple as a near necessity to perform well on the exams.

All homework and take-home exam solutions were submitted electronically to the colleges' course management platform, Sakai [8]. Each submission had to be in the form of a single PDF document. While students were encouraged to try to use the Maple's capabilities to represent answers to all questions and to solve all problems, hand-written solutions by the students were accepted. Solutions written by hand would have to be "imaged" and added to their homework document.

At the end of the semester, the Maple-based capabilities of the students were evaluated by noting the frequency with which each student used a Maple-based skillset or one of the problem-solving capabilities of Maple on their last homework set submission and the "open book, open note, open internet, full computational calculation" final exam submission. This evaluation was performed by the instructor. In addition, the students were surveyed anonymously to measure their self-assessed Maple capabilities and usage in the course both before and post course.

Importantly, during the course, the students were told that their Maple skillsets would not be evaluated. Technically, all the problems could have been solved either by traditional means or via another software package. Hence, their ability to use Maple, or lack of, did not affect their grade. This allowed the instructor to ascertain the degree to which the students found Maple to be an effective, implementable problem-solving tool.

To separate a student's understanding of quantum principles and their capabilities to apply the principles to systems from their ability to use Maple, both a midterm exam and an additional separate final exam were given for which all electronic devices were disallowed. On this type of exam, students could bring in a single page "reference sheet" to remind themselves of equations and functions. These type of exams were monitored while the Maple-permitted exams were not. An honor code of having them not share answers was applied for the computationally based exams.

Finally, the textbook was a relatively common textbook which was chosen because it introduced Dirac notation early in the book [12]. It contains no code nor computational type problems. Problems both from the textbook and those generated by the instructor where used in the exams for the homework assignments.

## 3 Results

The course was taught in the Spring semester of 2019. There were 14 students enrolled in the course. No student left the course. Of the 14 students, 11 self identified as female, three as male.

### 3.1 Maple Skillsets and Problem-Solving Capabilities Evaluated

The formulism of quantum mechanics deals with states and operators that belong to an abstract Hilbert space. Hence, in an undergraduate upper division level quantum mechanics course, there are three forms of representation of a physical system: (1) a meta-physical representation for which Dirac notation linear algebra is used; (2) a discrete based outcome representation for which vectors, matrices and properties of linear algebra are used; (3) a continuous based outcome representation for which functions and differential equations are used. These representations, along with Maple documentation and graphics presentation skillsets, formed the basis for the rubric used to evaluate the capabilities of the students using Maple. See Table 1.

**Table 1.** Rubric for characterizing a student's ability to use Maple, organized by Maple skillset or mathematical representation of quantum mechanical systems.

| Skillset, mathematical representation | Advanced | Skilled | Novice |
|---|---|---|---|
| Dirac notation representation | Uses the *Physics* library, can represent state, generate orthonormal-sets, apply operators such as creation/annihilation, apply commutation relationship algebra, generate outcomes using purely Dirac notation | Uses the *Physics* library, can represent states, generate orthonormal sets, generate some outcomes using Dirac notation | Using instructor created worksheets, can modify worksheets to perform some of the calculations |
| Vector & matrix representation | Uses *LinearAlgebra* library to create vectors, matrices, solve eigenvalue/eigenvector problems, normalize eigenvectors, perform vector and matrix multiplication to create matrix-based operators | Uses *LinearAlgebra* library to create matrices, solve for the eigenvalue/eigenvectors of a matrix and can show results are eigenvectors and eigenvalues of the matrix | Uses *LinearAlgebra* library to create matrices, solve for the eigenvalue/eigenvectors of a matrix. Can read Maple output |
| Continuous function representation | Can solve ordinary differential equations both algebraically and numerically, demonstrate Maple output is correct, plot output, use solutions to determine weighted mean values. Can use "shooting method" and "variational method" to calculate numerical values for both eigenfunctions and eigenvalues | Can perform some substantial subset of the advanced skillset group | Can use Maple to calculate outcomes for only a couple types of problems |
| Graphics | Regardless of the format of the calculation, can generate a physically justifiable graph or type of animation. Can include titles, vary line styles and colors, and provide a legend. Near publishable quality output | Can generate graphs of continuous functions and some numerically based data (pointplot). Can include some useful graphical attributes | Can generate graphs of continuous functions |
| Document format of Maple text and math usage | Document is nearly a paper embedded with Maple code and output. Problems are separated for easy identification and readability. Any derivation requiring text plus mathematical characters is typed in using the math format in a text box. Page numbers | Problems are clearly identifiable. Can add text. Uploads images of handwritten solutions. No page numbers | Problems are barely labeled. Minimal Maple calculation (or placed at end.) Text + math answers are nearly always added as image of handwritten solutions |

Any student who did not fall into one of the three categories was characterized as having no skillset or capabilities in that area.

## 3.2  Fraction of Implementation

Using the rubric, the final exam and final homework submission, each student was evaluated for Maple usage capabilities. Table 2 shows the number of students who achieved a classification for each of the skillsets. Some students have been assessed as falling somewhere between two of the classifications, hence a half-point was given to each of the classifications.

**Table 2.** Number of students in each classification based on the rubric, organized by Maple skillset or mathematical representation of quantum mechanical system. $N = 14$.

| Skillset, mathematical representation | Advanced | Skilled | Novice | None |
|---|---|---|---|---|
| Dirac notation representation | 6 | 2.5 | 3.5 | 2 |
| Vector & matrix representation | 6.5 | 5 | 2.5 | 0 |
| Continuous function representation | 6 | 6.5 | 1.5 | 0 |
| Graphics | 4.5 | 4.5 | 5 | 0 |
| Document format of Maple text and math usage | 5.5 | 3.5 | 3.5 | 1 |

## 3.3  Student Self-reporting on Their Use of Maple

After the in-class final exam was completed, students filled out an anonymous survey about their use of Maple in the course. Outcomes of the survey are:

- After completing the lower division course, Modern Physics, where Maple was used for presentations, but was not required, four of the students state they ended up using Maple for virtually all the problems in that course. Two rarely used it. The rest claim they used it about half the time, but not all the time.
- Nine of the 14 consider themselves strongly familiar with Maple coming into the upper division course. The rest characterize themselves as somewhat familiar.
- Upon completing the upper division course, there are only a couple students who feel they became even more comfortable with Maple.
- For the upper division course, the percentage of the problems where Maple was used to perform the mathematics: three claim they used it 60–74% of the time, four used it 75–89% of the time, three used it 90–94% of the time and four used it 95% or more of the time.
- Of the possible mathematical areas, only the Dirac-notation based problems were ones where the students say they were less likely to use Maple.
- Six of the 14 students say they used a Maple document as the lead document and then imported images or text as necessary. This document was eventually exported as a single PDF document that was submitted as their work. The remaining eight created PDFs from several sources and then used a merging program or website to create the single PDF document.

- How do they see Maple as part of the problem-solving process in quantum mechanics? Nine of the 14 feel it was a natural part of executing the mathematical calculations. Four feel Maple required an additional effort beyond what they experience in most upper division quantitative courses, but it did solve problems faster, with fewer errors, than they could solve by hand. Only one student feels Maple is frustrating to use and thus an additional burden for the course.
- When asked to rank the value of Maple for learning quantum mechanics, eight of the 14 characterize it as essential, the remaining six rank it as useful, but not essential. None characterizes it as not useful.
- When compared to other mathematical software packages/platforms such as Mathematica™, MATLAB™, Python, etc., none of the students recommends switching away from Maple.
- Every student prefers the online submission method for submitting homework or take-home exams to the "print out and submit a physical copy of your work" method.
- The single most requested improvements of Maple by the students are more examples in the Help and easier ways to search online for help or examples, similar to what one experiences when learning and working with MATLAB™.

## 4  Analysis: Success and "Opportunities"

As the results show, by the end of the course, a minimum of five of the students (36%) wrote all their solutions using Maple. This included using all problem solving, writing out solutions where the problem required them to "solve by hand", and organizing as a logical document.

Between 60% to 75% of the students were able to use competently Maple to help solve the mathematics associated with the quantum mechanics problems. Oddly, fewer students than expected possessed strong graphics skills. Some could generate animations, but many were limited to one-line commands. This may be a consequence of the instructor not emphasizing this skillset sufficiently. This deficiency will have to be addressed for the next iteration of the course.

Not surprisingly, the students overstated their use of Maple in solving problems on the homework. For example, based on the instructor's knowledge of the students' performance in the lower division course, coming into the upper division course, at most two students could use Maple for all their homework assignment. Indeed, most students did not use the software as often or as extensively as they claimed.

From the perspective of the students, the online submission process is a hit. Students do not have to print out and run over to the instructor's office to submit. In addition, for the instructor, it provides a timestamp of when the document was submitted which is important since the course contains a no-late-homework-is-accepted policy. However, it is clear the instructor needed to perform a better job of demonstrating how to make the Maple document the key document and importing files or images as required. One can create a highly organized and readable document with embedded material even without using the Workbook system of Maple.

Finally, based on the survey, the instructor sees no reason to move away from Maple to another platform as the mathematical instructional tool in the physics courses. This is true despite the popularity of some of the other systems, such as Mathematica, with fellow colleagues and in other mathematics and physics courses.

# 5  Conclusion

I had taught this course two other times where I often used Maple to solve problems, including Dirac-notation based problems. However, until this semester, I had never explicitly required students to use it as a substitute for solving problems by hand. Of the three groups of students, this one was probably the weakest academically. In addition, while the goal was for them to understand the principles of quantum mechanics and apply them, the reality is that I never expected them to become quantum mechanics experts. Rather, the realistic hope was that they would appreciate the experience of thinking through the abstract environment known as Hilbert space using a new mathematical language, that of Dirac. Moreover, while I wanted them to be able to present all material and solve all problems using Maple, I never expected them to be Maple gurus. Rather, my more realistic goal was for them to come away from this course with a greater appreciation of the power of a mathematical tool like Maple. To this end, I believe the realistic goals were accomplished. The reason for the more realistic goals is that upon graduation, most of the students in our classes do not plan to enroll in physics graduate school. Rather, they move on to some type of health science, life science or engineering graduate or medical school. Hence, they have no reason to view quantum mechanics or Maple as knowledge that they will need to rely upon for the remainder of their lives.

Based on an informal survey of colleagues who teach this type of course at other institutions, including the textbook author himself, the Maple enhanced course contained a greater number of problems and more physics content than found in a typical upper division undergraduate quantum mechanics courses [13]. Overall, by using Maple to complete the numerous mathematical calculations assigned, the vast majority of the students developed a more reinforced understanding of the physical principles of quantum mechanics. As one student said, Maple's ability to simplify a complex mathematical statement allowed her to understand better, which properties of the system were important and how those properties affected the value of the measurement of the system she was investigating. Yet, because Maple code is not a complete black box, she could understand all the steps in the calculations. The enhanced pace was an unexpected positive outcome of this experiment.

# References

1. Baggott, J.: The Meaning of Quantum Theory. Oxford University Press, Oxford (1992)
2. Dirac, P.A.M.: The Principles of Quantum Mechanics, 4th edn. Oxford University Press, Oxford (1981)
3. Feagin, J.M.: Quantum Methods with Mathematica. Springer, New York (1994)

4. Griffiths, D., Schroeter, D.: Introduction to Quantum Mechanics, 3rd edn. Cambridge University Press, Cambridge (2018)
5. Horbatsch, M.: Quantum Mechanics Using Maple®. Springer, Heidelberg (1995). https://doi.org/10.1007/978-3-642-79538-1
6. Maple™: Maplesoft, Waterloo, Ontario, Canada (2018)
7. MATLAB: Mathworks, Nantick, MA, USA (2018a)
8. Sakai. www.sakailms.org
9. Steeb, W.-H., Hardy, Y.: Quantum Mechanics Using Computer Algebra, 2nd edn. World Scientific Publishing, Singapore (2010)
10. Townsend, J.: A Modern Approach to Quantum Mechanics, 2nd edn. University Science Books, Sausalito (2012)
11. Wolfram, S.: Mathematica 12, Champaign, IL, USA (1999)
12. Zettili, N.: Quantum Mechanics, Concepts and Applications, 2nd edn. Wiley, West Sussex (2010)
13. Zettili, N., et al.: Private conversations

# The Fermat-Torricelli Problem
# of Triangles on the Sphere with Euclidean
# Metric: A Symbolic Solution with Maple

Xiaofeng Guo[1,3], Tuo Leng[2], and Zhenbing Zeng[1]($\boxtimes$) [ID]

[1] Department of Mathematics, Shanghai University,
Shanghai 200444, China
{gxf16720010,zbzeng}@shu.edu.cn
[2] School of Computer Engineering and Science, Shanghai University,
Shanghai 200444, China
tleng@shu.edu.cn
[3] School of Mathematical Sciences, East China Normal University,
Shanghai 200062, China
52195500015@ecnu.edu.cn

**Abstract.** The Fermat-Torricelli problem of triangles on the sphere under Euclidean metric asks to find the optimal point $P$ on the sphere $S^2$ for three given points $A, B, C$ on $S^2$, so that the sum of the Euclidean distances $L = PA + PB + PC$ from that point $P$ to the three vertices is minimal (or maximal). In this paper we introduce a solution to this problem done with help of the symbolic computation software MAPLE and interpolation of implicit function, where the minimal and the maximal sum of the distances are expressed by same polynomial $f(L, a, b, c)$ of degree 12 with $a = BC, b = CA, c = AB$.

**Keywords:** Fermat-Torricelli problem · Elimination · Sylvester resultant · Dixon resultant · Implicit function interpolation · Symbolic-Numeric hybrid computation

## 1 Introduction

On the plane, the Fermat point of a triangle, also called the Torricelli point or Fermat-Torricelli point, is a point such that the total distance from the three vertices of the triangle to the point is the minimum possible. The original question was proposed by Pierre de Fermat in his book *Methodus ad disquirendam maximam et minimam* (manuscript written in 1629, and published in 1636 and 1979, cf. [31]) as a challenge problem. This problem also appeared in Fermat's book *Œuvres* [11] (published in the 1890s). It is easily seen that the optimal point must be contained in the interior or on the edges of the given triangle. Mersenne had introduced the problem to Italy. Evangelista Torricelli gave the first solution

Support by the Chinese National Natural Science Foundation (11471209 and 11501352).

© Springer Nature Switzerland AG 2020
J. Gerhard and I. Kotsireas (Eds.): MC 2019, CCIS 1125, pp. 263–278, 2020.
https://doi.org/10.1007/978-3-030-41258-6_20

to Fermat's problem in 1640 (published in 1659 by Viviani, cf. [16,31]). Some literature said that Fermat told the optimal problem as a challenge to Torricelli in private communication. Torricelli proved a nice property that if all angles of the triangle are less than $2\pi/3$, then the Fermat point $P$ of $ABC$ can be constructed by drawing equilateral triangles on the outside of the given triangle and connecting opposite vertices, and in this case, the intersection point $P$ is contained in the interior of $ABC$, and that from the point $P$, each side subtends an angle of $2\pi/3$, i.e.,

$$\angle APB = \angle BPC = \angle CPA = \frac{2\pi}{3},$$

In case that one of the angle of the triangle $ABC$ is larger than $2\pi/3$, Torricelli proved that the point $P$ that minimizes $PA+PB+PC$ is one of the vertices of the triangle. From this solution it is easy to derive that $L = \min_P (PA + PB + PC)$ satisfies

$$L^2 = \frac{1}{2}(a^2 + b^2 + c^2) + 2\sqrt{3} S_{\triangle ABC},$$

or

$$(L - a - b)(L - b - c)(L - c - a) = 0,$$

where the area $S_{\triangle ABC}$ of the triangle can be found in terms of $a, b, c$ through Heron's formula. Fasbender proved that the Fermat point problem is dual to construction of the maximal regular triangle that circumscribes to the given triangle $ABC$ in 1846 [12]. More solutions of the Fermat-Torricelli problem can be seen at the web page *Cut The Knot* of Alexander Bogomoly [6].

The Fermat-Torricelli problem has many application in economics and engineering fields. In 1902, Weber [29] investigated the following allocation problem: Given $n$ locations $P_1, P_2, \cdots, P_n$ and the price indices (products of the transportation price and the quantity of products) $w_1, w_2, \cdots, w_n$, find the best location $X$ of the factory, so that the total transportation cost

$$W(X) = \sum_{i=1}^{n} w_i \|X - P_i\| \tag{1}$$

is minimal. For $n = 3$, a general solution to this problem was given by Launhardt in [24]. For $n = 4$ and $w_1 = w_2 = \cdots = w_4 = 1$, a solution was given by G. Fagnano (cf. [23]).

Kupitz and Martini studied the isogonal property of the Fermat-Rorricelli point of the four vertices of the simplex in $\mathbb{R}^3$. They proved in [22] that likes to the plane case, the Fermat-Torricelli point determined by any simplex must be inside (or on the surface) of the vertex, and the four solid angles formed by the Fermat-Rorricelli point with the four surfaces of the vertex are equal. Nevertheless, Dalla [8] proved that this property is not valid anymore for $n \geq 4$. Here, the solid angle formed by a point $P \in \mathbb{R}^n$ with the $n - 1$ simplex $A_1 A_2 \ldots A_n$ (with $P \notin A_2 A_2 \ldots A_n$) is defined as the volume of the geometry object intersected by the cone$(P, A_1 A_2 \ldots A_n)$ (spanned ny point $P$ and the $n - 1$ simplex) and the unit sphere $S^{n-1}(P)$ in $\mathbb{R}^n$ (centered at $P$).

Alexandrescu proved in [1] that for any $n$ points in a Hilbert space, $W(X)$ defined by (1) is a continuously differentiable function, and the set of points that accesses the global minimum is the convex hull formed by $n$ points when $w_1 = w_2 = \cdots = w_n = 1$. In [35], Zuo and Lin investigated the existence and the uniqueness of the Fermat-Torricelli point for finite points in general metric space (including hyperbolic and Banach spaces).

For the numerical solution of the general form of the Fermat-Torricelli problem in $\mathbb{R}^n$, Weiszfeld (named as Andrew Vázsonyi after 1936, and also called *Zepartzatt Gozinto* in certain occasion since 1956) gave a gradient descent algorithm in 1936 (cf. [9,30]), called as *Weiszfeld Algorithm*. More results related to the distribution of the optimal solution to Eq. 1) can be found in [21].

The minimal Steiner tree problem is also connected with Fermat-Torricelli problem. Given a set of finite points in the space, find the minimal network that connects the all given points. It can be derived from the property of the Fermat-Torricelli point that in the minimal Steiner tree, the degree of every non-leaf vertex is 3, and that any two of the three edges associated to the non-leaf vertex formed an angle of $2\pi/3$. See [10,14] for more information. For this reason, the Fermat-Torricelli problem is also called as Steiner-Weber problem in some literatures.

Fermat problems regarding geometric manifolds have been studied by many people. In 1979, Katz and Cooper studied the Fermat problem on a sphere with spherical distance, Euclidean distance and squared Euclidean distance metric; in addition, for $N$ given points, the Weiszfeld algorithm and its convergence property were given to compute their Fermat point [20]. Using this result, Ghalieh and Hajja, studying the sphere with spherical distance metric, showed in [13] that the Fermat point $P$ of a spherical triangle $ABC$ is unique when the length of each side is less than $\pi/2$. This is the same as the case on the plane. In 2011, Chen [4] proved that the above result also holds for any three points on a regular surface with the geodesic metric. Furthermore, Zachos studied the Fermat problem on a surface of rotation and hyperbolic plane $H^2$ with a negative constant Gauss curvature; this was completed in [32] and [33].

In [15], the authors of this paper studied the Fermat-Torricelli problem on the unit sphere with Euclidean metric: Given a triangle $\triangle ABC$ whose sides are of length $a = BC, b = CA, c = AB$, find a point $P$ on that sphere such that $PA + PB + PC$ reaches its minimum, where all distances between points on the sphere are measured by the Euclidean metric. Let $u = PA$, $v = PB$, $w = PC$ and $L = u + v + w$ (Fig. 1). Then, the problem there can be written into a non-linear programming problem as below:

$$\min\ L = u + v + w$$
$$\text{s.t. }\ u = PA, v = PB, w = PC,$$
$$u^2 = (x - x_1)^2 + (y - y_1)^2 + (z - z_1)^2,$$
$$v^2 = (x - x_2)^2 + (y - y_2)^2 + (z - z_2)^2,$$
$$w^2 = (x - x_3)^2 + (y - y_3)^2 + (z - z_3)^2,$$
$$x^2 + y^2 + z^2 = 1,$$

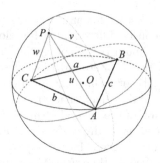

**Fig. 1.** A triangle $ABC$ on the unit sphere $S^2$ and the Fermat-Torricelli point $P$ so that $PA + PB + PC$ is minimal. All distances $a, b, c, u, v, w$ are measured in the Euclidean metric space $\mathbb{R}^3$.

$$x_i^2 + y_i^2 + z_i^2 = 1 \,(i = 1, 2, 3),$$
$$a^2 = (x_2 - x_3)^2 + (y_2 - y_3)^2 + (z_2 - z_3)^2,$$
$$b^2 = (x_3 - x_1)^2 + (y_3 - y_1)^2 + (z_3 - z_1)^2,$$
$$c^2 = (x_1 - x_2)^2 + (y_1 - y_2)^2 + (z_1 - z_2)^2.$$

It is natural that one expects that in the final result the minimal value $L = u + v + w$ would be a function determined by $a, b, c$. Applying the distance geometry knowledge, Guo et al. transformed the above problem to an optimization problem that involves only the lengths $a, b, c, u, v, w$ as follows:

$$\min \; L = u + v + w$$

$$\text{s.t.} \; V(a, b, c, u, v, w) = \begin{vmatrix} 0 & 1 & 1 & 1 & 1 & 1 \\ 1 & 0 & u^2 & v^2 & w^2 & 1 \\ 1 & u^2 & 0 & c^2 & b^2 & 1 \\ 1 & v^2 & c^2 & 0 & a^2 & 1 \\ 1 & w^2 & b^2 & a^2 & 0 & 1 \\ 1 & 1 & 1 & 1 & 1 & 0 \end{vmatrix} = 0. \tag{2}$$

here the determinant $V(a, b, c, u, v, w)$ is called the Cayley-Menger determinant (cf. [2,3]). Let

$$G = u + v + w + \lambda \cdot V(a, b, c, u, v, w).$$

Then applying the Lagrange multiplier method, it is easy to know that the minimal value $L$ is determined by the following polynomial equations

$$\begin{cases} L - u - v - w = 0 \\ \dfrac{\partial G}{\partial \lambda} = V(a, b, c, u, v, w) = 0, \\ \dfrac{\partial G}{\partial u} = 1 + \lambda \dfrac{\partial V}{\partial u} = 0, \\ \dfrac{\partial G}{\partial v} = 1 + \lambda \dfrac{\partial V}{\partial v} = 0, \\ \dfrac{\partial G}{\partial w} = 1 + \lambda \dfrac{\partial V}{\partial w} = 0. \end{cases} \tag{3}$$

It is also clear that if we consider the maximal of $L = u + v + w$ in (2), we would obtained the same equation system as (3). With the combination of the Sylvester resultant, Dixon resultant (cf. [19, 25] and the interpolation of implicit equation as developed in [27, 28], Guo et al. conclude the following result.

**Theorem 1.** *Suppose that $ABC$ is a triangle on the unit sphere $S$ such that $a = BC$, $b = CA$, $c = AB$ respective to the Euclidean metric. Assume that $P$ is a point on $S$. Let $L = PA + PB + PC$. Then, the inequality $r_{min} \leq L \leq r_{max}$ holds, where $r_{min}$ and $r_{max}$ are the minimal and maximal real root of the equation*

$$f(L, a, b, c) = (L - a - b)(L - a - c)(L - b - c)g(L, a, b, c) = 0 \tag{4}$$

*where*

$$g(L, a, b, c) = q_{12} \cdot L^{12} + q_{10} \cdot L^{10} + \cdots + q_2 \cdot L^2 + q_0,$$

*and the coefficients $q_{12}, q_{10}, \cdots, q_0$ are symmetric polynomials of $a, b, c$, expressed in a compact form by*

$$k = a^2 b^2 c^2, \quad m = a^2 + b^2 + c^2, \quad n = a^2 b^2 + b^2 c^2 + c^2 a^2$$

*as follows:*

$q_{12} = k - 4n + 16m - 64$

$q_{10} = -32m^2 + 4mn - 72k - 704m + 256n + 3072$

$q_8 = 16m^3 + 48km + 1984m^2 - 368mn - 4n^2 + 1584k + 2560m - 4800n - 30720$

$q_6 = -1728m^3 + 128m^2 n - 1728km - 14848m^2 + 7040mn + 192n^2 - 11520k$
$\quad + 18432m + 30720n + 114688$

$q_4 = 512m^4 + 768km^2 + 16640m^3 - 3712m^2 n - 128mn^2 + 11520km - 1152kn$
$\quad + 14336m^2 - 53760mn + 2688n^2 + 34560k - 110592m - 15360n - 147456$
$\quad + 30720n + 114688$

$$q_2 = -11264m^4 + 1024m^3n - 9216km^2 - 29696m^3 + 62464m^2n - 3072mn^2$$
$$- 46080km + 27648kn + 98304m^2 + 82944mn - 70656n^2 + 55296k$$
$$+ 147456m - 294912n$$

$$q_0 = 4096m^5 + 4096km^3 + 23552m^4 - 30720m^3n - 1024m^2n^2 + 55296km^2$$
$$- 18432kmn - 8192m^3 - 141312m^2n + 55296mn^2 + 4096n^3 + 27648k^2$$
$$- 165888kn - 147456m^2 + 36864mn + 211968n^2 - 110592k + 442368n.$$

As [15] is written in Chinese, we will give a general description to the method used in that paper in English for convenience of the non-Chinese-speaking readers. For sake of the page limitation, we will not show the details of complicated computation in the Dixon resultant elimination and the implicit interpolation as done in [15]. In the introduction (Sect. 1) we add more materials on the history of Fermat-Torrichelli problem. In the Sect. 2 the metric equation has been rewritten with emphasis on its application in algebraic representation for four points on the unit sphere. The Sect. 3 shows the symbolic computation part we have used in elimination. In Sect. 4 we will show the interpolation computation for computing large determinant, namely, the Dixon resultant. Section 5 is added for a brief discussion on some geometric properties of the Fermat-Torricelli points.

## 2    The Metric Equation and the Algebraic Representation of Fermat-Torricelli Problem

It is well-known that in $\mathbb{R}^n$ the volume $V$ of the simplex formed by $n+1$ points $P_0, P_1, P_2, \cdots, P_n$ with $d_{i,j} = P_iP_j = \|P_i - P_j\|$ in this space can be given by the determinant (the Cayley-Menger determinant) of an $(n+1) \times (n+1)$ matrix as below:

$$V^2 = \frac{(-1)^{n+1}}{2^n \cdot (n!)^2} \begin{vmatrix} 0 & 1 & 1 & 1 & \cdots & 1 \\ 1 & 0 & d_{01}^2 & d_{02}^2 & \cdots & d_{0,n}^2 \\ 1 & d_{10}^2 & 0 & d_{12}^2 & \cdots & d_{1,n}^2 \\ 1 & d_{20}^2 & d_{21}^2 & 0 & \cdots & d_{2,n}^2 \\ \vdots & \vdots & \vdots & \vdots & \ddots & \vdots \\ 1 & d_{n,0}^2 & d_{n,1}^2 & d_{n,2}^2 & \cdots & 0 \end{vmatrix}. \tag{5}$$

Thus, for four points $A, B, C, P$ on the unit sphere $S^2$, centered on $O$, in the three-dimension space $\mathbb{R}^3$ with

$$a = BC, b = CA, c = AB, \quad u = PA, v = PB, c = PC, \tag{6}$$

the five points $O, A, B, C, P$, as a subset in $\mathbb{R}^3$, form a degenerate simplex in $\mathbb{R}^4$, and therefore, its volume equals zero, which implies that

$$V(a,b,c,u,v,w) := \begin{vmatrix} 0 & 1 & 1 & 1 & 1 & 1 \\ 1 & 0 & u^2 & v^2 & w^2 & 1 \\ 1 & u^2 & 0 & c^2 & b^2 & 1 \\ 1 & v^2 & c^2 & 0 & a^2 & 1 \\ 1 & w^2 & b^2 & a^2 & 0 & 1 \\ 1 & 1 & 1 & 1 & 1 & 0 \end{vmatrix} = 0. \tag{7}$$

Below we prove that this condition is also a sufficient condition for $P, A, B, C$ are on the same unit sphere. Namely, we have the following result.

**Lemma 1.** *Assume that $P, A, B, C$ are points in $\mathbb{R}^3$, $a, b, c, u, v, w$ are distances shown as (6), so that the radius of the circumscribed circle of $ABC$ is not greater than 1. Then, (7) implies that $P, A, B, C$ are contained in some unit sphere in $\mathbb{R}^3$.*

*Proof.* Let $O_1$ be the center of the circle that circumscribed $ABC$ and $O$ be the point (any one of the two points) so that $OO_1$ is perpendicular to the plane determined by $ABC$ and $OA = OB = OC = 1$. Let $S^2$ be the unit sphere in $\mathbb{R}^3$ with center at $O$. In the case that $O, A, B, C$ are not coplanar, the existence of a point $P$ on $S^2$ is guaranteed by the Lemma 42.1 in [2]. In the case that $O, A, B, C$ are coplanar, we may construct a sequence of points $\{A_n\}$ that are not coplanar with $B$, $C$ with $A_n B = c, CA_n = b$, and $A_n \to A$. Let $P_n$ be the points on the sphere $S^2$ determined by $A_n BC$. As the unit sphere in the Euclidean space is compact, we conclude that $P_n$ converges to a point on $P$ that satisfies (6). $\quad\square$

Lemma 1 guarantees that the searching of the point $P$ on $S^2$ with minimal $PA + PB + PC$ for given points $A, B, C \in S^2$ can be transformed to the non-linear programming problem (2) as claimed in the Sect. 1. Furthermore, it is easily seen that $\lambda \neq 0$ in the equation system (3) and therefore, the equation can be simplified to

$$
\begin{cases}
f_1 := L - u - v - w = 0 \\
f_2 := V(a, b, c, u, v, w) = \underbrace{-a^4 u^4 + 2a^2 b^2 u^2 v^2 + \cdots + 4c^4 w^2 + 4a^2 b^2 c^2}_{28 \text{ terms}} = 0, \\
f_3 := \frac{1}{4}\left(\frac{\partial f_2}{\partial u} - \frac{\partial f_2}{\partial v}\right) = \underbrace{-a^4 u^3 - a^2 b^2 u^2 v + \cdots - 2b^4 v + 2b^2 c^2 v}_{26 \text{ terms}} = 0, \\
f_4 := \frac{1}{4}\left(\frac{\partial f_2}{\partial u} - \frac{\partial f_2}{\partial w}\right) = \underbrace{-a^4 u^3 + a^2 b^2 uv^2 + \cdots - 2b^2 c^2 w - 2c^4 w}_{26 \text{ terms}} = 0.
\end{cases}
\tag{8}
$$

With MAPLE it is easy to check that $f_2, f_3, f_4$ are irreducible polynomials.

Now the task of computing the optimal point $P$ becomes to elimination. The best result would be a triangular form of equations (called as *an ascending chain*) as follows.

$$
\begin{cases}
g_1(a, b, c, u) = 0, \\
g_2(a, b, c, u, v) = 0, \\
g_3(a, b, c, u, v, w) = 0, \\
g_4 = -u - v - w + L = 0,
\end{cases}
\tag{9}
$$

where $g_1$ is a polynomial which takes $u$ as the variable and $a, b, c$ as parameters in its coefficients, and $g_2$ a polynomial which takes $v$ as the main variable and $a, b, c, u$ as parameters, and so on in $g_3$. It is clear that (9) can be considered as a symbolic solution of the Fermat-Torricelli problem. However, the computation with MAPLE shows that when taking $a, b, c$ as symbolic parameters, both the time complexity and the space complexity for getting the ascending chain are too high. If taking $a, b, c$ as specific rational numbers, there is no difficulty to do elimination. For example, if we take

$$a = \frac{1}{2}, b = \frac{3}{10}, c = \frac{2}{5},$$

then, we would have

> hh1 := factor(resultant(resultant(f2, f3, w), resultant(f2, f4, w), v));
> hh1 := primpart(hh1, u)
> $= u^{32}(2 - 5u)^8(2 + 5u)^8(3 - 10u)^{16}(3 + 10u)^{16} \cdot h_{20,1}(u)^2 \cdot h_{12}(u)^2,$

and

> hh2 := factor(resultant(resultant(f2, f3, v), resultant(f2, f4, v), w));
> hh2 := primpart(hh2, u)
> $= u^{32}(2 - 5u)^8(2 + 5u)^8(3 - 10u)^8(3 + 10u)^8 \cdot h_{20,2}(u)^2 \cdot h_{12}(u)^2,$

where $h_{20,1}(u), h_{20,2}(u)$ are irreducible polynomials of degree 20, with

$$\gcd(h_{20,1}(u), h_{20,2}(u)) = 1$$

and

$$h_{12}(u) = 9636074153000000000u^{12} - 14797674532025000000u^{10} + \cdots$$
$$-256923876528228240000u^2 + 2261669941382971392$$

This experiment indicates that the final result of $L$ is possibly a polynomial of degree 12 in $L$ with $a, b, c$ as parameters, or

$$(L - a - b)(L - b - c)(L - c - a) = 0$$

for certain situation, and we may do further computation via accurate numeric computation with MAPLE and then recover the final result. As we have not found successful work on interpolation for a system likes the ascending chain, we will show how to recover the final relation that connects $L$ and $a, b, c$ in the next section.

## 3    Symbolic Elimination via Sylvester Resultant and Dixon Resultant

In this section we describe the symbolic computation we have done with MAPLE for eliminating variables $u, v, w$ from the equation system (8). As the Sylvester resultant has been used for eliminating one variable from two polynomials, we will not display its definition here. It worths to indicate that immediately after using the built-in Maple command $\texttt{resultant}(f, g, x)$, we always use a sub-procedure that factorize the result of the resultant computation, and remove all duplicate factors:

$$\texttt{resultant}(f, g, x) \xrightarrow{\texttt{factor}} h_1^{d_1} \cdot h_2^{d_2} \cdots h_j^{d_j} \xrightarrow{\texttt{square-free}} h_1 \cdot h_2 \cdots h_j.$$

Next is the procedure that we designed to eliminate variables $u, v, w$ with the Sylvester resultant.

$$\begin{cases} P_1 = \texttt{resultant}(f_1, f_2, u), & (86, [4, 4, 4, 4, 0, 4, 4]) \\ P_2 = \texttt{resultant}(f_1, f_3, u), & (56, [3, 4, 4, 2, 0, 3, 3]), \\ P_3 = \texttt{resultant}(f_1, f_4, u), & (56, [3, 4, 2, 4, 0, 3, 3]); \\ P_4 = \texttt{primpart}(\texttt{factor}(\texttt{resultant}(P_1, P_2, v)), w), (2133, [12, 14, 14, 10, 0, 0, 12]), \\ P_5 = \texttt{primpart}(\texttt{factor}(\texttt{resultant}(P_1, P_3, v)), w); (4354, [12, 18, 12, 14, 0, 0, 12]), \\ P_6 = \texttt{resultant}(P_4, P_5, w). \end{cases}$$

$$(10)$$

The numbers shown after each result are the number of terms in the result, and the degree of the result respect to $a, b, c, u, v, w$. Thus, $P_4$ has 2133 monomials, in which the degrees of $L, p, q, r, w$ of $P_4, P_5$ are $[12, 14, 14, 10, 12]$, while $P_5$ has 4354 monomials, the degrees are $[12, 18, 12, 14, 12]$. The computation was implemented with Maple 18 on a personal computer with Inter(R)Core(TM)i7CPU and 8 GB memory. The total time used for $P_1, P_2, P_3, P_4, P_5$ is 0.422 s in the above mentioned machine. And for computing $P_6$, the allocated memory had been exceeded after 5970 s, so it failed to return any result.

While there are many alternative ways, including Gröbner base and the pseudo-remainder (i.e., $\texttt{prem}$), could be used for elimination, we have selected the Dixon resultant for further experiment according to the practical experience from past works. Before explaining our computation, we shall give a brief introduction to the definition and the basic property of the Dixon result.

Consider the polynomial equation system

$$\mathcal{F} = \{p_1(x_1, \ldots, x_n), \ldots, p_{n+1}(x_1, \ldots, x_n)\} \subset \mathcal{K}[x_1, \ldots, x_n].$$

For $1 \le i \le n$, denote

$$d_i = \max\{\deg(p_1, x_i), \deg(p_2, x_i), \cdots, \deg(p_{n+1}, x_i)\}$$

Now construct the $(n + 1) \times (n + 1)$ determinant as follows:

$$\Delta(x_1, \ldots, x_n, \alpha_1, \ldots, \alpha_n) = \begin{vmatrix} p_1(x_1, x_2, \ldots, x_n) & \cdots & p_{n+1}(x_1, x_2, \ldots, x_n) \\ p_1(\alpha_1, x_2, \ldots, x_n) & \cdots & p_{n+1}(\alpha_1, x_2, \ldots, x_n) \\ \cdots & \cdots & \cdots \\ p_1(\alpha_1, \alpha_2, \ldots, \alpha_n) & \cdots & p_{n+1}(\alpha_1, \alpha_2, \ldots, \alpha_n) \end{vmatrix} \quad (11)$$

where $\alpha_1, \cdots, \alpha_n$ represent the new variables and $p_i(\alpha_1, \cdots, \alpha_k, x_{k+1}, \cdots, x_n)$ stands for $x_j$ in $p_i$ being substituted by $\alpha_j$ for all $1 \le j \le k$. Next, $\Delta$ can be divided by $\prod_{i=1}^{n}(x_i - \alpha_i)$, therefore, we have

$$\delta(x_1, \ldots, x_n, \alpha_1, \ldots, \alpha_n) = \frac{\Delta(x_1, \ldots, x_n, \alpha_1, \ldots, \alpha_n)}{(x_1 - \alpha_1) \cdots (x_n - \alpha_n)}. \tag{12}$$

and call $\delta$ as the Dixon polynomial of $\mathcal{F}$.

We may view $\delta$ as a polynomial of $\alpha_1 \cdots \alpha_n$ and denote different coefficients of power products of $\alpha_1, \cdots, \alpha_n$ by

$$c_i(x_1, x_2, \cdots, x_n) \quad (i = 1, 2, \ldots, s_1). \tag{13}$$

Since $\delta$ vanishes at all common zero of $\mathcal{F}$, we conclude that

$$c_i(x_1, x_2, \cdots, x_n) = 0 \tag{14}$$

for all $1 \le i \le s_1$. We call (14) as the derived equations of the original system $\mathcal{F}$. Furthermore, we may use $v_1, \cdots, v_k$ to represent all different power products of $x_1, x_2, \cdots, x_n$

$$v_1 = 1, \quad v_2 = x_1, \quad \cdots, \quad v_k = \prod_{i=1}^{n} x_i^{i \cdot d_i - 1}, \tag{15}$$

and assume that there are $s_2$ variables among $v_1, \cdots, v_k$ appearing in (14), without loss of generality, we may denote them as $v_1, \cdots, v_{s_2}$. Then, (14) can be rewritten as the following linear equation systems:

$$D(v_1, \cdots, v_{s_2})^T = \mathbf{0}, \tag{16}$$

where $D$ is the coefficient matrix of (14), related to the power product $v_1, \cdots, v_{s_2}$, and called as *the Dixon matrix* of $\mathcal{F}$.

It is apparent that if the Dixon matrix is nonsingular, that is, $\det(D)$ is not a zero polynomial, then the expression $\det(D) = 0$ could be taken as a necessary condition for the existence of a nontrivial common zero of $\mathcal{F}$. And therefore, $\det(D)$ is called as *the Dixon resultant* of the equation system $\mathcal{F}$. For fast computation of the Dixon resultant, see [5].

In case that the Dixon matrix is singular, i.e., $\det(D) \equiv 0$, we cannot acquire any useful information then. Moreover, when the original polynomial system $\mathcal{F}$ is sparse, $s_1$ in (13) may not equals to $s_2$ in (16), so that the obtained Dixon matrix $D$ is not a square one, and the Dixon resultant is not well-defined anymore. To overcome the above problems, Kapur et al. (see [19,25]) proposed the following KSY method.

Denote $r = \text{rank}(D)$ and $m_i$ the $i$-th column of D. Let $monom(m_i)$ be the power product $v_i$ corresponding to $m_i$. In addition, let

$$\mathcal{R} = \{Y : r \times r \text{ nonsingular submatrix}\},$$

$$nvcol(\mathcal{C}) = \{m_i : \mathcal{C} \implies monom(m_i) \neq 0, 1 \le i \le s_2\},$$

for given constraints $\mathcal{C}$, and $\mathcal{D}_1$ be the set of all $s_1 \times (s_2 - 1)$ sub-matrices obtained by deleting all columns from $nvcol(\mathcal{C})$.

Now we may view the coefficients of the system $\mathcal{F}$ as parameters $a_1, \cdots, a_m$, and define the map

$$\phi : \{a_1, \cdots, a_m\} \mapsto \bar{\mathcal{K}}$$

in which the values $a_1, \cdots, a_m$ are given from $\bar{\mathcal{K}}$, and the values $\phi(\mathcal{F})$, $\phi(D)$, $\phi(X)$ are obtained by substituting the parameters in $\mathcal{F}$, $D$, $X$ by the particular values in $\bar{\mathcal{K}}$, here $\phi(\mathcal{D}_1)$ is obtained by mapping all elements $X$ in $\mathcal{D}_1$ to $\phi(X)$. Then, we have the following theorem.

**Theorem 2.** *If the following* KSY *condition*

$$\exists X \in \mathcal{D}_1 \ such \ that \ \mathrm{rank}(X) < \mathrm{rank}(D), \tag{17}$$

*holds, then* $\phi(\det(Y)) = 0$ *for all* $Y \in \mathcal{R}$, *provides that* $\phi(\mathcal{F})$ *has common affine zero that satisfies* $\mathcal{C}$.

It is easy to see that the essentials of the KSY condition in the Theorem 2 is the existence of a column in the Dixon matrix $D$ that is linear independent to all other columns in $D$. The theorem implies that if the KSY-condition is verified, then the determinant of any maximal non-singular square-matrix can be taken as a necessary condition for the original equations to have a common affine zero. A practical way of checking the KSY condition is to solve the linear equation $Dw = \mathbf{0}$ in which $w = (w_1, \cdots, w_{s_2})^T$. It is not difficult to verify that the KSY condition is equivalent to the fact that there exists $w_i = 0 \ (1 \leq i \leq s_2)$ in $w$ and that $\mathcal{C} \implies monom(m_i) \neq 0$.

If the validity of KSY condition has been verified, it remains to compute the determinant of a $r \times r$ submatrix of $D$. For this, Kapur et al. [19] proved that one can execute Gaussian elimination on the Dixon matrix $D$ to construct $D_{row}$ and take the product of all its pivots as $\det(Y)$ as below.

**Lemma 2.** *There exists some* $Y \in \mathcal{R}$ *such that the product of all the pivots of* $D_{row}$ *is equal to* $\det(Y)$.

Since the KSY condition is satisfied in most situations, the KSY method expands the range of application of the Dixon resultant effectively. We shall denote the determinant of the sub-matrix that the KSY method produces by $\varrho_{\mathcal{C}}$ and also call it the Dixon resultant below in this paper.

Return to the equation system (8), we have constructed an $85 \times 85$ Dixon matrix $D$ in $5\,\mathrm{s}$ in our computer, and verified the KSY-condition very quickly, too. However, when we applied the Maple command `gausselim` to do row-reduce to the Dixon matrix $D$, again the overflow happened for allocating memory.

## 4 Reconstruct the Dixon Resultant Using Implicit Equation Interpolation

In this section, we present a concise description to the interpolation for reconstructing the Dixon resultant in the Fermat-Torricelli problem. We consider the interpolation problem as follows. Given a multivariate implicit polynomial function

$$r(x_1, \cdots, x_n, u_s) = \sum_{k=0}^{d} p_k(x_1, \cdots, x_n) u_s^k = 0,$$

where $p_k \in \mathcal{K}[x_1, \cdots, x_n]$ for $0 \leq k \leq d$, $p_d \neq 0$. If we take an $n$-tuple $(\xi_1, \cdots, \xi_n) \in \mathcal{K}^n$ to substitute the variables $(x_1, \cdots, x_n)$ in the function, we would get a univariate polynomial of variable $u_s$

$$c \cdot r(\xi_1, \cdots, \xi_n, u_s), \quad c \neq 0. \tag{18}$$

in which $c \in \mathcal{K}$. Given this information, we want to reconstruct the polynomial coefficients $p_k(x_1, \cdots, x_n)$ of $r(x_1, \cdots, x_n, u_s)$. This type of interpolation problem is called implicit function interpolation. Notice that $c$ may not be the same for different value $(\xi_1, \cdots, \xi_n)$, As a consequence, the common methods for multivariate polynomial interpolation may not be applied directly to solve the implicit interpolation problem here. According to [17,18], if the constant term of one of $p_k(x_1, \cdots, x_n)$ $(k = 0, 1, \cdots, d)$ is not zero, say,

$$p_d(x_1, \cdots, x_n) = c_0 + \sum_{d_1 + \cdots + d_n \geq 1} c_{d_1, \cdots, d_n} x_1^{d_1} \cdots x_n^{d_n}, \quad c_0 \neq 0,$$

then $r(x_1, \cdots, x_n, u_s)$ can be reconstructed by using the normalized multivariate rational function interpolation on

$$\frac{p_0(x_1, \cdots, x_n)}{p_d(x_1, \cdots, x_n)}, \cdots, \frac{p_{d-1}(x_1, \cdots, x_n)}{p_d(x_1, \cdots, x_n)},$$

and the number of the interpolation points $(\xi_1, \cdots, \xi_n)$ we need is $(D_1 + 1) \times \cdots \times (D_n + 1)$, where

$$D_1 = \max_{k=0, \cdots, d} \deg(p_k, x_1), \quad \cdots, \quad D_n = \max_{k=0, \cdots, d} \deg(p_k, x_n).$$

Otherwise, a more complicated general rational function interpolation method (see [7,27,28]) is needed. A practical method for checking whether or not the constant term of $p_k(x_1, \cdots, x_n)$ is zero, is computing $r(x_1, \cdots, x_n, u_s)$ by taking $x_1 = \xi_1 z, \cdots, x_n = \xi_n z$, where $\xi_1, \cdots, \xi_n$ are numbers and $z$ is a letter. It is clear that if an $n$-tuple $(\xi_1, \cdots, \xi_n) \in \mathcal{K}^n$ is found (indeed, this can be done randomly with high probability in view of [26] and [34]) so that

$$p_k(\xi_1 z, \cdots, \xi_n z) = c_0 + c_1 z + \cdots + c_{N_k} z^{N_k}$$

for certain $k = 0, \cdots, d-1, d$, then we claim that $p_k(0, \cdots, 0) \neq 0$, and moreover, the upper bounds $D_1, \cdots, D_n$ can be obtained by

$$D_j = \max_{k=0, \cdots, d} \deg(p_k, x_j) \leq \max_{k=0, \cdots, d} (p_k(\xi_1 z, \cdots, \xi_n z), z) = \max_{k=0, \cdots, d} N_k$$

for $j = 1, \cdots, n$.

To the implicit interpolation of the Dixon resultants for the Fermat-Torricelli problem, we have used

$$p = a^2, q = b^2, r = c^2$$

to reduce computation complexity since we have seen that only $a^2, a^4, b^2, b^4, c^2, c^4$ appeared in the (8). We have tried to construct the Dixon matrix in two ways. The first way is starting from the equation system (8), as indicated in the end of the Sect. 3, an $85 \times 85$ Dixon matrix $D$ (in symbolic expression) has been constructed in 5 s, and the KSY-condition is also verified in seconds. Then the experiment on computing the Dixon resultant for interpolation points $p = \xi_1 z, q = \xi_2 z, r = \xi_3 z$, where $\xi_1, \xi_2, \xi_3$ are positive rational numbers, shows $d = 24$, the constant term of $p_{24}(p, q, r)$ is not zero, and

$$D_1 = \max_{k=0,\cdots,24} (p_k, p) \leq 12, \quad D_2 = \max_{k=0,\cdots,24} (p_k, q) \leq 12, \quad D_3 = \max_{k=0,\cdots,24} (p_k, r) \leq 12.$$

For instance, taking $p = 10z, q = 3z, r = 6z$, we get

$$p_{24}(10z, 3z, 6z)L^{24} + p_{22}(10z, 3z, 6z)L^{22} + \cdots + p_2(10z, 3z, 6z)L^2 + p_0(10z, 3z, 6z)$$

where

$$p_{24}(10z, 3z, 6z) = 45z^3 - 108z^2 + 76z - 16,$$
$$p_{22}(10z, 3z, 6z) = -3420z^4 + 7020z^3 - 1752z^2 - 2128z + 768,$$

$$\vdots$$

$$p_2(10z, 3z, 6z) = 86617423872z^{11} - 91866968064z^{10} + \cdots + 48545464320z^7,$$
$$p_0(10z, 3z, 6z) = -12745506816z^{12} + 14703575040z^{11} + \cdots - 9624158208z^8.$$

It takes about 10 s to generate an interpolation instance from the symbolic $85 \times 85$ Dixon matrix $D$. In the second way, we use the polynomial $P_1, P_2, P_3$ generated in (10) to construct the Dixon matrix $D$. Again, we use substitution $p = a^2, q = b^2, r = c^2$ for shorter expression. Then, we have

$$\begin{cases} P_1(L, p, q, r, v, w) := -p^2 L^4 + 4p^2 r L^3 + \cdots + 4r^2 w^2 + 4pqr \ (86 \text{ terms}), \\ P_2(L, p, q, r, v, w) := p^2 L^3 - 3p^2 v L^2 + \cdots + -2prw + 2q^2 v \ (56 \text{ terms}), \quad (19) \\ P_3(L, p, q, r, v, w) := p^2 L^3 - 3p^2 v L^2 + \cdots - 2qrw + 2r^2 w \ (56 \text{ terms}), \end{cases}$$

here the highest degrees of $L, p, q, r, v, w$ in $P_1, P_2, P_3$ are $[4, 2, 2, 2, 4, 4]$, $[4, 2, 2, 1, 3, 3]$, $[4, 2, 1, 2, 3, 3]$. Then the Dixon matrix $D_1$ for (10) or (19) is an $18 \times 18$ matrix in symbolic expression. It is also easy to check the KSY condition of $D_1$. Now it takes only 2 s to generate an instance $r(\xi_1 z, \xi_2 z, \xi_3 z, L)$ for $\xi_1, \xi_2, \xi_3$ selected randomly and uniformly from 0 to 100. Again, we found that $d = 24$, $p_{24}(0, 0, 0) \neq 0$, and the degree bounds for $p, q, r$ are 12. Therefore, it is suffice to reconstruct the $p_{2k}(p, q, r)/p_{24}(p, q, r)$ for $k = 0, 1, \cdots, 11$, and therefore, the implicit equation

$$\rho(L, p, q, r) = p_1(p, q, r)L^{24} + p_2(p, q, r)L^{22} + \cdots + p_{12}(p, q, r)L^2 + p_{13}(p, q, r), \quad (20)$$

on $13^3 = 2197$ instances, or on the grid $[\xi_1 z, \xi_2 z, \xi_3 z]$ for $\xi_1, \xi_2, \xi_3 \in \{0, 1, 2, \cdots, 12\}$.

Finally, factorizing the obtained Dixon resultant, we obtained the following polynomial:

$$\rho(L, p, q, r) = \rho_0(L, q, r)\rho_0(L, r, p)\rho_0(L, p, q)\rho_4(L, p, q, r) \qquad (21)$$

where

$$\rho_0(L, x, y) = L^4 - 2(x + y)L^2 + (x - y)^2,$$

and

$$\rho_4(L, p, q, r) = q_{12}L^{12} + q_{10}L^{10} + \cdots + q_2 L^2 + q_0,$$

with $q_{12}, q_{10}, \cdots, q_2, q_0 \in \mathbb{Q}[p, q, r]$, same as polynomials displayed in the Theorem 1, when taking $k = pqr, m = p + q + r, n = pq + qr + rp$ and $p = a^2, q = b^2, r = c^2$ for shorter in expression, i.e., $\rho_4(L, a^2, b^2, c^2)$ equals to the polynomial $g(L, a, b, c)$ defined in (4).

Notice that

$$L^4 - 2(x + y)L^2 + (x - y)^2 = (L - \sqrt{x} - \sqrt{y})(L - \sqrt{x} + \sqrt{y})(L + \sqrt{x} - \sqrt{y})(L + \sqrt{x} + \sqrt{y}),$$

and

$$L + a + b > 0, \quad L - a + b > 0, \quad L + a - b > 0$$

for any four points $P, A, B, C \in S^2$ with $a = BC, b = CA, c = AB, L = PA + PB + PC$, so we proved that $a, b, c$ are edges of a triangle on $S^2$ and $L$ be the minimal or maximal of $PA + PB + PC$ for $P \in S^2$, then it satisfies

$$(L - a - b)(L - b - c)(L - c - a)\rho_4(L, a^2, b^2, c^2) = 0,$$

as claimed in the Theorem 1.

## 5  Geometric Properties of the Fermat-Torricelli Points

To conclude the paper, we list two geometric properties of the Fermat-Torricelli points.

**Proposition 1.** *If $A, B, C \in S^2$ and one of the $\angle CAB, \angle ABC, \angle BCA$ is larger than $2\pi/3$, then the point $P \in S^2$ such that $PA + PB + PC$ is the minimal coincides to one of the vertices. In reverse, $(L - a - b)(L - b - c)(L - c - a) = 0$ implies that one of the angles of the triangle is larger than or equals to $2\pi/3$.*

*Proof.* Without loss of generality, we may assume that

$$A = (x_1, y_1, z_0), \quad B = (x_2, y_2, z_0), \quad C = (x_3, y_3, z_0),$$

and $z_0 \geq 0$. Let $P = (x, y, z)$ be any point on $S^2$ and $P_0 = (x, y, z_0)$. Then we have

$$\min(a + b, b + c, c + a) \leq P_0 A + P_0 B + P_0 C \leq PA + PB + PC$$

whenever $\max(\angle ABC, \angle BCA, \angle CAB) \geq 2\pi/3$. $\qquad \square$

**Proposition 2.** *Assume $A, B, C \in S^2$ and $\Pi_1$ be the plane determined by $A, B, C$, and $\Pi_0$ the plane that parallels to $\Pi_1$, and passes the center $O$ of $S^2$. Assume that $\Pi_1 /\!/ \Pi_0$. Then $\Pi_1$ divides the sphere $S^2$ into two parts. Let $\Sigma_1$ be the minor part. Let $\Sigma_0$ be the hemisphere of $S^2$ divided by plane $\Pi_0$ so that $\Sigma_0 \cap \Sigma_1 = \emptyset$. Let $P_1$ be the point on $S^2$ satisfies $P_1A + P_1B + P_1C$ is minimal, and $P_0$ the point on $S^2$ satisfies that $P_0A + P_0B + P_0C$ is maximal. Then $P_0 \in \Sigma_0$, $P_1 \in \Sigma_1$.*

*Proof.* Without loss of generality, assume that $A = (x_1, y_1, z_0), B = (x_2, y_2, z_0), C = (x_3, y_3, z_0)$, and $z_0 \geq 0$. Then

$$\Sigma_0 = \{(x, y, z) | x^2 + y^2 + z^2 = 1, z \leq 0\}, \quad \Sigma_1 = \{(x, y, z) | x^2 + y^2 + z^2 = 1, z \geq z_0\},$$

Then it is easy to prove that any points $P_1 \in \Sigma_1, P_0 \in \Sigma_0$ satisfy that $P_1A + P_1B + P_1C \leq P_0A + P_0B + P_0C$ if $P_0P_1$ is perpendicular to $ABC$. For point $P \in S^2 \setminus (\Sigma_0 \cup \Sigma_1)$, we can prove that there exist points $P_0' \in \Sigma_0$ and $P_1' \in \Sigma_1$ so that $P_1'A + P_1'B + P_1'C \leq PA + PB + PC \leq P_0'A + P_0'B + P_0'C$. We leave the proof of this fact to readers for saving pages.   □

# References

1. Alexandrescu, D.-O.: A characterization of the Fermat point in Hilbert spaces. Mediterr. J. Math. **10**(3), 1509–1525 (2013)
2. Blumenthal, L.F.: Theory and Application of Distance Geometry, 2nd edn. Chelsea Publishing Company, New York (1970)
3. Cayley, A.: A theorem in the geometry of position. Camb. Math. J. **2**, 267–271 (1841)
4. Chen, Z.: The Fermat-Torricelli problem on surfaces. Appl. Math. J. Chin. Univ. **31**(3), 362–366 (2016)
5. Chionh, E.-W., Zhang, M., Goldman, R.: Fast computation of the Bezout and Dixon resultant matrices. J. Symb. Comput. **33**(1), 13–29 (2002)
6. Cut The Knot. https://www.cut-the-knot.org/Generalization/fermat_point.shtml. Accessed 6 Sept 2019
7. Cuyt, A., Lee, W.S.: Sparse interpolation of multivariate rational functions. Theor. Comput. Sci. **412**(16), 1445–1456 (2011)
8. Dalla, L.: A note on the Fermat-Torricelli point of a d-simplex. J. Geom. **70**, 38–43 (2001)
9. Drezner, Z., Plastria, F.: In Memoriam Andrew (Andy) Vazsonyi: 1916–2003. Ann. Oper. Res. **167**, 1–6 (2009). https://doi.org/10.1007/s10479-009-0523-6
10. Du, D.-Z., Hwang, F.K.: A proof of the Gilbert-Pollak conjecture. Algorithmica **7**, 121–135 (1992)
11. Œuvres de Fermat. Paris: Gauthier-Villars et fils (1891–1896)
12. Fasbender, E.: Über die gleichseitigen Dreiecke, welche um ein gegebenes Dreieck gelegt werden können. J. Reine Angew. Math. **30**, 230–231 (1846)
13. Ghalich, K., Hajja, M.: The Fermat point of a spherical triangle. Math. Gazette **80**(489), 561–564 (1996)
14. Gilbert, E., Pollak, H.: Steiner minimal trees. SIAM J. Appl. Math. **16**, 1–29 (1968)

15. Guo, X., Leng, T., Zeng, Z.: The Fermat-Torricelli problem on sphere with Euclidean metric. J. Syst. Sci. Complex. **38**(12), 1376–1392 (2018). (in Chinese)
16. Johnson, R.A.: Modern Geometry: An Elementary Treatise on the Geometry of the Triangle and the Circle, pp. 221–222. Houghton Mifflin, Boston (1929)
17. Kai, H.: Rational interpolation and its Ill-conditioned property. In: Wang, D., Zhi, L. (eds.) Symbolic-Numeric Computation. Trends in Mathematics, pp. 47–53. Birkhäuser, Basel (2007)
18. Kaltofen, E., Yang, Z.: On exact and approximate interpolation of sparse rational functions. In: Proceedings of the International Symposium on Symbolic and Algebraic Computation, pp. 203–210 (2007)
19. Kapur, D., Saxena, T., Yang, L.: Algebraic and geometric reasoning using Dixon resultants. In: Proceeding ISSAC 1994 (Proceedings of the International Symposium on Symbolic and Algebraic Computation), pp. 99–107 (1994)
20. Katz, I., Cooper, L.: Optimal location on sphere. Comput. Math. Appl. **6**(2), 175–196 (1980)
21. Kuhn, H.: Steiner's problem revisited. In: Dantzig, G.B., Eaves, B.C. (eds.) Studies in Optimization. Studies in Mathematics, vol. 10, pp. 52–70. Mathematical Association of America, Washington, DC (1974)
22. Kupitz, Y., Martini, H.: The Fermat-Torricelli point and isosceles tetrahedra. J. Geom. **49**(1–2), 150–162 (1994)
23. Kupitz, Y., Martini, H.: Geometric aspects of the generalized Fermat-Torricelli problem. In: Básrásny, I., Böröczky, K. (eds.) Intuitive Geometry. Bolyai Society Mathematical Studies 1995, vol. 6, pp. 55–127. János Bolyai Mathematical Society, Budapest (1995)
24. Launhardt, W.: Kommercielle Tracirung der Verkehrswege. Hannover (1872)
25. Saxena, T.; Efficient variable elimination using resultants. Ph.D. thesis, State University of New York at Albany, Albany (1996)
26. Schwartz, J.: Fast probabilistic algorithms for verification of polynomial identities. J. ACM **27**(4), 701–717 (1980)
27. Tang, M.: Polynomial algebraic algorithms and their applications based on sparse interpolation. Ph.D. thesis, East China Normal University, Shanghai (2017)
28. Tang, M., Yang, Z., Zeng, Z.: Resultant elimination via implicit equation interpolation. J. Syst. Sci. Complex. **29**(5), 1411–1435 (2016)
29. Weber, A.: Über den Standort der Industrien, Teil I: Reine Theorie des Standorts. J.C.B. Mohr, Tübingen (1909). English ed. by C.J. Friedrichs, University of Chicago Press (1929)
30. Weiszfeld, E.: Sur le point pour lequel la somme des distances de n points donnés est minimu. Tôhoku Math. J. **43**, 355–386 (1937)
31. Wikipedia: Fermat point - Wikipedia. https://en.wikipedia.org/wiki/Fermat_point. Accessed 4 Apr 2019
32. Zachos, A.: Exact location of the weighted Fermat-Torricelli point on flat surfaces of revolution. Results Math. **65**(1–2), 167–179 (2014)
33. Zachos, A.: Hyperbolic median and its applications. In: International Conference "Differential Geometry and Dynamical Systems", pp. 84–89 (2016)
34. Zippel, R.: Probabilistic algorithms for sparse polynomials. In: Ng, E.W. (ed.) Symbolic and Algebraic Computation. LNCS, vol. 72, pp. 216–226. Springer, Heidelberg (1979). https://doi.org/10.1007/3-540-09519-5_73
35. Zuo, Q., Lin, B.: The Fermat point of finite points in the metric space. Math. J. **17**(3), 359–364 (1997). (in Chinese)

# Using Leslie Matrices as the Application of Eigenvalues and Eigenvectors in a First Course in Linear Algebra

Michael Monagan[✉]

Department of Mathematics, Simon Fraser University, Burnaby, Canada
mmonagan@cecm.sfu.ca

**Abstract.** Leslie matrices may be used to model the age distribution of a population as well as population growth. The dominant eigenvalue tells us the long term population growth and the corresponding eigenvector tells us the long term age distribution. Because the model is so simple, and it does not require any knowledge of physics or chemistry or biology, it's ideal for presenting in a first course on Linear Algebra as the main application of eigenvalues and eigenvectors.

In this paper we present the Leslie age distribution model and provide accompanying exercises suitable for students. We use Maple for both numerical calculations and symbolic calculations. We include some data for real populations that instructors may use for classroom presentation or for assignments.

## 1 Introduction

Linear algebra is my favourite subject to teach. Like most lower division mathematics courses, it is packed with topics that someone wants to be covered. So there's not much room for applications and certainly no room for applications that require additional mathematics to be introduced first. An application needs to fit in one lecture or less. For linear systems, I like to use Markov matrices as the application as they also introduce a family of matrices. What application should we use to illustrate eigenvalues and eigenvectors? I have 19 linear algebra texts on my office shelf. The most common application for eigenvalues and eigenvectors is to solving linear systems of first order differential equations. A problem with this application is that many students will not yet have seen first order linear systems of differential equations. And for those who have, their understanding will likely be superficial. So this is not a good choice.

We need an application where the model is simple to understand and there are interesting questions that the student can easily explore. I propose that we use Leslie matrices and the Leslie age distribution model. This model is popular in ecology and demographics. It takes 10 to 15 min to understand the model and see how to express it as a linear transformation. It takes 15 to 20 min to compute the dominant eigenvalue $\lambda^+$ and corresponding eigenvector $v^+$ for an

© Springer Nature Switzerland AG 2020
J. Gerhard and I. Kotsireas (Eds.): MC 2019, CCIS 1125, pp. 279–291, 2020.
https://doi.org/10.1007/978-3-030-41258-6_21

example and give a physical interpretation of what $\lambda^+$ and $v^+$ mean for the population. That leaves time to pose some interesting questions and exercises. For the reader, a dominant eigenvalue is an eigenvalue, possibly complex, of largest magnitude. Leslie matrices are non-negative matrices with a unique positive dominant eigenvalue $\lambda^+$.

Four of my texts, Anton and Rorres [1], Poole [8], Lay et al. [5], Boyd and Vandenberghe [2] use Leslie matrices as an application of eigenvalues and eigenvectors. The latter two do so under the label "linear dynamical systems". Anton and Rorres give a longer theoretical treatment, data from actual populations, and study what happens when we harvest from the population.

In Sect. 2 I develop the Leslie age distribution model, calculate the dominant eigenvalue and eigenvector for an example in Maple and give a physical interpretation of them. In Sect. 3 I explore the Leslie matrix from an algebraic viewpoint using Maple to do some of calculations. In Sect. 4 I explore some questions about controlling the growth of a population. For Sects. 2–4, I have included some exercises that can be used for student assignments. I have also gathered some data from real populations from the literature in the Appendix.

## 2    The Leslie Population Distribution Model

Leslie matrices model the age distribution of a population over time. They model births, the aging process and deaths of a human or animal population. I think the best way to introduce this subject is to present an actual example and show how the model leads to a matrix times a vector before presenting the general case. A real example will focus the attention of the students. I use the example of the grey seal population on Sable island, an island off the coast of Nova Scotia. The data for the example is taken from [7]. The model is presented in Fig. 1.

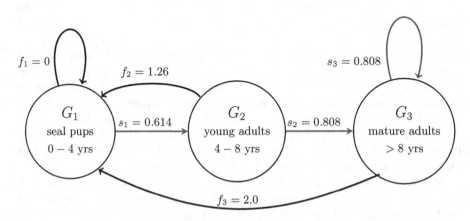

**Fig. 1.** Leslie model for grey seal population. The $G_k$ are age groups, $f_k$ are fertility rates and $s_k$ are survival probabilities.

We divide the females of the population into $n$ age groups $G_1, G_2, \ldots, G_n$. In Fig. 1 we have divided the seal population into three age groups: $G_1$, seal pups, ages 0–4 years, $G_2$, young seal adults, ages 4–8 years, and $G_3$, mature seal adults aged over 8 years. We model the fertility rates $f_1, f_2, \ldots, f_n$ which are the average number of female births per female in the time period. In our seal population we have $f_1 = 0$ which means seal pups are not mature enough to reproduce, $f_2 = 1.26$ for young adult female seals and $f_3 = 2.0$ for mature female seals which means each seal has on average two female seal pups in 4 years, so one seal pup per year until they die. We also model average survival rates for each age group. In our seal population these are $s_1 = 0.614$, $s_2 = s_3 = 0.808$ meaning over 60% of seal pups survive to be 4 years old. These numbers (see [7]) are estimates based on that has been gathered by scientists over a long period.

In [5] Lay et al. study the northern spotted owl population. They also divide the owl population into three age groups with fertility rates $f_1 = 0.0, f_2 = 0.0, f_3 = 0.33$ and survival rates $s_1 = 0.18, s_2 = 0.71, s_3 = 0.94$.

Let $p_i^t$ be the number of females in age group $i$ at time $t$. So the population vector at time $t$ is $P^{(t)} = [p_1^t, p_2^t, \ldots, p_n^t]$. It is the female population at time $t$. According to the model the population at time $t + 1$ is given by

$$P^{(t+1)} = \begin{bmatrix} f_1 p_1^t + f_2 p_2^t + f_3 p_3^t \\ s_1 p_1^t \\ s_2 p_2^t + s_3 p_3^t \end{bmatrix}$$

The key observation is that the model is a linear transformation so we may write $P^{(t+1)} = L P^{(t)}$ for some $n \times n$ matrix $L$, a Leslie matrix. One should spend some time constructing the matrix $L$ here so the student can see where the matrix comes from and why Linear Algebra is involved.

$$\begin{bmatrix} p_1^{(t+1)} \\ p_2^{(t+1)} \\ p_3^{(t+1)} \end{bmatrix} = \begin{bmatrix} f_1 & f_2 & f_3 \\ s_1 & 0 & 0 \\ 0 & s_2 & s_3 \end{bmatrix} \begin{bmatrix} p_1^t \\ p_2^t \\ p_3^t \end{bmatrix}$$

The Leslie matrices for the grey seal population and northern spotted owl population are given below in Fig. 2. For the reader, what we will eventually find is that both matrices have a dominant eigenvalue $\lambda^+$. For the grey seals $\lambda^+ = 1.49$ which means the seal population is growing rapidly (it is exploding) and because

$$\begin{bmatrix} 0 & 1.26 & 2.0 \\ 0.614 & 0 & 0 \\ 0 & 0.808 & 0.808 \end{bmatrix} \begin{bmatrix} 0 & 0 & 0.33 \\ 0.18 & 0 & 0 \\ 0 & 0.71 & 0.94 \end{bmatrix}$$

**Fig. 2.** Leslie matrices for the grey seals (left) and northern spotted owls (right)

of this there is an effort to stop the population growing. For the owl population $\lambda^+ = 0.91$ which means the owl population is dying and there is a concerted effort to save it. For teaching we proceed as follows.

We are interested in the distribution of the population among the $n$ age groups. Let us define population distribution vector $D^{(t)}$ to be $P^{(t)}/p_t$ where $p_t = \sum_{i=1}^{n} p_i^t$ is the total female population at time $t$.

Suppose the current seal population is $P^{(0)} = [1, 1, 1]$ thousands. We calculate $P^{(1)}$ using Maple as follows.

```
> L := Matrix([[0.0,1.26,2.00],[0.614,0,0],[0,0.808,0.808]]);
```

$$L := \begin{bmatrix} 0.0 & 1.26 & 2.0 \\ 0.614 & 0 & 0 \\ 0 & 0.808 & 0.808 \end{bmatrix}$$

```
> P[0] := <1,1,1>: P[1] := L.P[0];
```

$$P_1 := \begin{bmatrix} 3.260000000 \\ 0.6140000000 \\ 1.616000000 \end{bmatrix}$$

After 16 time periods (64 years) we get the following data.

| $P^{16}$ | $D^{16}$ | $P^{15}$ | $D^{15}$ |
|---|---|---|---|
| $\begin{bmatrix} 1115.126895 \\ 459.0148362 \\ 542.5083059 \end{bmatrix}$ | $\begin{bmatrix} 0.5268357429 \\ 0.2168591067 \\ 0.2563051504 \end{bmatrix}$ | $\begin{bmatrix} 747.5811664 \\ 307.7235768 \\ 363.6975940 \end{bmatrix}$ | $\begin{bmatrix} 0.5268357541 \\ 0.2168591050 \\ 0.2563051409 \end{bmatrix}$ |

Maple code to compute these vectors is

```
> local D; # by default D is the differential operator in Maple
> t := 15;
> pop := proc(v) local i; add(v[i],i=1..numelems(v)) end;
> for i to t do P[t] := L.P[t-1]; D[t] := P[t]/pop(P[t]); od;
```

Now we connect what has happened to the seal population with the eigenvalues and eigenvectors of $L$. First, the seal population has exploded! It has increased from 3 thousand to 2,116 thousand. Second, comparing $D^{(15)}$ and $D^{(16)}$, the population age distribution has stabilized at 52.7% seal pups, 21.7% young adults, and 25.6% mature adults. Consider the quantities

$$\frac{P_1^{16}}{P_1^{15}} = 1.491646586, \quad \frac{P_2^{16}}{P_2^{15}} = 1.491646630, \quad \text{and} \quad \frac{P_3^{16}}{P_3^{15}} = 1.491646673.$$

This means $\lambda = 1.4916466$ and $v = D^{15}$ satisfy $Lv = \lambda v$ to 7 decimal places. Thus the sequence $D_1, D_2, D_3, \ldots$ is converging to an eigenvector of $L$. We state the following Theorem for a Leslie matrix $L$.

**Theorem 1.** *For any non-zero initial population* $P^0 = [p_1^0, p_1^0, \ldots, p_n^0]$, *if at least one fertility rate* $f_i$ *is positive, the Leslie matrix* $L$ *has a unique positive eigenvalue* $\lambda^+$. *If* $v^+$ *is the corresponding eigenvector and at least two consecutive fertility rates are positive,* $\lambda^+$ *is dominant and the population distribution will converge to an eigenvector of* $L$, *that is* $\lim_{t\to\infty} D^{(t)}$ *exists and is a multiple of* $v^+$.

We also have the following physical interpretation for $\lambda^+$.

$\lambda^+ < 1$ means the population will decline exponentially.

$\lambda^+ > 1$ means the population will grow exponentially.

$\lambda^+ = 1$ means the population is stable, it does not change.

Below we calculate the eigenvalues of $L$ and the dominant eigenvector of $L$ using Maple. This is too difficult to do by hand. In exercise 3 below, I've constructed a Leslie matrix $L$ with $\lambda^+ = 7/6$ so that a hand calculation is easy.

```
> with(LinearAlgebra):
> E := Eigenvalues(L);
```

$$E := \begin{bmatrix} -0.341823317441679 + 0.359549749028222\,i \\ -0.341823317441679 - 0.359549749028222\,i \\ 1.49164663488336 + 0.\,i \end{bmatrix}$$

```
> lambda := Re(E[3]);
> I3 := IdentityMatrix(3):
> v := NullSpace(L-lambda*I3)[1]:
> v/pop(v);
```

$$\begin{bmatrix} 0.526835747502870 \\ 0.216859101480196 \\ 0.256305151016934 \end{bmatrix}$$

**Exercises**

1. For a population with two age groups with $f_1 = 1$, $f_2 = 1$, $s_1 = 0.75$ and $s_2 = 0$. Write down the Leslie matrix. Calculate the eigenvalues and eigenvectors. Is the population growing or declining? What is the long term population distribution?

2. Calculate the positive eigenvalue of $L$ for the spotted owl population. See Fig. 2. Is the population growing or dying?

3. For the Leslie matrix $L$ below, what does $L_{22} = 0.5$ mean? Calculate the eigenvalues by hand. For the positive eigenvalue, determine the corresponding eigenvector. What is the long term population distribution vector?

$$L = \begin{bmatrix} 0.5 & 0.75 \\ 0.5 & 0.75 \end{bmatrix}$$

4. For the Leslie matrix below calculate the eigenvalues. You should find that one is 0 and one is positive. For the positive eigenvalue, determine the corresponding eigenvector. What is the long term population distribution vector?

$$L = \begin{bmatrix} 0 & 7/6 & 7/6 \\ 1/2 & 0 & 0 \\ 0 & 2/3 & 2/3 \end{bmatrix}$$

5. For the northern spotted owl population (see Fig. 2), starting with $P^0 = [0.2, 0.1, 0.7]$, calculate $P^4 = L^4 P^0$ and $P^5 = L^5 P^0$ and determine the age distribution. To how may decimal places has the population distribution converged. Estimate the corresponding eigenvalue.

6. This exercise is taken from Poole [8]. Woodland caribou are found primarily in western Canada and the American northwest. The fertility rates and survival rates are given in the table below. The data shows that caribou cows do not give birth during their first two years and the survival rate for caribou calves is low.

| Age | 0–2 | 2–4 | 4–6 | 6–8 | 8–10 | 10–12 | 12–14 |
|-----|-----|-----|-----|-----|------|-------|-------|
| $f_i$ | 0.0 | 0.2 | 0.9 | 0.9 | 0.9 | 0.8 | 0.3 |
| $s_i$ | 0.3 | 0.7 | 0.9 | 0.9 | 0.9 | 0.6 | 0.0 |
| $P_i^{(0)}$ | 10 | 2 | 8 | 5 | 12 | 0 | 1 |

Construct the Leslie matrix. Shown also in the last row is the female caribou population in Jasper National park in 1990. Predict the female population in 1992, 1994, 1996, 1998 and 2000. What do you conclude will happen to the population in the long term? Use a computer to compute the eigenvalues of $L$. What is $\lambda^+$? What does this tell you about the population?

## 3   The Leslie Matrix

A Leslie matrix is an $n$ by $n$ matrix of the form

$$L = \begin{bmatrix} f_1 & f_2 & \cdots & f_{n-1} & f_n \\ s_1 & 0 & \cdots & 0 & 0 \\ 0 & s_2 & \cdots & 0 & 0 \\ \vdots & \vdots & & \vdots & \vdots \\ 0 & 0 & \cdots & s_{n-1} & 0 \end{bmatrix} \tag{1}$$

where $n \geq 2$, the survival rates $s_i > 0$ and fertility rates $f_i \geq 0$ with at least one $f_i > 0$. Thus the Fibonacci matrix $[[1,1],[1,0]]$ is a Leslie matrix. Notice that we have $s_n = L_{nn} = 0$. If $s_n > 0$, as was the case for the grey seals and spotted owls, we say $L$ is a generalized Leslie matrix. Here we assume $s_n = 0$. One of the exercises in [1] is to show that the characteristic polynomial of $L$ is

$$c(x) = x^n - f_1 x^{n-1} - s_1 f_2 x^{n-2} - s_2 s_1 f_3 x^{n-3} - \cdots - s_{n-1} \ldots s_3 s_2 s_1 f_n \quad (2)$$

The difficulty is that the matrix has an arbitrary dimension. To do this we would suggest that the student first calculate $c(x)$ for $n = 2$ and $n = 3$. Using Maple it is easy to do this. We will do it first for $n = 3$ then for $n = 4$.

```
> with(LinearAlgebra):
> L := Matrix([[f[1],f[2],f[3]],[s[1],0,0],[0,s[2],0]]);
```

$$\begin{bmatrix} f_1 & f_2 & f_3 \\ s_1 & 0 & 0 \\ 0 & s_2 & 0 \end{bmatrix}$$

```
> CharacteristicPolynomial(L,x);
```

$$x^3 - x^2 f_1 - x f_2 s_1 - s_2 s_1 f_3$$

```
> L := Matrix([[f[1],f[2],f[3],f[4]],[s[1],0,0,0],
              [0,s[2],0,0],[0,0,s[3],0]]):
> C := CharacteristicMatrix(L,x);
```

$$C := \begin{bmatrix} x - f_1 & -f_2 & -f_3 & -f_4 \\ -s_1 & x & 0 & 0 \\ 0 & -s_2 & x & 0 \\ 0 & 0 & -s_3 & x \end{bmatrix}$$

```
> Determinant(C);
```

$$x^4 - f_1 x^3 - s_1 f_2 x^2 - s_2 s_1 f_3 x - s_3 s_2 s_1 f_4$$

Anton and Rorres [1] give the following formula for the eigenvector of $L$ where $\lambda$ is an eigenvalue.

$$v = \left[ 1 \quad \frac{s_1}{\lambda} \quad \frac{s_1 s_2}{\lambda^2} \quad \frac{s_1 s_2 s_3}{\lambda^3} \quad \cdots \quad \frac{s_1 s_2 \ldots s_{n-1}}{\lambda^{n-1}} \right]^T \quad (3)$$

How would we check this? The right way is simply to calculate $Lv$ and $\lambda v$ and try to show $Lv = \lambda v$. Another not so clever way, which I confess to trying at first, is to try to calculate the eigenvector, that is, solve $(L - \lambda I)z = 0$ for $z$ in terms of $\lambda$ and try to show that $v = sz$ for some scalar $s$. Again, using Maple, we can only do this for a fixed $n$. Let us try the first way for $n = 4$.

```
> v := <1,s[1]/x,s[1]*s[2]/x^2,s[1]*s[2]*s[3]/x^3> :
> y := L.v:
> y, x*v;
```

$$
\begin{bmatrix} f_1 + \dfrac{s_1 f_2}{x} + \dfrac{s_2 s_1 f_3}{x^2} + \dfrac{s_3 s_2 s_1 f_4}{x^3} \\[2ex] s_1 \\[2ex] \dfrac{s_2 s_1}{x} \\[2ex] \dfrac{s_3 s_2 s_1}{x^2} \end{bmatrix} , \quad \begin{bmatrix} x \\[2ex] s_1 \\[2ex] \dfrac{s_2 s_1}{x} \\[2ex] \dfrac{s_3 s_2 s_1}{x^2} \end{bmatrix}
$$

It seems that all I have to do is check that $y_1 = x$. It is tempting to try to manipulate $y_1$ to get $x$. It is better to show $y_1 - x$ equals zero, that is, to show that $y_1 - x \mod c(x) = 0$. Simplifying to zero is always the best approach if you are using a computer algebra system. It is also often true for a hand calculation. How do we tell Maple to simplify $y_1 - x$ using the constraint $c(x) = 0$? One way to do this is to use the `simplify` command directly as follows

```
> simplify(y[1]-x,{c = 0});
```

$$0$$

If the second input to the `simplify` command is a set of algebraic equations, they are treated as constraints. Alternatively one could use division. First multiply $y_1 - x$ by $x^3$ to clear the denominators so that $x^3(y_1 - x)$ is a polynomial in $x$ then and divide $x^3(y_1 - x)$ by $c(x)$ to get the remainder.

```
> zero := numer(y[1]-x);
```

$$f_1 x^3 + f_2 s_1 x^2 + f_3 s_1 s_2 x + f_4 s_1 s_2 s_3 f_4 - x^4$$

This is just the negative of the characteristic polynomial.

```
> rem(zero,c,x);
```

$$0$$

The remainder command treats the inputs as polynomials in $x$.

Another way is tell Maple that $c(x) = 0$ directly. I will use a Maple RootOf to do this. The way to read the following command is that $\lambda$ is one of the roots of $c(x) = 0$ and $\lambda$ is how this root will be displayed. Then we use Maple's `evala` facility to evaluate algebraic expressions.

```
> alias( lambda=RootOf(c,x) ) ;
> evala( lambda^4-f[1]*lambda^3 );
```

$$s_1 \left( f_2 \lambda^2 + f_3 \lambda s_2 + f_4 s_2 s_3 \right)$$

```
> evala( subs(x=lambda,y[1]) );
```

$$\lambda$$

Now for the not so smart way. We've just told Maple that $\lambda$ is a root of the characteristic polynomial. Let's calculate the eigenvector the way we teach a student to do it by solving $(L - \lambda I)u = 0$ for $u$ in terms of the $f_i$ and $s_i$. Since the system is homogeneous it should have a free parameter. In the Maple code below I tell Maple to use $t$ for the parameter and, to save space, I print $u^T$.

```
> I4 := IdentityMatrix(4):
> u  := LinearSolve(L-lambda*I4,<0,0,0,0>,free=t):
> Transpose(u);
```

$$\left[ \frac{\lambda^3 t_4}{s_3 s_2 s_1} \quad \frac{\lambda^2 t_4}{s_3 s_2} \quad \frac{\lambda t_4}{s_3} \quad t_4 \right]$$

Let's try $t = v_4 = s_1 s_2 s_3 / \lambda^3$.

```
> u := subs( t[4]=v[4], u ):
> Transpose(u);
```

$$\left[ 1 \quad \frac{s_1}{\lambda} \quad \frac{s_1 s_2}{\lambda^2} \quad \frac{s_1 s_2 s_3}{\lambda^3} \right]$$

Well that worked with no further simplification required.

To show that the Leslie matrix $L$ has one positive eigenvalue $\lambda_1$ we introduce

$$q(x) = \frac{f_1}{x} + \frac{f_2 s_1}{x^2} + \frac{f_3 s_1 s_2}{x^3} + \cdots + \frac{f_n s_1 s_2 \cdots s_{n-1}}{x^n} \tag{4}$$

and claim $q(\lambda) = 1$ where $\lambda$ is a non-zero eigenvalue of $L$. We leave this as an exercise. Now since $f_i \geq 0$ and $s_i > 0$ the function $q(x)$ is monotonically decreasing and $\lim_{x \to \infty} q(x) = 0$. Consequently there is only one $\lambda$, say $\lambda = \lambda^+$ such that $q(\lambda^+) = 1$. That is, $L$ has a unique positive eigenvalue $\lambda^+$. Exercise 3 below shows that $\lambda^+$ has multiplicity 1.

## Exercises

1. Show that characteristic polynomial for an $n$ by $n$ Leslie matrix given by Eq. (1) is (2).
2. Show that $q(\lambda) = 1$.
3. Show that the positive eigenvalue $\lambda_1$ of a Leslie matrix has algebraic multiplicity 1. Hint: a root $\lambda_1$ of a polynomial $q(x)$ has multiplicity 1 if and only if $q'(\lambda_1) \neq 0$.
4. For a generalized Leslie matrix

$$L = \begin{bmatrix} f_1 & f_2 & f_3 \\ s_2 & 0 & 0 \\ 0 & s_2 & s_3 \end{bmatrix}$$

   use Maple to calculate the characteristic polynomial $c(x)$. Now try to find a formula for the $c(x)$ for an $n$ by $n$ generalized Leslie matrix.
5. The net reproduction rate of a population is defined as

$$r = f_1 + f_2 s_1 + f_3 s_1 s_2 + \cdots + f_n s_1 s_2 \ldots s_{n-1}.$$

   Explain why $r$ can be interpreted as the average number of daughters born to a female over her lifetime. It follows that if $r > 1$ the population will grow but if $r < 1$ it will decline. Calculate $r$ for caribou population in Sect. 2 Exercise 5.

# 4  Population Stabilization and Harvesting

Consider the Leslie matrix for the grey seal population.

$$\begin{bmatrix} 0 & 1.26 & 2.0 \\ 0.614 & 0 & 0 \\ 0 & 0.808 & 0.808 \end{bmatrix}$$

We have determined that the dominant eigenvalue $\lambda^+ = 1.49$ which means the seal population is growing by almost 50% every four years. How can we stabilize the population so that it is neither growing nor declining? The idea is to change the fertility rates $f_1, f_2, f_3$ or the survival rates $s_1, s_2, s_3$ to force $\lambda^+ = 1$. We consider two possibilities.

1. Reduce $s_1$ by culling the seal pups every 4 years.
2. Reduce all $f_i$ by shooting all seals with infertility darts.

I do the first option in class by hand and leave the second as an exercise. Here I will run the both experiments using Maple. These calculations can easily be be done by hand but one will worry about errors. I use Maple here to check my calculations.

```
> L := Matrix([[0.0,1.26,2.0],[s[1],0,0],[0,0.808,0.808]]);
```

$$L := \begin{bmatrix} 0 & 1.26 & 2.0 \\ s_1 & 0 & 0 \\ 0 & 0.808 & 0.808 \end{bmatrix}$$

Now force the eigenvalue $\lambda = 1$ and solve for $s_1$.

```
> I3 := IdentityMatrix(3):
> C := L - 1*I; # lambda=1
```

$$C := \begin{bmatrix} -1. & 1.26 & 2.0 \\ s_1 & -1. & 0 \\ 0 & 0.808 & -0.192 \end{bmatrix}$$

```
> c := Determinant(C);
```

$$c := -0.192 + 1.85792\, s_1$$

```
> s[1] = solve( c=0, s[1] );
```

$$s_1 = 0.1033413710$$

So we must reduce $s_1$ from 61% to 10% to stop the population growing. Such a huge reduction indicates how healthy the population is. Continuing with the second option.

```
> L := Matrix([[0,s*1.26,s*2.0],[0.614,0,0],[0,0.808,0.808]]);
```

$$L := \begin{bmatrix} 0 & 1.26\,s & 2.0\,s \\ 0.614 & 0 & 0 \\ 0 & 0.808 & 0.808 \end{bmatrix}$$

```
> c := Determinant(L-I3);
```

$$c := -0.192 + 1.14076288\,s$$

```
> s = solve(c=0,s);
```

$$s = 0.1683084218$$

Again, a drastic reduction in the fertility rates is needed to stabilize the population.

Another kind of question that one can ask is, what is the maximal sustainable harvest rate $h$. That is what value of $h$ can we use such that the matrix

$$L = \begin{bmatrix} 0 & 1.26(1-h) & 2.0(1-h) \\ 0.614(1-h) & 0 & 0 \\ 0 & 0.808(1-h) & 0.808(1-h) \end{bmatrix}$$

has an eigenvalue 1? The answer below is one third.

```
> L := Matrix([[0,1.26,2.0],[0.614,0,0],[0,0.808,0.808]]):
> c := Determinant( (1-h)*L - I3 );
```

$$c := 0.94876288 - 3.45664864h + 1.87500864h^2 - 0.36712288h^3$$

```
> h = fsolve( c=0, h );
```

$$h = 0.3295999357$$

## Exercises

1. Consider the following Leslie matrix

$$L = \begin{bmatrix} 0.5 & f \\ 0.75 & 0 \end{bmatrix} .$$

For $f = 0.25$ is the population growing or declining? What must $f$ be to stabilize the population?
2. For the grey seal population in Fig. 2, what is the maximum sustainable harvesting rate assuming we do not harvest seal pups. Why might it be risky to harvest at this rate?
3. For the northern spotted owl population in Fig. 2, what must $s_1$ be so that the owl population stabilizes? Comment on the stability of the population.
4. If the government tries to eradicate northern owl predators so that all $s_1, s_2, s_3$ increase, what rate must they increase by to stabilize the population?

## 5   Conclusion

I am indebted to Carl Schwarz of Simon Fraser University for introducing me to Leslie matrices and the Leslie age distribution model and showing me the Sable island grey seal data in [7]. I have taught Linear Algebra at Simon Fraser University many times. I now use Leslie matrices as the sole application for eigenvalues and eigenvectors in the course. The main advantage is that one does not require any new mathematics nor any understanding of physics, chemistry or biology to understand the model.

One difficulty with using data from real applications is that the characteristic polynomials will not have a simple real root. Students would need a computer to compute $\lambda^+$. Furthermore, calculating the corresponding eigenvector by solving $(L - \lambda^+ I)v+ = 0$ for $v+$ is also difficult to do correctly by hand.

One can obtain a good estimate for the dominant eigenvalue $\lambda^+$ and corresponding eigenvector $v^+$ using the power method. If you teach the power method, then that is a reasonable approach. It will require only a few matrix multiplications and the Leslie matrices are sparse. If not, then one needs Leslie matrices which are suitable for hand calculations. I've provided some in the exercises in this paper.

## Appendix

The following data is taken from [1,3]. It is for a sheep population. The sheep have a lifespan of 12 years so the age groups are 1 year each. The dominant eigenvalue is 1.176.

| Age | 0–1 | 1–2 | 2–3 | 3–4 | 4–5 | 5–6 | 6–7 | 7–8 | 8–9 | 9–10 | 10–11 | 11–12 |
|-----|-----|-----|-----|-----|-----|-----|-----|-----|-----|------|-------|-------|
| $f_i$ | .000 | .045 | .391 | .472 | .484 | .546 | .543 | .502 | .468 | .459 | .433 | .421 |
| $s_i$ | .845 | .975 | .965 | .950 | .926 | .895 | .850 | .786 | .691 | .561 | .370 | .000 |

The following data is taken from [1]. The data is for Canadian females in 1965. Because few women over 50 bear children, we ignore those older than 50. The age groups are 5 years each so 10 age groups. The dominant eigenvalue is 1.076.

| Age | 0–5 | 5–10 | 10–15 | 15–20 | 20–25 | 25–30 | 30–35 | 35–40 | 40–45 | 45–50 |
|-----|-----|------|-------|-------|-------|-------|-------|-------|-------|-------|
| $f_i$ | .0000 | .00024 | .0586 | .2861 | .4479 | .3640 | .2226 | .1046 | .0283 | .0024 |
| $s_i$ | .9965 | .9982 | .9980 | .9973 | .9969 | .9962 | .9946 | .9918 | .9780 | – |

The following data is taken from [7]. The data is for the grey seal population on Sable island, an island in the Atlantic off Nova Scotia. If one expands the first age group to 4 age groups of 1 year each with $s_i = \sqrt[4]{0.614} = 0.8852$ and $f_i = 0.0$, I get $\lambda^+ = 1.114$. To compress the model into three 4 year age groups, for $G_2$ (4–8 yrs) use $f_2 = .142 + .948(.347 + .948(.436 + .948(.468))) = 1.26$ and $s_2 = .948^4 = 0.808$.

| Age | 0–4 | 4–5 | 5–6 | 6–7 | 7–8 | 8–9 | 9–10 |
|---|---|---|---|---|---|---|---|
| $f_i$ | 0.000 | 0.142 | 0.347 | 0.436 | 0.468 | 0.491 | 0.500 |
| $s_i$ | 0.614 | 0.948 | 0.948 | 0.948 | 0.948 | 0.948 | 0.948 |

The following data it taken from [9]. The data is for northern fur seals. The birth rates include female and male seal pups. I calculate $\lambda^+ = 1.333$.

| Age | 0–2 | 2–4 | 4–6 | 6–8 | 8–10 | 10–12 | 12–14 |
|---|---|---|---|---|---|---|---|
| $f_i$ | 0.00 | 0.02 | 0.70 | 1.53 | 1.67 | 1.65 | 1.56 |
| $s_i$ | 0.91 | 0.88 | 0.85 | 0.80 | 0.74 | 0.67 | 0.59 |
| Age | 14–16 | 16–18 | 18–20 | 20–22 | 22–24 | 24–26 | |
| $f_i$ | 1.45 | 1.22 | 0.91 | 0.70 | 0.22 | 0.00 | |
| $s_i$ | 0.49 | 0.38 | 0.27 | 0.17 | 0.15 | 0.00 | |

# References

1. Anton, H., Rorres, C.: Elementary Linear Algebra with Applications. Wiley, Hoboken (1987)
2. Boyd, S., Vandenberghe, L.: Introduction to Applied Linear Algebra. Cambridge University Press, Cambridge (2018)
3. Caughley, G.: Parameters for seasonally breeding populations. Ecology **48**, 834–839 (1967)
4. Laberson, R.H., McKelvey, R., Noon, B.R., Voss, C.: A dynamic analysis of the viability of the northern spotted owl in a fragmented forest environment. J. Conserv. Biol. **6**, 505–512 (1992)
5. Lay, D.C., Lay, S.R., McDonald, J.J.: Linear Algebra and its Applications, 5th edn. Pearson, London (2016)
6. Leslie, P.H.: The use of matrices in certain population mathematics. Biometrika **33**(3), 183–212 (1945)
7. Manske, M., Schwarz, C.J., Stobo, W.T.: The estimation of the rate of population change of Grey Seals (Halichoerus grypus) on Sable Island using a Leslie projection matrix with Capture-Recapture data. Unpublished manuscript
8. Poole, D.: Linear Algebra, A Modern Introduction. Brooks/Cole, Pacific Grove (2003)
9. York, A.E., Hartley, J.R.: Pup production following harvest of female northern fur seals. Can. J. Fish. Aquat. Sci. **38**, 84–90 (1981)

# Transforming Maple into an Intelligent Model-Tracing Math Tutor

Dimitrios Sklavakis[✉][iD]

European School Brussels III,
135 Boulevard du Triomphe, 1050 Brussels, Belgium
sklavadi@teacher.eursc.eu
http://www.ontomath.com/

**Abstract.** This article describes an intelligent, model-tracing system for tutoring expansion and factoring of algebraic expressions. The system is implemented as a set of procedures in a Maple document that tutor a breadth of 18 top-level mathematical skills (algebraic operations). Twelve (12) skills for expansion (monomial multiplication, monomial division and power of monomial, monomial-polynomial and polynomial-polynomial multiplication, parentheses elimination, collection of like terms, identities square of sum and difference, product of sum by difference, cube of sum and difference) and six (6) skills for factoring (common factor, identities square of sum and difference, product of sum by difference, quadratic form by sum and product, quadratic form by roots). These skills are further decomposed in simpler ones giving a deep domain expertise model of 68 primitive skills. The tutor has two novel features: (a) it exhibits *intelligent task recognition* by identifying all skills present in any expression through intelligent parsing, and (b) for each identified skill, the tutor traces all the sub-skills, a feature called *deep model tracing*. Furthermore, based on these features, the tutor achieves *broad knowledge monitoring* by recording student performance for all skills present in any expression.

**Keywords:** Intelligent Tutoring Systems · Model-Tracing tutors · Assessment systems

## 1 Introduction

One-to-one tutoring has proven to be one of the most effective ways of teaching. It has been shown [4] that the performance of the average student under an expert tutor is about two standard deviations above the average performance of the conventional class (30 students to one teacher). That is, 50% of the tutored students scored higher than 98% of students in the conventional class. However, it is also known that one-to-one tutoring is the most expensive form of education. Due to this cost, we are still in the era of mass education, struggling to raise the teacher to student ratio. The problem of designing and implementing educational

© Springer Nature Switzerland AG 2020
J. Gerhard and I. Kotsireas (Eds.): MC 2019, CCIS 1125, pp. 292–306, 2020.
https://doi.org/10.1007/978-3-030-41258-6_22

environments as effective as individual tutoring was termed by Bloom as "the two sigma problem", named after the mathematical symbol of standard deviation, $\sigma$.

The implementation of the one-to-one tutoring model by Intelligent Tutoring Systems (ITSs) has motivated researchers to develop ITSs that provide the same tutoring quality as a human tutor [17]. Model Tracing Tutors (MTTs) [3] have shown significant success in domains like mathematics [7], computer programming [5] and physics [18]. These tutors are based on a domain expertise model that solves the problem under tutoring and produces the correct step(s). At each step, the model-tracing algorithm matches the solution(s) produced by the model to that provided by the student and gives positive or negative feedback, hints or/and help messages.

However, the domain models of MTTs are hard to author [1]. The main reason for this is the *knowledge acquisition bottleneck*: extracting the knowledge from the domain experts and encoding it into a MTT. Knowledge reuse has been proposed as a key factor to overcome this obstacle [8,10]. Since expert knowledge and, particularly, tutoring knowledge is so hard to create, re- using it is of paramount importance. A good example of knowledge reuse is the Mass Production mechanism provided by Carnegie Mellon's Cognitive Tutors Authoring Tools (CTAT). This mechanism allows the creation of new tutors from existing ones for isomorphic problems, that is problems having nearly the same solution steps [1].

The problem of knowledge reuse permeates the whole domain of digital education, from the development of digital learning resources to the creation of online assessment questions and tests. While these resources demand considerable time and effort from domain experts, their reuse is almost impossible. The main reason for this is that the domain and tutoring models for these systems are developed from scratch and therefore they cannot reach the breadth and depth necessary to build on top of them and extend them. Despite the research efforts in the field of Artificial Intelligence in Education aimed in using semantic technologies (ontologies) to overcome this problem [6,9,14,16], reuse of digital learning resources is still limited.

A considerable effort in developing a model-tracing tutor, with a broad and deep domain model and knowledge reuse as its primary design principle, is the MATHESIS Algebra Tutor [12,13,15] and the MATHESIS meta-knowledge engineering framework for intelligent tutoring systems in mathematics [14,16]. The experience gained from these efforts, showed the paramount importance of a broad, deep and fine-grained domain (mathematical) model. Maple, a research and development effort of more than 30 years, provides one of the most advanced and sophisticated mathematical models, both in terms of computational power and in parsing mathematical expressions. This power of Maple in mathematical computation and parsing is the main research motive in redeveloping the MATHESIS Algebra Tutor. The result is the model-tracing tutor described in this article.

Based on the computational power of Maple, the tutoring system has a broad domain model of eighteen (18) top-level skills which are analysed - after elaborate cognitive task analysis - to a depth of sixty-eight (68) sub-skills. On the same time, the development of such a broad and deep domain model gives rise to the *scaling-up* problem: if a problem contains more than one (sub)tasks to be performed then a more complex task arises, that of identifying and ordering the (sub)tasks to perform! For example, the task of expanding the expression

$$3x \cdot (2x + 5) - (5x^2 - 1) \cdot (3x - 2)$$

contains the (sub)tasks of monomial by polynomial multiplication, $3x \cdot (2x + 5)$ and polynomial multiplication $(5x^2 - 1) \cdot (3x - 2)$ which contain the (sub)tasks of monomial multiplications. After the execution of these tasks follow the tasks of parentheses elimination and collection of like terms. The solution to this tutoring problem was to equip the tutor with *intelligent task recognition* through sophisticated parsing of the algebraic expressions based on the powerful parsing functions of Maple. Another, rather positive, consequence of adopting a broad and deep domain expertise model was the development of an equally detailed student model. Instead of simply keeping a percentage measure of the students' skill performance, the student model was extended to keep full records of the interactions between the tutor and the student for each solution step.

This article describes the intelligent model-tracing tutor for expanding and factoring algebraic expressions, developed as a set of procedures in a Maple document. The rest of the article is structured as follows: Sect. 2 describes the three models of the tutor, the *mathematical domain expertise model*, the *tutoring or pedagogical model* and the *student model*. Section 3 presents the performance of the expanding tutoring procedures of the tutor, while Sect. 4 presents the factoring tutoring procedures. Section 5 concludes the article with a discussion of the implementation and future directions of development.

## 2   The Tutor's Domain, Tutoring and Student Models

The tutor consists of three models:

(a) The domain expertise or mathematical model, which produces the correct solutions for each task. this is implemented through the *expand* and *factor* commands of Maple.

(b) The tutoring or pedagogical model, which parses the student answer, compares it with the correct answer produced by the domain model and gives appropriate feedback in the form of messages and guides the student to the correct solution. It must be noted that the most difficult part of the tutoring model is parsing the student's answer and finding any errors made. This is feasible due to the powerful Maple commands *op* and *type*.

(c) The student model, which records the skill/sub-skill to be performed, the student answer, and whether it is correct or not. The student model is implemented as a *Table* of *Records*. Each *Record* has the following structure

*Record*(STUDENT-ID, SESSION-ID, DATE, SKILL, SUB-SKILL, TASK,
ANSWER, CORRECT)

There are three procedures that manage the student model. These are:

- *initializeStudentModel(student-id)*: Sets the global variable STUDENT-ID and gets the next session number, SESSION-ID
- *showStudentModel(student-id,[sub-skill1,sub-skill-2,...,sub-skill-N],[true,false])*: Displays all records of student with student-id which contain the skills contained in the list of the second argument and the student's answer is correct ([true]), wrong ([false]) or both ([true,false]).
- *showStudentModelSkills(student-id)*: Displays the skills contained in the student model for the student with student-id

The development of the domain expertise model was based on deep cognitive task analysis in the paradigm of Carnegie-Mellon's cognitive tutors [3]. The tutor can teach a breadth of 18 top-level cognitive mathematical skills (algebraic operations). Twelve (12) skills for expansion:

- Monomial multiplication,
- Monomial division,
- Power of monomial
- Monomial-Polynomial multiplication
- Polynomial multiplication,
- Parentheses elimination,
- Collection of like terms,
- Identity square of sum
- Identity square of difference
- Identity product of sum by difference,
- Identity cube of sum,
- Identity cube of difference

Six (6) Skills for factoring:

- Common factor,
- Identity square of sum,
- Identity square of difference,
- Identity difference of squares,
- Factoring quadratic form by sum S and product P,
- Factoring quadratic form by roots

For each one of these skills, the corresponding Maple procedure checks the student's answer for correct steps as well as for common mistakes. If the student's answer is correct, positive feedback is given. In the case of mistakes, the tutor produces a series of messages and guidance to the correct solution. The following two sections present in detail the performance of the tutoring procedures, using illustrative examples of erroneous student answers.

# 3    The Tutoring Processes for Expanding

## 3.1    Procedure *multiplyMonomials(expression, answer)*

$multiplyMonomials((3x^2y) \cdot (-4xz^3), 12x^3yz^3)$:
SUB-SKILL:multiplyCoefficients
FEEDBACK:The coefficient of your answer is not correct. You must multiply the coefficients of the monomials.

$multiplyMonomials((3x^2y) \cdot (-4xz^3), -12x^3z^3)$:
SUB-SKILL:multiplyVariables
FEEDBACK:You omitted variable, y.

$multiplyMonomials((3x^2y) \cdot (-4xz^3), -12x^2z^3)$:
SUB-SKILL:multiplyVariables
FEEDBACK:The exponent, 2, of variable, x, is not correct. You must add the exponents of the common variables.
Hint: when a variable does not have an exponent, the exponent is 1.

## 3.2    Procedure *divideMonomials(expression, answer)*

$divideMonomials(\frac{-12x^3yz^3}{3x^2y}, 4xz^3)$:

SUB-SKILL:divideCoefficients
FEEDBACK:The coefficient of your answer is not correct. You must divide the coefficients of the monomials.

$divideMonomials(\frac{-12x^3yz^3}{3x^2y}, -4z^3)$:

SUB-SKILL:divideVariables
FEEDBACK:You omitted variable, x.

$divideMonomials(\frac{-12x^3yz^3}{3x^2y}, -4x^2z^3)$:

SUB-SKILL:divideVariables
FEEDBACK:The exponent, 2, of variable, x, is not correct. You must subtract the exponents of the common variables.
Hint: when a variable does not have an exponent, the exponent is 1.

## 3.3    Procedure *monomialPower(expression, answer)*

$monomialPower((-2x^2yz^3)^3, -6x^6y^3z^9)$:
SUB-SKILL:raiseCoefficient
FEEDBACK:The coefficient of your answer is not correct. You must raise the coefficient, -2, to the exponent, 3, of the power

$monomialPower((-2x^2yz^3)^3, -8x^6z^9)$:
SUB-SKILL:raiseVariable
FEEDBACK:You omitted variable, y.

$monomialPower((-2x^2yz^3)^3, -8x^6y^3z^6)$:
SUB-SKILL:raiseVariable
FEEDBACK:The exponent, 6, of variable, z, is not correct. You must multiply the exponent, 3, of variable, z, by the exponent, 3

### 3.4   Procedure $multiplyMonomialPolynomial(expression, answer)$

$multiplyMonomialPolynomial(2x^2y^3 \cdot (3x + 5x^2y - 2xy^2), 6x^2y^3 + 10x^3y^4 - 4x^3y^5)$:
SUB-SKILL:partialProducts
FEEDBACK:For product, $2x^2y^3 \cdot 3x$, your answer is not correct.
For product, $2x^2y^3 \cdot 5x^2y$, your answer is not correct.
For product, $2x^2y^3 \cdot -2xy^2$, your answer, $-4x^3y^5$, is correct.
For products, $[2x^2y^3 \cdot 3x, 2x^2y^3 \cdot 5x^2y]$, you must check your wrong answers, $[6x^2y^3, 10x^3y^4]$ with the command multiplyMonomials.

$multiplyMonomialPolynomial(2x^2y^3 \cdot (3x + 5x^2y - 2xy^2), 6x^3y^3 + 10x^3y^4 - 4x^3y^5)$:
SUB-SKILL:partialProducts
FEEDBACK:For product, $2x^2y^3 \cdot 3x$, the answer,$6x^3y^3$, is correct.
For product, $2x^2y^3 \cdot 5x^2y$, your answer is not correct.
For product, $2x^2y^3 \cdot -2xy^2$, your answer, $-4x^3y^5$, is correct.
(At this point, the tutor automatically calls procedure multiplyMonomials)
Expression: $10x^4y^4$, Answer: $10x^3y^4$
The exponent, 3, of variable, x, is not correct. You must add the exponents of the common variables.
Hint: when a variable does not have an exponent, the exponent is 1.

### 3.5   Procedure $multiplyPolynomials(expression, answer)$

$multiplyPolynomials((3xy - 2x^2) \cdot (2x^2y^2 - 4xy), 6x^2y^3 - 12x^2y^2 - 4x^4y^2 - 8x^3y)$:
SUB-SKILL:partialProducts
FEEDBACK:For product, $3xy \cdot 2x^2y^2$, your answer is not correct.
For product, $3xy \cdot -4xy$, your answer, $-12x^2y^2$, is correct
For product, $-2x^2 \cdot 2x^2y^2$, your answer, $-4x^4y^2$, is correct.
For product, $-2x^2 \cdot -4xy$, your answer is not correct.
For products, $[3xy \cdot 2x^2y^2, -2x^2 \cdot -4xy]$, you must check your wrong answers, $[6x^2y^3, -8x^3y]$ with the command multiplyMonomials.

### 3.6   Procedure $expandSquareOfSum(expression, answer)$

$expandSquareOfSum((ax - by)^2, a^2x - 2axby + b^2y^2)$
SUB-SKILL:squareTerm

FEEDBACK:The first term, $a^2x$, is not correct because...
(At this point, the Tutor automatically calls procedure
$monomialPower((ax)^2, a^2x)$, to check the error)
The exponent, 1, of variable, x, is not correct.
You must multiply the exponent, 1 of variable, x, by the exponent, 2, of the power.

$expandSquareOfSum((x^2 - 1)^2, x^4 + 2x^2 + 1)$
SUB-SKILL:doubleProduct
FEEDBACK:The second term, $2x^2$, is not correct because...
(At this point, the Tutor automatically calls procedure
$multiplyMonomials(2 \cdot x^2 \cdot -1, 2x^2)$, to check the error)
The coefficient of your answer is not correct. You must multiply the coefficients of
the monomials.
You must calculate the product of terms, $2, x^2 and - 1$

### 3.7   Procedure *expandSumByDifference(expression, answer)*

$expandSumByDifference((2y + x) \cdot (2y - x), 4y - x^2)$
SUB-SKILL:squareTerm
FEEDBACK:The first term, $4y$, is not correct because...
(At this point, the Tutor automatically calls procedure
$monomialPower((2y)^2, 4y)$, to check the error)
The exponent, 1, of variable, y, is not correct.
You must multiply the exponent, 1 of variable, y, by the exponent,2, of the power.

$expandSumByDifference((ax + by) \cdot (ax - by), a^2x^2 + b^2y^2)$
SUB-SKILL:recallIdentity
FEEDBACK:The second term, $b^2y^2$, is not correct because...
The second term of your answer must have a NEGATIVE sign

### 3.8   Procedure *expandCubeOfSum(expression, answer)*

$expandCubeOfSum((2x - 3)^3, 2x^3 + 36x^2 + 54x - 27)$
SUB-SKILL:cubeTerm
FEEDBACK:The first term, $2x^3$, is not correct because...
(At this point, the Tutor automatically calls procedure
$monomialPower((2x)^3, 2x^3)$, to check the error)
The coefficient, 2, of your answer is not correct.
You must raise the coefficient of the monomial, 2, to the exponent, 3, of the power.

$expandCubeOfSum((2x - 3)^3, 8x^3 - 12x^2 + 54x - 27)$
SUB-SKILL:doubleProduct
FEEDBACK:The second term, $-12x^2$, is not correct because...
(At this point, the Tutor automatically calls procedure

*multiplyMonomials*$(3 \cdot 4x^2 \cdot -3, -12x^2)$, to check the error)
The coefficient of your answer is not correct. You must multiply the coefficients of the monomials.
You must calculate the product of terms, $3, 4x^2 and - 3$

### 3.9   Procedure *addPolynomials(expression, answer)*

*addPolynomials*$\left(\left(2x^2 - x\right) - \left(x^3 - 5x^2 + x - 1\right), 2x^2 - x - x^3 + 5x^2 - x - 1\right)$
SUB-SKILL:ALL
(Maple simplifies automatically both the expression and the answer, so the tutor cannot identify the source of errors)
FEEDBACK:Expression: $\left[-x^3, 7x^2, -2x, 1\right]$, Answer: $\left[-x^3, 7x^2, -2x, -1\right]$
Correct terms in your answer: $\left[-x^3, 7x^2, -2x\right]$
Incorrect or Missing terms: $[1]$

*addPolynomials*$\left(-3x^2y - \left(2xy - yx^2\right) + \left(3xy - y^3\right), -3x^2y - 2xy - yx^2 + 3xy - y^3\right)$
SUB-SKILL:ALL
(Maple simplifies automatically both the expression and the answer, so the tutor cannot identify the source of errors)
FEEDBACK:Expression: $\left[-2x^2y, -y^3, xy\right]$, Answer: $\left[-4x^2y, -y^3, xy\right]$
Correct terms in your answer: $\left[-y^3, xy\right]$, Incorrect or Missing terms: $\left[-2x^2y\right]$

### 3.10   Scaling Up: Procedure *expandTutor(expression, answer)*

Procedure *expandTutor* combines the expansion procedures described in the previous subsections. Using the powerful parsing commands of Maple, like *op* and *type*, it uses intelligent parsing to identify the expansion tasks present in both the expression to be expanded and the student's answer. By comparing the two representations, the tutor identifies the task(s) executed by the student and calls the corresponding procedures to check for errors and guide the student. Below follow some illustrative examples:

*expandTutor*$\left(-5x \cdot (2x - 3) - 3x \cdot (2 - 3x), 10x^2 + 15x - 3x \cdot (2 - 3x)\right)$
FEEDBACK:Your answer is not correct...
(The tutor identifies that only the first monomial-polynomial multiplication was performed and calls *myltiplyMonomialPolynomial*)
For product, $-5x \cdot 2x$, your answer is not correct
For product, $-5x \cdot -3$, your answer, $-15x$, is correct
(Procedure *myltiplyMonomialPolynomial* now calls *myltiplyMonomials*)
Expression: $-10x^2$, Answer: $10x^2$
The coefficient of your answer is not correct. You must multiply the coefficients of the monomials.

*expandTutor*$\left(-5x \cdot (2x - 3) - 3x \cdot (2 - 3x), -10x + 15x - 6x + 9x^2\right)$
FEEDBACK:Your answer is not correct...

(The tutor identifies that both monomial–polynomial multiplications were performed and prompts the student to check them separately)
More than one operations performed. You must check them one by one:
Procedure to Call:multiplyMonomialPolynomial($-5x \cdot (2x - 3)$, your-answer
Procedure to Call:multiplyMonomialPolynomial($-3x \cdot (2 - 3x)$, your-answer

$expandTutor(3x^2 \cdot (-2x + 3) \cdot (5 - x), 3x^2 \cdot (10x + 2x^2 + 15 - 3x))$
FEEDBACK:Your answer is not correct...
(The tutor identifies that only the polynomials' multiplication was performed and calls *multiplyPolynomials*)
Expression:$(-2x + 3) \cdot (5 - x)$, Answer: $[2x^2, 7x, 15]$ For product, $-2x \cdot 5$, you have not given a correct answer.
For product, $-2x \cdot -x$,your answer, $2x^2$, is correct
For product, $3 \cdot 5$,your answer, 15, is correct.
For product, $3 \cdot -x$, your answer, $-3x$, is correct.
You must check your answers, $[7x]$, with *multiplyMonomials*

The student model, after these three calls of *expandTutor* looks like this:
*showStudentModel(1,[],[true,false])*
Record(STUDENT-ID = 1, SESSION-ID = 1,DATE = Date: 2019-05-24, SKILL = "expand", SUBSKILL = "ALL", TASK = $-5x \cdot (2x-3) - 3x \cdot (2-3x)$, ANS = $10x^2 + 15x - 3x \cdot (2-3x)$, CORRECT = false)
Record(STUDENT-ID = 1, SESSION-ID = 1,DATE = Date: 2019-05-24, SKILL = "multiplyMonomialPolynomial", SUBSKILL = "partialProducts", TASK = $-5x(2x - 3)$ , ANS = $[10x^2, 15x]$, CORRECT = false)
Record(STUDENT-ID = 1, SESSION-ID = 1,DATE = Date: 2019-05-24, SKILL = "multiplyMonomials", SUBSKILL = "multiplyCoefficients", TASK = $-10x$, ANS = $10x$, CORRECT = false)
Record(STUDENT-ID = 1, SESSION-ID = 2,DATE = Date: 2019-05-24, SKILL = "expand", SUBSKILL = "ALL", TASK = $-5x(2x - 3) - 3x(2 - 3x)$, ANS = $9x^2 - x$, CORRECT = false)
Record(STUDENT-ID = 1, SESSION-ID = 3,DATE = Date: 2019-05-24, SKILL = "expand", SUBSKILL = "ALL", TASK = $3x^2(-2x + 3)(5 - x)$, ANS = $3x^2 (2x^2 + 7x + 15)$, CORRECT = false)
Record(STUDENT-ID = 1, SESSION-ID = 3,DATE = Date: 2019-05-24, SKILL = "multiplyPolynomials", SUBSKILL = "partialProducts", TASK = $(-2x + 3)(5 - x)$, ANS = $[2x^2, 7x, 15]$, CORRECT = false)

## 4 The Tutoring Procedures for Factoring

### 4.1 Procedure *factorByCommonFactor(expression, answer)*

*factorByCommonFactor($2a^3 - 4a^2 + 6a^2b, 2a \cdot (a - 2 + 3b)$)*
SUB-SKILL:commonFactor
FEEDBACK:Your answer is not correct...

Common Factor: $2a^2$, Partial quotients: $a - 2 + 3b$
Answer Common Factor: $2a$, Answer Partial Quotients:$a - 2 + 3b$
Your common factor, $2a$, is not correct
You must find the Greatest Common Divisor of the coefficients and the smallest exponents of the common variables.

$factorByCommonFactor(2a^3 - 4a^2 + 6a^2b, 2a^2 \cdot (a - 2 + 3ab))$
SUB-SKILL:partialQuotients
FEEDBACK:Your answer is not correct...
Common Factor: $2a^2$, Partial quotients: $a - 2 + 3b$
Answer Common Factor: $2a^2$, Answer Partial Quotients:$a - 2 + 3ab$
Your partial quotient, $3ab$, is not correct
(The tutor automatically calls $divideMonomials(\frac{6a^2b}{2a^2}, 3ab)$)
The exponent, 1, of variable, a, is not correct. You must subtract the exponents of common variables.

## 4.2   Procedure $factorBySquareOfSum(expression, answer)$

$factorBySquareOfSum(25a^2 - 10ab + b^2, (25a - b)^2)$
SUB-SKILL:rootTerm
FEEDBACK:Your answer is not correct...
Term A: $5a$, Term B: $-b$
Your Term A: $25a$, Your Term B:$-b$
Your term A, $25a$, is not correct. It must be the square root of, $25a^2$, or,$b^2$

$factorBySquareOfSum(25a^2 - 10ab + b^2, (5a + b)^2)$
SUB-SKILL:doubleProductSign
FEEDBACK:Your answer is not correct...
Term A: $5a$, Term B: $-b$
Your Term A: $5a$, Your Term B:$b$
The sign inside the parenthesis must be the same as that of the double product, $-10ab$.

## 4.3   Procedure $factorByDifferenceOfSquares(expression, answer)$

$factorByDifferenceOfSquares(x^4 - 16, (x^2 + 2) \cdot (x^2 - 2))$
SUB-SKILL:rootTerm
FEEDBACK:Your answer is not correct...
Term A: $x^2$, Term B: $4$
Your Term A: $x^2 + 2$, Your Term B: $x^2 - 2$
Your term A, $x^2 + 2$, is not correct. It must be the sum or difference of, $x^2$, and, $4$

$factorByDifferenceOfSquares(16x^2 - 1, (4x + 1) \cdot (2x - 1))$
SUB-SKILL:rootTerm
FEEDBACK:Your answer is not correct...
Term A: $4x$, Term B: $1$
Your Term A: $4x + 1$, Your Term B: $2x - 1$
Your term B, $2x - 1$, is not correct. It must be the difference of, $4x$, and, $1$

## 4.4 Procedure *factorByTrinomialSumProduct(expression, answer)*

*factorByTrinomialSumProduct(*$w^2 + 5w + 6, (w + 1) \cdot (w + 6))$

SUB-SKILL:findSum

FEEDBACK:Your answer is not correct...

Roots: $-2, -3$

A: 1, B: 6

The numbers A, 1, and B, 6, that you gave do not have the right sum S.
The sum, $S = 5$ is the coefficient of the $1^{st}$ degree term.

*factorByTrinomialSumProduct(*$w^2 + 5w - 6, (w + 1) \cdot (w + 6))$

SUB-SKILL:findProduct

FEEDBACK:Your answer is not correct...

Roots: $1, -6$

A: 1, B: 6

The numbers A, 1, and B, 6, that you gave do not have the right product P.
The product $P = -6$ is the constant term of the trinomial.

## 4.5 Procedure *factorByTrinomialRoots(expression, answer)*

*factorByTrinomialRoots(*$x^2 + 3x + 2, (x - 1) \cdot (x - 3))$

SUB-SKILL:findRoot

FEEDBACK:Your answer is not correct...

The root, $x_1 = 1$, that you gave is not correct.
The roots of the trinomial $ax^2 + bx + c$ are given by the formulae:

$$x_1 = \frac{-b - \sqrt{b^2 - 4ac}}{2a}, x_1 = -1 \text{ and } x_2 = \frac{-b + \sqrt{b^2 - 4ac}}{2a}, x_2 = -2$$

The factored form of the trinomial is $a \cdot (x - x_1) \cdot (x - x_2)$

*factorByTrinomialRoots(*$4x^2 - 4x + 1, 4 \cdot (x - 1)^2)$

SUB-SKILL:findRoot

FEEDBACK:Your answer is not correct...

The root, $x_1 = 1$, that you gave is not correct.
The roots of the trinomial $ax^2 + bx + c$ are given by the formulae:

$$x_1 = \frac{-b - \sqrt{b^2 - 4ac}}{2a}, x_1 = \frac{1}{2} \text{ and } x_2 = \frac{-b + \sqrt{b^2 - 4ac}}{2a}, x_2 = \frac{1}{2}$$

The factored form of the trinomial is $a \cdot (x - x_1) \cdot (x - x_2)$

## 4.6 Scaling Up: Procedure *factorTutor(expression, answer)*

Procedure *factorTutor* combines the factoring procedures described in the previous subsections. Using the powerful parsing commands of Maple, like *op* and *type*, it uses intelligent parsing to identify the factoring tasks present in both the expression to be expanded and the student's answer. By comparing the two representations, the tutor identifies the task(s) executed by the student and calls the corresponding procedures to check for errors and guide the student. Below follow some illustrative examples:

$factorTutor(2a^3 - 4a^2 + 6a^2b, 2a^2 \cdot (a^2 - 2 + 3b))$
SUB-SKILL:factorByCommonFactor, partialQuotients
FEEDBACK:Your answer is not correct...
Common Factor: $[2a^2, \text{"NoIdentity"}]$,
Partial Quotients: $[a - 2 + 3b, \text{"NoIdentity"}]$
Factor $a - 2 + 3b$ is not correct...
There is a common factor and it seems that you did not find correctly the partial quotients.
You must first find correctly the partial quotients.
(Tutor calls $factorByCommonFactor(2a^3 - 4a^2 + 6a^2b, 2a^2 \cdot (a^2 - 2 + 3b)))$
Common Factor: $2a^2$, Partial quotients: $a - 2 + 3b$
Answer Common Factor: $2a^2$, Answer Partial Quotients:$a^2 - 2 + 3b$
Your partial quotient, $a^2$, is not correct
(The tutor automatically calls $divideMonomials(\frac{2a^3}{2a^2}, a^2))$
The exponent, 2, of variable, a, is not correct. You must subtract the exponents of common variables.

$factorTutor(5ax^2 - 80a, 5a \cdot (x^2 - 16))$
SUB-SKILL:factorByCommonFactor, factorByDifferenceOfSquares
FEEDBACK:Correct!
Common Factor: $[5a, \text{"NoIdentity"}]$,
Partial Quotients: $[x^2 - 16, \text{"factorByDifferenceOfSquares"}]$
Your answer is correct but you can factor it further.
Use command factorByDifferenceOfSquares($x^2 - 16$,your-answer)

$factorTutor(5ax^2 + 30ax + 25a, 5a \cdot (x + 2) \cdot (x + 3))$
SUB-SKILL:factorByCommonFactor, factorByDifferenceOfSquares
FEEDBACK:Your answer is not correct...
Common Factor: $[5a, \text{"NoIdentity"}]$,
Partial Quotients: $[(x + 5) \cdot (x + 1), \text{"factorByTrinomialRoots"}]$
Factor $x + 2$ is not correct in your answer.
There is a common factor and it seems that you did not find correctly the partial quotients.
You must first find correctly the partial quotients.
You must use the command factorByCommonFactor($5ax^2 + 30ax + 25a$,your-answer)

The student model, after these three calls of *expandTutor* looks like this:
*showStudentModel(1,[],[true,false])*
*Record(STUDENT-ID = 1, SESSION-ID = 1,DATE = Date: 2019-05-24,*
*SKILL = "factor", SUBSKILL = "ALL", TASK = $2a^3 + 6a^2b - 4a^2$, ANS*
*= $2a^2(a^2 + 3b - 2)$, CORRECT = false)*
*Record(STUDENT-ID = 1, SESSION-ID = 1,DATE = Date: 2019-05-24,*

$SKILL$ = *"factorByCommonFactor"*, $SUBSKILL$ = *"partialQuotients"*, $TASK$ = $2a^3 + 6a^2b - 4a^2$, $ANS$ = $2a^2\left(a^2 + 3b - 2\right)$, $CORRECT$ = *false)*
$Record(STUDENT\text{-}ID$ = 1, $SESSION\text{-}ID$ = 1, $DATE$ = *Date: 2019-05-24*,
$SKILL$ = *"divideMonomials"*, $SUBSKILL$ = *"divideVariables"*, $TASK$ = $a$,
$ANS$ = $a^2$, $CORRECT$ = *false)*
$Record(STUDENT\text{-}ID$ = 1, $SESSION\text{-}ID$ = 2, $DATE$ = *Date: 2019-05-24*,
$SKILL$ = *"factorByCommonFactor"*, $SUBSKILL$ = *"ALL"*, $TASK$ = $5ax^2 -$
$80a$, $ANS$ = $5a\left(x^2 - 16\right)$, $CORRECT$ = *true)*
$Record(STUDENT\text{-}ID$ = 1, $SESSION\text{-}ID$ = 2, $DATE$ = *Date: 2019-05-24*,
$SKILL$ = *"factor"*, $SUBSKILL$ = *"factorPartialQuotients"*, $TASK$ = $5ax^2 - 80a$,
$ANS$ = $5a\left(x^2 - 16\right)$, $CORRECT$ = *false)*
$Record(STUDENT\text{-}ID$ = 1, $SESSION\text{-}ID$ = 3, $DATE$ = *Date: 2019-05-24*,
$SKILL$ = *"factorByCommonFactor"*, $SUBSKILL$ = *"partialQuotients"*, $TASK$ = $5ax^2 + 30ax + 25a$, $ANS$ = $5a \cdot (x+2) \cdot (x+3)$, $CORRECT$ = *false)*

## 5    Discussion and Further Work

The last example of $factorTutor\left(5ax^2 + 30ax + 25a, 5a \cdot (x+2) \cdot (x+3)\right)$ presented, illustrates clearly that the success of Model-Tracing Tutors as tutoring and/or assessment systems lies on the development of *broad and deep domain*, *tutoring* and *student* models. In this example, when the tutor identifies that the factoring $(x+2) \cdot (x+3)$ of $x^2 + 6x + 5$ is not correct, it must *backtrack* to locate where the student made the error. There are two possible choice points in this backtracking: (a) The student may have calculated erroneously the partial quotients, $x^2 + 6x + 5$ or (b) The student calculated erroneously the roots $x_1 = -5$ and $x_2 = -1$ of the trinomial $x^2 + 6x + 5$. Based on tutoring experience, point (a) must be considered first since it precedes in the solution path. However, if the student model suggests a good performance in calculating the partial products, point (b) could be considered. Of course, this knowledge of the possible backtracking points is based on the intelligent parsing of the student answer using Maple's *op* and *type* commands as well as the computational power of Maple's *expand* and *factor* commands.

The tutor presented here is in an experimental stage of development. Testing and evaluation is planned in real-world conditions at the European School Brussels III from the coming school year. However, the tutor performance per se is far ahead of most tutoring and assessment systems to the best of the author's knowledge. This performance is clearly due to the two models of the tutor: (a) The mathematical model implemented by Maple's powerful computational, *expand* and *factor*, and parsing, *op* and *type*, commands, and (b) The detailed tutoring model based on cognitive task analysis of the author's tutoring expertise. It is the author's belief, based on previous research [12–16], that the powerful, broad and deep mathematical domain model of Maple, extended with adequately deep tutoring models can transform Maple into a new paradigm of tutoring and assessment system. A system that can surpass the two main and interconnected obstacles that inhibit the widespread and effective use of these systems, the *knowledge acquisition bottleneck* and the *scaling up* problem.

Towards this direction, the next research and development steps are:

(a) Adding more factoring methods,
   - Factoring by groups: $ax+by+ay+bx = a\cdot(x+y)+b\cdot(x+y) = (x+y)\cdot(a+b)$
   - Factoring by Cube of Sum/Difference: $a^3 \pm 3a^2b + 3ab^2 \pm b^3 = (a \pm b)^3$
   - Factoring by Sum of Cubes $(a^3 + b^3 = (a + b) \cdot (a^2 - ab + b^2)$
   - Factoring by Difference of Cubes $(a^3 - b^3 = (a - b) \cdot (a^2 + ab + b^2)$
(b) Elaborating the tutoring model in terms of backtracking and student guidance,
(c) Using a graphical user interface instead of the default command execution interface of a Maple document,
(d) Extending the tutor to handle operations (addition, subtraction, multiplication, division and powers) of rational expressions.

# References

1. Aleven, V., McLaren, B.M., Sewall, J., Koedinger, K.R.: The cognitive tutor authoring tools (CTAT): preliminary evaluation of efficiency gains. In: Ikeda, M., Ashley, K.D., Chan, T.-W. (eds.) ITS 2006. LNCS, vol. 4053, pp. 61–70. Springer, Heidelberg (2006). https://doi.org/10.1007/11774303_7
2. Aleven, V., McLaren, B.M., Sewall, J.: Scaling up programming by demonstration for intelligent tutoring systems development: an open-access web site for middle school mathematics learning. IEEE Trans. Learn. Technol. **2**(2), 64–78 (2009)
3. Anderson, J.R., Corbett, A.T., Koedinger, K.R., Pelletier, R.: Cognitive tutors: lessons learned. J. Learn. Sci. **4**(2), 167–207 (1995)
4. Bloom, B.S.: The 2 sigma problem: the search of methods for group instruction as effective as one-to- one tutoring. Educ. Res. **13**(6), 4–16 (1984)
5. Corbett, A.: Cognitive computer tutors: solving the two-sigma problem. In: Bauer, M., Gmytrasiewicz, P.J., Vassileva, J. (eds.) UM 2001. LNCS (LNAI), vol. 2109, pp. 137–147. Springer, Heidelberg (2001). https://doi.org/10.1007/3-540-44566-8_14
6. Dicheva, D., Mizoguchi, R., Greer, J. (eds.): Semantic Web Technologies for e-Learning, The Future of Learning, vol. 4. IOS Press, Amsterdam (2009)
7. Koedinger, K., Corbett, A.: Cognitive tutors: technology bringing learning science to the classroom. In: Sawyer, K. (ed.) The Cambridge Handbook of the Learning Sciences, pp. 61–78. University Press, Cambridge (2006)
8. Mizoguchi, R., Bourdeau, J.: Using ontological engineering to overcome common AI-ED problems. Int. J. Artif. Intell. Educ. **11**(2), 107–121 (2000)
9. Mizoguchi, R., Hayasi, Y., Bourdeau, J.: Inside a theory-aware authoring system. In: Dicheva, D., Mizoguchi, R., Greer, J. (eds.) Semantic Web Technologies for e-Learning: The Future of Learning, vol. 4, pp. 59–76. IOS Press, Amsterdam (2009)
10. Murray, T.: Principles for pedagogy-oriented knowledge based tutor authoring systems. In: Murray, T., Ainsworth, S., Blessing, S. (eds.) Authoring Tools for Advanced Technology Learning Environments, pp. 439–466. Kluwer Academic Publishers, Netherlands (2003)

11. Sklavakis, D.: Implementing problem solving methods in CYC. MSc dissertation, Department of Artificial Intelligence, University of Edinburgh (1998)
12. Sklavakis, D., Refanidis, I.: An individualized web-based algebra tutor based on dynamic deep model tracing. In: Darzentas, J., Vouros, G.A., Vosinakis, S., Arnellos, A. (eds.) SETN 2008. LNCS (LNAI), vol. 5138, pp. 389–394. Springer, Heidelberg (2008). https://doi.org/10.1007/978-3-540-87881-0_38
13. Sklavakis, D., Refanidis, I.: The MATHESIS algebra tutor: web-based expert tutoring via deep model tracing. Interactive Event. Proceedings of the 14th International Conference on Artificial Intelligence in Education (AIED 2009), p. 795. IOS Press, Amsterdam (2009)
14. Sklavakis, D., Refanidis, I.: Ontology-based authoring of intelligent model-tracing math tutors. In: Dicheva, D., Dochev, D. (eds.) AIMSA 2010. LNCS (LNAI), vol. 6304, pp. 201–210. Springer, Heidelberg (2010). https://doi.org/10.1007/978-3-642-15431-7_21
15. Sklavakis, D., Refanidis, I.: MATHESIS: an intelligent web-based algebra tutoring school. Int. J. Artif. Intell. Educ. **22**(2), 191–218 (2013)
16. Sklavakis, D., Refanidis, I.: The MATHESIS meta-knowledge engineering framework: ontology-driven development of intelligent tutoring systems. Appl. Ontol. **9**(3–4), 237–265 (2014)
17. VanLehn, K.: The behavior of tutoring systems. Int. J. Artif. Intell. Educ. **16**(3), 227–265 (2006)
18. VanLehn, K., Lynch, C., Schulze, K., Shapiro, J., Shelby, R.: The andes physics tutoring system: lessons learned. Int. J. Artif. Intell. Educ. **15**(3), 147–204 (2005)

# A Heilbronn Type Inequality for Plane Nonagons

Zhenbing Zeng[1] , Jian Lu[1](✉), Lydia Dehbi[1](✉), Liangyu Chen[2],
and Jianlin Wang[3](✉)

[1] Department of Mathematics, Shanghai University, Shanghai 200444, China
{zbzeng,lydia_dehbi}@shu.edu.cn, {zbzeng,lujian}@picb.ac.cn
[2] East China Normal University, Shanghai 200062, China
lychen@sei.ecnu.edu.cn
[3] Henan University, Henan 475001, China
jlwang@henu.edu.cn

**Abstract.** In this paper, we present a proof of the property that for any convex nonagon $P_1 P_2 \ldots P_9$ in the plane, the smallest area of a triangle $P_i P_j P_k (1 \le i < j < k \le 9)$ is at most a fraction of $4 \cdot \sin^2(\pi/9)/9 = 0.05199\ldots$ of the area of the nonagon. The problems is transformed into an optimization problem with bilinear constraints and solved by symbolic computation with Maple.

**Keywords:** Heilbronn problem · Convex nonagon · Computer algebra · Lagrange multipliers

## 1 Introduction

The general form of the Heilbronn triangle problem of $n$ points in a given compact set $K$ in the plane with the unit area can be written as the following max-min problem

$$H_n(K) := \max\{\min\{\text{Area}(P_i P_j P_k), 1 \le i < j < k \le n\} | P_1, P_2, \cdots, P_n \in K\}. \tag{1}$$

The original question posed by Heilbronn was to find a function $f(n, K)$, and constants $c_1$ and $c_2$, so that

$$c_1 f(n, K) < H_n(K) < c_2 f(n, K). \tag{2}$$

Notice that for any convex compact set $K$ in the plane, there exist a unique ellipse $E_i$ of the largest area inscribed to $K$ and a unique ellipse $E_c$ of the smallest area circumscribed to $K$, as proven in [1], and that $E_i, E_c$ satisfy the inequalities

$$\text{Area}(E_i) \ge \frac{\pi}{3\sqrt{3}} \text{Area}(K), \quad \text{Area}(E_c) \le \frac{4\pi}{3\sqrt{3}} \text{Area}(K), \tag{3}$$

Supported by the National Natural Science Foundation of China (No. 11471109).

J. Gerhard and I. Kotsireas (Eds.): MC 2019, CCIS 1125, pp. 307–323, 2020.
https://doi.org/10.1007/978-3-030-41258-6_23

where in both cases equality holds for triangles. This fact implies that the shape of $K$ may affect the exact values of $H_n(K)$, but only by a constant factor. Thus the function $f(n, K)$ in (2) can be chosen as independent of $K$ through adjusting $c_1, c_2$ when $K$ is limited to convex bodies. For a general $K$, the best lower and upper bounds satisfy the following form

$$\frac{\log n}{n^2} \ll H_n(K) \ll \frac{\exp\left(\sqrt{\log n}\right)}{n^{8/7}}, \tag{4}$$

proven by Komlós, Pintz and Szemerédi in [10,11].

For specific convex sets and numbers, Goldberg [9] had investigated the optimal arrangements of $n$ points in the unit square and disc, for $n$ up to 16. In [6], Comellas and Yebra improved Goldberg's bounds for $n = 7, 8, 9, 10, 12$ points in the square, as shown in Table 1. Better configurations were found for $n = 13, 15$ by Peter Karpov in 2011 and for $n = 14, 16$ by Mark Beyleveld in 2006 (see [8,12]).

**Table 1.** Heilbronn number of the unit square proposed by Goldberg and the new bounds found by Comellas et al.

| $n$ | $H_n$ suggested by Goldberg | $H_n$ suggested by Comellas et al. | $n$ | $H_n$ suggested by Goldberg | $H_n$ suggested by Comellas et al. |
|---|---|---|---|---|---|
| 5 | $1/(3+\sqrt{5}) = 0.1909\cdots$ | $\sqrt{3}/9 = 0.1924\cdots$ | 11 | 0.037037 | |
| 6 | $1/8$ | $1/8$ | 12 | 0.030303 | 0.032599 |
| 7 | 0.079416 | 0.083859 | 13 | | $1/27$ |
| 8 | 0.066987 | 0.072376 | 14 | | 0.0243 |
| 9 | 0.047619 | 0.054876 | 15 | | 0.0211 |
| 10 | 0.042791 | 0.046537 | 16 | | $7/341 = 0.0205$ |

The following results for the unit square were proven in [5,14,15,17] by Yang, Zhang, Zeng, and Chen.

$$H_5 = \frac{\sqrt{3}}{9}, \quad H_6 = \frac{1}{8}, \quad H_7 = (1 - 14x + 12x^2 + 152x^3)_2 = 0.083859\cdots, \tag{5}$$

$$\frac{9\sqrt{65} - 55}{320} = 0.054875999\cdots \leq H_9 < 0.054878314. \tag{6}$$

Here, the notation $(P(x))_k$ indicates the $k$-th smallest positive root of the polynomial $P(x)$. For triangles $\triangle$, the following results were proven in [3,4,16,18] by Yang, Zhang, Zeng, Cantrell, Chen, and Zhou.

$$H_5(\triangle) = 3 - 2\sqrt{2}, \ H_6(\triangle) = \frac{1}{8}, \ H_7(\triangle) = \frac{7}{72}, \ 0.067789 \leq H_8(\triangle) < 0.067816. \tag{7}$$

And for a general convex body $K$ in the plane, the following results were proven in [7,13] by Dress, Yang, and Zeng.

$$H_6(K) \leq \frac{1}{6}, \quad H_7(K) \leq \frac{1}{9}. \tag{8}$$

A very interesting problem that remains unsolved for $n \geq 9$ encountered in computing Heilbronn optimal configurations is as follows:

**Open Problem 1.** *Let $P_1 P_2 \cdots P_n$ be a convex polygon in the plane satisfying $Area(P_i P_j P_k) \geq a$ for $1 \leq i < j < k \leq n$. Determine the minimal area of the polygon.*

Solutions to this problem for $n \leq 8$ can be found in [7,13,19]. The proven results can be expressed as follows.

**Theorem 1.** *Let $4 \leq n \leq 8$, $P_1 P_2 \cdots P_n$ be any convex polygon in the plane, and*

$$a = \min\{Area(P_i P_j P_k) | 1 \leq i < j < k \leq n\}.$$

*Then*

$$Area(P_1 P_2 \cdots P_n) \geq \frac{n}{4 \sin^2(\pi/n)} \cdot a, \tag{9}$$

*and the equality holds if and only if $P_1 P_2 \cdots P_n$ is an affine regular polygon.*

In this paper, we solve the above open problem for $n = 9$. Namely, we will prove the following inequality.

**Theorem 2.** *For any convex nonagon $P_1 P_2 \cdots P_9$ in the plane, the inequality*

$$Area(P_1 P_2 \cdots P_9) \geq \frac{9}{4 \sin^2(\pi/9)} \cdot \min\{Area(P_i P_j P_k) | 1 \leq i < j < k \leq 9\} \tag{10}$$

*is valid, and the equality holds if and only if $P_1 P_2 \cdots P_9$ is an affine regular nonagon.*

The strategy of the proof is as follows: in the first step we prove a property (as stated in the Theorem 2) for the convex polygon $P_1 P_2 \cdots P_n$ of smallest area that satisfies $Area(P_i P_j P_k) \geq a$ for $1 \leq i < j < k \leq n$ for general $n$, in the second step we transform the problem of determining the convex polygon $P_1 P_2 \cdots P_n$ of smallest area to a global optimization problem, and in the third step we solve the optimization problem via elimination using symbolic computation.

The rest of the paper is organized as follows:

- In Sect. 2 we prove some geometric properties of polygons that satisfy the extremal condition in the Theorem 2 and represent the Heilbronn type inequality for nonagons as a global optimization problem using the Lagrange multiplier method.

- In Sect. 3 we use an elimination procedure to solve the derived system of polynomial equations. The symbolic computations are run with MAPLE on a notebook computer with Intel Core i7 CPU and 8 GB RAM. Since the main Maple functions we have used are polynomial manipulations including `factor`, `gcd` and `resultant`, and the symbolic computations follows the text description in Sect. 3, we have not included the Maple code for saving space.

The authors are grateful to Prof. Dr. Ilias Kotsireas and the anonymous referees, who have helped to improve this paper substantially.

## 2    The Convex Polygon with Equal Peripheral Triangles

Let $n > 4$, $P_1 P_2 \cdots P_n$ be any simple convex polygon, and

$$a = \min\{\text{Area}(P_i P_j P_k) | 1 \leq i < j < k \leq n\}.$$

Then, $\text{Area}(P_i P_j P_k) = a$ implies that $P_i, P_j, P_k$ are three consecutive points in the list $P_1, P_2, \cdots, P_n, P_1, P_2$. Otherwise, let $Q$ be the intersect point of $P_{j-1}P_{j+1}$ and $P_k P_j$, as shown in Fig. 1, we would have

$$
\begin{aligned}
\text{Area}(P_i P_j P_k) &> \text{Area}(P_i Q P_k) \\
&\geq \min\{\text{Area}(P_i P_{j-1} P_k), \text{Area}(P_i P_{j+1} P_k)\} \geq a. \quad (11)
\end{aligned}
$$

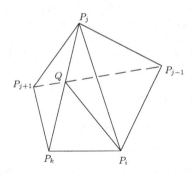

**Fig. 1.** $P_i, P_j, P_k$ are not consecutive implies that $\text{Area}(P_i P_j P_k) > \text{minimal}$.

For convenience, we define the triangle $P_i P_j P_k$ formed by three consecutive vertices of a convex polygon $P_1 P_2 \cdots P_n$ as a *peripheral triangle*. Therefore, we have the following property.

**Proposition 1.** *Let $P_1 P_2 \cdots P_n$ be any convex polygon in the plane, $P_{n+1} = P_1, P_{n+2} = P_2$, and $a$ the minimal area of triangles $P_i P_j P_k$ for $1 \leq i < j < k \leq n$. Then*

$$a = \min\{\text{Area}(P_i P_{i+1} P_{i+2}) | i = 1, 2, \cdots, n\},$$

*and $\text{Area}(P_i P_j P_k) > a$ for any non-peripheral triangle $P_i P_j P_k$.*

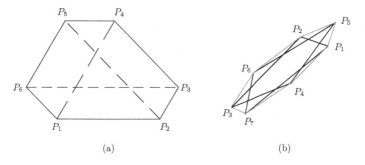

$$P_5 \quad P_4$$
$$P_6 \quad P_3$$
$$P_1 \quad P_2$$
(a)

$$P_2 \quad P_5$$
$$P_1$$
$$P_6$$
$$P_4$$
$$P_3 \quad P_7$$
(b)

**Fig. 2.** (a) An equi-peripheral hexagon $P_1 P_2 \cdots P_6$ that is not affine regular. (b) A self-intersecting heptagon $P_1 P_2 \cdots P_7$ that satisfies Area($P_i P_{i+1} P_{i+2}$) are equal.

*Proof.* For $n \geq 5$ this is guaranteed by the inequality (11), and for $n < 5$, all triangles formed by vertices of a convex polygon are peripheral ones.     □

If all peripheral triangles of a convex polygon are of equal area, we call the polygon as an *equi-peripheral polygon*. It is easy to prove that any equi-peripheral pentagon is an affine regular pentagon. The figures in Fig. 2 show that an equi-peripheral polygon is not necessarily an affine regular polygon for $n \geq 6$, and that a polygon $P_1 P_2 \cdots P_n$ satisfying the condition Area($P_i P_{i+1} P_{i+2}$) = constant is not necessarily a simple polygon. See Fig. 8(b) in the page 161 of [2] for drawing (non-regular) simple nonagon $ABCDEFGHI$ with equal peripheral triangles starting from an equi-peripheral (called *right equiangular*) hexagon $ABDEGH$. Figure 3 shows a way to construct an equi-peripheral polygon through solving a polynomial equation system. Note that for any three points

$$P_i = (x_i, y_i), \quad P_j = (x_j, y_j), \quad P_k = (x_k, y_k)$$

in the plane, the *oriented area* of the triangle $P_i P_j P_k$ can be expressed as a determinant

$$\text{Area}(P_i, P_j, P_k) = \Delta(x_i, y_i, x_j, y_j, x_k, y_k) := \frac{1}{2} \cdot \begin{vmatrix} x_i & y_i & 1 \\ x_j & y_j & 1 \\ x_k & y_k & 1 \end{vmatrix}. \quad (12)$$

Therefore, an equi-peripheral polygon with

$$P_1 = (0,0), P_2 = (1,0), P_i = (x_i, y_i)(i = 2, \cdots, n-1), \text{ and } P_n = (0,1)$$

can be constructed by solving the bi-linear polynomial equations system

$$\begin{cases} f_1 = \Delta(x_1, y_1, x_2, y_2, x_3, y_3) - 1/2 = 0, \\ f_2 = \Delta(x_2, y_2, x_3, y_3, x_4, y_4) - 1/2 = 0, \\ \quad \vdots \\ f_{n-1} = \Delta(x_{n-1}, y_{n-1}, x_n, y_n, x_1, y_1) - 1/2 = 0, \\ (x_1 = 0, y_1 = 0, x_2 = 1, y_2 = 0, x_n = 0, y_n = 1). \end{cases} \quad (13)$$

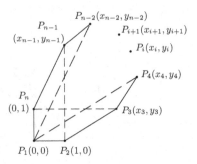

**Fig. 3.** Construction of an equi-peripheral convex $n$-gon.

To prevent self-intersection in the polygon $P_1 P_2 \cdots P_n$, it also requires that $(x_1, y_1, x_2, y_2, \cdots, x_n, y_n)$ are satisfying the *four-point-condition* as follows

$$\text{Area}(P_i, P_j, P_k) > 0, \quad \text{Area}(P_j, P_k, P_l) > 0, \quad \text{Area}(P_k, P_l, P_i) > 0, \qquad (14)$$

for all combinations $(i, j, k, l)$ with $1 \leq i < j < k < l \leq n + 3$ and $P_{n+1} := P_1, P_{n+2} := P_2, P_{n+3} := P_3$.

Moreover, we can prove that for $n \geq 6$, any convex polygon $P_1 P_2 \cdots P_n$ contains an equi-peripheral polygon $P_1' P_2' \cdots P_n'$ in its inside so that

$$\min\{P_i' P_j' P_k' | 1 \leq i < j < k \leq n\} \geq \min\{P_i P_j P_k | 1 \leq i < j < k \leq n\}.$$

Namely, we have

**Proposition 2.** *Let* $n \geq 6$, $P_1 P_2 \cdots P_n$ *be any convex polygon, and* $a$ *be the minimal area of triangles* $P_i P_j P_k$ *for* $1 \leq i < j < k \leq n$. *Then there exists a convex polygon* $P_1' P_2' \cdots P_n'$ *contained in* $P_1 P_2 \cdots P_n$ *such that* $\text{Area}(P_i' P_{i+1}' P_{i+2}') = a$ *for* $i = 1, 2, \cdots, n$ *with* $P_{n+1}' = P_1'$, $P_{n+2}' = P_2'$.

*Proof.* If $P_1 P_2 \cdots P_n$ is an equi-peripheral polygon, just take $P_1' P_2' \cdots P_n' = P_1 P_2 \cdots P_n$. Otherwise, we may assume that

$$\text{Area}(P_1 P_2 P_3) > a := \min\{P_i P_j P_k | 1 \leq i < j < k \leq n\}.$$

Then, taking

$$P_2' = P_2'(\epsilon) = (1 - \epsilon) P_2 + \epsilon P_5 \qquad (15)$$

as shown in Fig. 4. It is clear that

$$\text{Area}(P_2' P_3 P_4) > \min\{\text{Area}(P_2 P_3 P_4), \text{Area}(P_3 P_4 P_5)\} \geq a,$$

$$\text{Area}(P_n P_1 P_2') > \min\{\text{Area}(P_n P_1 P_2), \text{Area}(P_n P_1 P_5)\} \geq a,$$

for any $0 < \epsilon < 1$, and

$$\text{Area}(P_1 P_2' P_3) = (1 - \epsilon) \cdot \text{Area}(P_1 P_2 P_3) - \epsilon \cdot \text{Area}(P_5 P_1 P_3) > a \qquad (16)$$

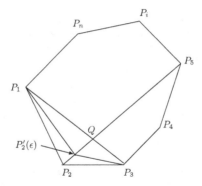

**Fig. 4.** If the triangle $P_1P_2P_3$ is not the smallest triangle, then moving $P_2'$ along to $P_2Q$ to $P_2'(\epsilon)$ so that Area($P_1P_2'P_3$) is still larger than the smallest triangle.

when $\epsilon > 0$ is sufficiently small so that $P_2'(\epsilon) \in P_2Q$, where $Q$ is the intersection of $P_1P_3$ and $P_2P_5$. We make a similar perturbation for all vertices $P_k$ for which Area($P_{k-1}P_kP_{k+1}$) $> a$ to $P_k'$ recursively, until all peripheral triangles of $P_1'P_2'\cdots P_n'$ are of equal area. It is clear that $P_1'P_2'\cdots P_n' \subset P_1P_2\cdots P_n$, and

$$\min\{P_i'P_j'P_k'|1 \le i < j < k \le n\} \ge a.$$

$\square$

Proposition 2 implies that if a convex polygon is a solution to Open Problem 1, then it must be an equi-peripheral polygon. We have the following result.

**Theorem 3.** *Let $k_n$ be the solution to the following problem*

$$\begin{cases} \min \, Area(P_1P_2\cdots P_n), \\ P_1P_2\cdots P_n \text{ is a convex polygon in the plane, and} \\ Area(P_iP_jP_k) \ge 1/2 \text{ for all } 1 \le i < j < k \le n. \end{cases} \quad (17)$$

*Then $k_n$ is monotonically increasing with $n$ with $k_{n+1} \ge k_n + 1/2$, and*

$$k_n \le \frac{n}{8\sin^2(\pi/n)}. \quad (18)$$

*Proof.* Assume $P_1P_2\cdots P_nP_{n+1}$ is an optimal polygon corresponding to $k_{n+1}$. Then Area($P_nP_{n+1}P_1$) $\ge 1/2$, and Area($P_1P_2\cdots P_n$) $\ge k_n$. Thus,

$$\begin{aligned} k_{n+1} &= \text{Area}(P_1P_2\cdots P_nP_{n+1}) \\ &= \text{Area}(P_1P_2\cdots P_n) + \text{Area}(P_nP_{n+1}P_1) \ge k_n + \frac{1}{2}. \end{aligned} \quad (19)$$

To prove the inequality (18), notice that the affine regular $n$-gon $P_1P_2\cdots P_n$ inscribed in the unit circle can be represented by

$$P_k = (\cos(\frac{(k-1)\cdots 2\pi}{n}), \sin(\frac{(k-1)\cdots 2\pi}{n})), \ k = 1, 2, \cdots, n-1,$$

so we have

$$\text{Area}(P_1 P_2 \cdots P_n) = n \cdot \frac{1}{2} \sin(\frac{2\pi}{n}),$$

and

$$\text{Area}(P_1 P_2 P_3) = 2 \cdot \frac{1}{2} \sin(\frac{2\pi}{n}) - \frac{1}{2} \sin(\frac{4\pi}{n}) = \sin(\frac{2\pi}{n}) \cdot (1 - \cos(\frac{2\pi}{n})),$$

therefore,

$$k_n \leq \frac{1}{2} \frac{\text{Area}(P_1 P_2 \cdots P_n)}{\text{Area}(P_1 P_2 P_3)} = \frac{n}{4(1 - \cos(\frac{2\pi}{n}))} = \frac{n}{8 \sin^2(\pi/n)},$$

as claimed. □

Together with Theorem 1, the above inequality immediately implies that

$$k_5 = 5/(8 \sin^2(\pi/5)) = (5 + \sqrt{5})/4 = 1.8090 \cdots, \quad k_6 = 3,$$
$$k_7 = 7/(8 \sin^2(\pi/7)) = 4.6479 \cdots, \quad k_8 = 4 + 2\sqrt{2} = 6.8284 \cdots,$$

Notice that $k_7$ is the largest real root of the equation $8 z^3 - 56 z^2 + 98 z - 49 = 0$.

## 3    The Smallest Equi-Peripheral Nonagon

This section is devoted to proving Theorem 2. In view of Proposition 2 and Theorem 3, we need only to prove that the smallest equi-peripheral nonagon is an affine image of the regular nonagon. We will prove this by solving the optimization problem

$$\begin{aligned}
\min S &= \text{Area}(P_1 P_2 \cdots P_9) \\
&= \tfrac{1}{2} (y_3 + x_3 y_4 - x_4 y_3 + x_4 y_5 - x_5 y_4 + x_5 y_6 - x_6 y_5 \\
&\quad + x_6 y_7 - x_7 y_6 + x_7 y_8 - x_8 y_7 + x_8),
\end{aligned} \tag{20}$$

subject to the constraint conditions (13) and (14) for $n = 9$. For convenience we only write the equality part of the condition as follows.

$$\begin{cases}
f_1 = -1 + y_3 = 0, \\
f_2 = x_3 y_4 - x_4 y_3 + y_3 - y_4 - 1 = 0, \\
f_3 = x_3 y_4 - x_3 y_5 - x_4 y_3 + x_4 y_5 + x_5 y_3 - x_5 y_4 - 1 = 0, \\
f_4 = x_4 y_5 - x_4 y_6 - x_5 y_4 + x_5 y_6 + x_6 y_4 - x_6 y_5 - 1 = 0, \\
f_5 = x_5 y_6 - x_5 y_7 - x_6 y_5 + x_6 y_7 + x_7 y_5 - x_7 y_6 - 1 = 0, \\
f_6 = x_6 y_7 - x_6 y_8 - x_7 y_6 + x_7 y_8 + x_8 y_6 - x_8 y_7 - 1 = 0, \\
f_7 = x_7 y_8 - x_8 y_7 - x_7 + x_8 - 1 = 0, \\
f_8 = -1 + x_8 = 0.
\end{cases} \tag{21}$$

Regarding $\{f_1, f_2, f_3, f_6, f_7, f_8\}$ as linear equations of $\{y_3, x_4, y_4, x_7, y_7, y_8\}$, we have

$$
\begin{cases}
x_4 = \dfrac{(x_3 - 1)(x_3\, y_5 - x_5 + 1)}{x_3\, y_5 - x_5 - y_5 + 1}, & y_4 = \dfrac{x_3\, y_5 - x_5 + 1}{x_3\, y_5 - x_5 - y_5 + 1}, \\[2ex]
x_7 = \dfrac{x_6\, y_8 - y_6 + 1}{x_6\, y_8 - x_6 - y_6 + 1}, & y_7 = \dfrac{(y_8 - 1)(x_6\, y_8 - y_6 + 1)}{x_6\, y_8 - x_6 - y_6 + 1}, \\[2ex]
x_8 = 1, & y_3 = 1.
\end{cases}
\tag{22}
$$

Substituting this solution into (20) and the constraint polynomial equations $f_4 = 0, f_5 = 0$, we have transformed the optimization problem related to the smallest equi-peripheral nonagon as follows:

$$
\begin{aligned}
&\min S \\
&\text{s.t. } g_0 = x_3\, y_5 + x_5\, y_6 - x_6\, y_5 + x_6\, y_8 - 2\,S - x_5 - y_6 + 4 = 0, \\
&\qquad g_1(x_3, x_5, y_5, x_6, y_6, y_8) = 0, \\
&\qquad g_2(x_3, x_5, y_5, x_6, y_6, y_8) = 0, \\
&\qquad \text{Ineq}(i, j, k, l) \text{ for all } 1 \le i < j < k < l \le n + 3.
\end{aligned}
\tag{23}
$$

where

$$
\begin{aligned}
g_1 = &-1 + y_6 + x_6 - y_5\, x_6 - x_5\, x_6 + x_5\, y_5 + x_5{}^2 - x_3\, y_6 \\
&+ y_5{}^2 x_6 - x_5\, y_5\, y_6 + x_5\, y_5\, x_6 - x_5{}^2 y_6 + x_3\, y_5\, y_6 + x_3\, y_5\, x_6 - x_3\, y_5{}^2 \\
&+ x_3\, x_5\, y_6 - 2\, x_3\, x_5\, y_5 - x_3\, y_5{}^2 x_6 + x_3\, x_5\, y_5\, y_6 - x_3{}^2 y_5\, y_6 + x_3{}^2 y_5{}^2,
\end{aligned}
\tag{24}
$$

$$
\begin{aligned}
g_2 = &-1 + y_5 + x_5 + y_6{}^2 + x_6\, y_6 - y_5\, y_6 - y_5\, x_6 - x_5\, y_8 \\
&- 2\, x_6\, y_6\, y_8 - x_6{}^2 y_8 + y_5\, x_6\, y_8 + y_5\, x_6\, y_6 + y_5\, x_6{}^2 + x_5\, y_6\, y_8 - x_5\, y_6{}^2 \\
&+ x_5\, x_6\, y_8 - x_5\, x_6\, y_6 + x_6{}^2 y_8{}^2 - y_5\, x_6{}^2 y_8 - x_5\, x_6\, y_8{}^2 + x_5\, x_6\, y_6\, y_8,
\end{aligned}
\tag{25}
$$

and Ineq$(i, j, k, l)$ refers to the inequalities defined in (14). Observing that $\{g_0, g_1\}$ are linear equations with respect to $\{x_6, y_6\}$, we obtain

$$
x_6 = \frac{P_1(S, x_3, x_5, y_5, y_8)}{Q(x_3, x_5, y_5, y_8)}, \qquad y_6 = \frac{P_2(S, x_3, x_5, y_5, y_8)}{Q(x_3, x_5, y_5, y_8)}
\tag{26}
$$

where

$$
\begin{aligned}
P_1 &= 2\,S - 3 - 2\,S x_3 + 4\,x_3 - 2\,S x_5\,y_5 + 3\,x_5\,y_5 - 2\,S x_5{}^2 + 3\,x_5{}^2 \\
&\quad + 2\,S x_3\,y_5 - 5\,x_3\,y_5 + 2\,S x_3\,x_5 - 5\,x_3\,x_5 + x_3\,y_5{}^2 + 2\,S x_3\,x_5\,y_5 \\
&\quad - x_3\,x_5\,y_5 + x_3\,x_5{}^2 - 2\,S x_3{}^2 y_5 + 5\,x_3{}^2 y_5 - 2\,x_3{}^2 y_5{}^2 - 2\,x_3{}^2 x_5\,y_5 + x_3{}^3 y_5{}^2, \\
P_2 &= -2\,S + 4 + y_8 + 2\,S y_5 - 5\,y_5 + 2\,S x_5 - 5\,x_5 - 2\,S y_5{}^2 + 4\,y_5{}^2 - 2\,S x_5\,y_5 \\
&\quad + 5\,x_5\,y_5 + x_5{}^2 - 2\,S x_3\,y_5 + 5\,x_3\,y_5 - x_5\,y_5\,y_8 - x_5{}^2 y_8 + 2\,S x_3\,y_5{}^2 \\
&\quad - 5\,x_3\,y_5{}^2 - 2\,x_3\,x_5\,y_5 + x_3\,y_5{}^2 y_8 + 2\,x_3\,x_5\,y_5\,y_8 + x_3{}^2 y_5{}^2 - x_3{}^2 y_5{}^2 y_8 \\
Q &= 1 + y_8 - 2\,y_5 - 2\,x_5 + y_5{}^2 + 2\,x_5\,y_5 + x_5{}^2 - x_3\,y_8 + 2\,x_3\,y_5 \\
&\quad - x_5\,y_5\,y_8 - x_5{}^2 y_8 + x_3\,y_5\,y_8 - 2\,x_3\,y_5{}^2 + x_3\,x_5\,y_8 - 2\,x_3\,x_5\,y_5 \\
&\quad + x_3\,x_5\,y_5\,y_8 - x_3{}^2 y_5\,y_8 + x_3{}^2 y_5{}^2.
\end{aligned}
$$

Therefore, the optimization problem (23) has been transformed to the following simple form:

$$
\begin{aligned}
&\min S, \\
&\text{s.t. } g(S, x_3, x_5, y_5, y_8) = 0, \\
&\qquad \text{Ineq}(i, j, k, l) \text{ fot all } 1 \le i < j < k < l \le n + 3,
\end{aligned}
\tag{27}
$$

where

$$
g = 3 - 2\,S + 15\,y_8 - 18\,y_5 + \text{ (135 terms) } - x_3{}^3 x_5\,y_5{}^2 y_8{}^2 - x_3{}^4 y_5{}^3 y_8
$$

is the polynomial obtained by substituting (26) into $g_2$ defined in (25), and Ineq$(i, j, k, l)$ are the inequalities determined by (14).

It is clear that the inequality constraint conditions in the problem (27) define an open subset in $\mathbb{R}^5$. Thus, a necessary condition for a nonagon $P_1 P_2 \cdots P_9$ with

$$
\begin{aligned}
&P_1 = (0, 0), \quad P_2 = (1, 0), \quad P_9 = (0, 1), \\
&P_3 = (x_3, 1), \quad P_i = (x_i, y_i)(4 \le i \le 7), \quad P_8 = (1, y_8),
\end{aligned}
\tag{28}
$$

to be an equi-peripheral nonagon with the smallest area is that $(S, x_3, x_5, y_5, y_8) \in \mathbb{R}^5$ satisfies the following system of equations:

$$
g = 0, \quad \frac{\partial g}{\partial x_3} = 0, \quad \frac{\partial g}{\partial x_5} = 0, \quad \frac{\partial g}{\partial y_5} = 0, \quad \frac{\partial g}{\partial y_8} = 0.
\tag{29}
$$

and $x_4, y_4, x_6, y_6, x_7, y_7$ satisfy (22) and (26). We list the information on the equation system in Table 2.

As we can see from Table 2, the equation $\frac{\partial g}{\partial y_8} = 0$ is a linear equation with respect to variable $y_8$, thus we have

$$
y_8 = \frac{A(x_3, x_5, y_5)\,S^2 + B(x_3, x_5, y_5)\,S + C(x_3, x_5, y_5)}{D(x_3, x_5, y_5)\,S + E(x_3, x_5, y_5)},
\tag{30}
$$

**Table 2.** The degree and number of terms of equation system (29)

| Polynomial equation | Degree w.r.t. variable | | | | | nops (number of terms) |
|---|---|---|---|---|---|---|
| | $S$ | $x_3$ | $x_5$ | $y_5$ | $y_8$ | |
| $g$ | 2 | 4 | 3 | 3 | 2 | 141 |
| $\partial g/\partial x_3$ | 2 | 3 | 3 | 3 | 2 | 91 |
| $\partial g/\partial x_5$ | 2 | 3 | 2 | 2 | 2 | 87 |
| $\partial g/\partial y_5$ | 2 | 4 | 2 | 2 | 2 | 94 |
| $\partial g/\partial y_8$ | 2 | 4 | 3 | 3 | 1 | 76 |

where $A, B, C, D, E \in \mathbb{Z}[x_3, x_5, y_5]$. Therefore, the system (29) can be transformed into a new system in variables $S, x_3, x_5$, and $y_5$ by substituting (30) into (29) as follows:

$$
\begin{aligned}
g \to h_1 &= (-3 + 2S - x_3 y_5) \cdot h_{1,20}^{(1,3,2,2)}(S, x_3, x_5, y_5) \cdot h_{1,240}^{(3,7,4,5)}(S, x_3, x_5, y_5), \\
\tfrac{\partial g}{\partial x_3} \to h_2 &= h_{2,923}^{(5,10,6,8)}(S, x_3, x_5, y_5), \\
\tfrac{\partial g}{\partial x_5} \to h_3 &= (-3 + 2S - x_3 y_5) \cdot h_{3,688}^{(4,10,6,7)}(S, x_3, x_5, y_5), \\
\tfrac{\partial g}{\partial y_5} \to h_4 &= h_{4,882}^{(5,11,6,7)}(S, x_3, x_5, y_5),
\end{aligned}
$$

$$(31)$$

here $h_{i,n}^{(d_1,d_2,d_3,d_4)}(S, x_3, x_5, y_5)$ represents a factor of $h_i$ with $n$ monomials, where the highest degrees of $S, x_3, x_5$, and $y_5$ are $d_1, d_2, d_3$, and $d_4$, respectively. For example,

$$
\begin{aligned}
d_{1,20}^{(1,3,2,2)}(S, x_3, x_5, y_5) &= 3 - 3\,x_3 - 2\,S - 3\,x_5\,y_5 - 3\,x_3^2 + 4\,x_3\,y_5 + 4\,x_3\,x_5 + 2\,Sx_3 \\
&+ 2\,x_3\,x_5\,y_5 - x_3\,x_5^2 - 4\,x_3^2 y_5 + 2\,Sx_5\,y_5 + 2\,S\,x_5^2 - 2\,Sx_3\,y_5 - 2\,Sx_3\,x_5 \\
&+ x_3^2\,y_5^2 + 2\,x_3^2 x_5\,y_5 - 2\,Sx_3\,x_5\,y_5 + 2\,S\,x_3^2 y_5 - x_3^3 y_5^2.
\end{aligned}
$$

For convenience, we will use the notation $h_{1,20}^{(1,3,2,2)}$ (or $h_{1,20}$) to represent the polynomial $h_{1,20}^{(1,3,2,2)}(S, x_3, x_5, y_5, y_8)$ for short.

The following property of a general equi-peripheral nonagon $P_1 P_2 \cdots P_9$ is very useful for simplifying the equation system (31).

**Proposition 3.** *For any equi-peripheral nonagon determined by the coordinates setting in (28), we have*

$$x_3, y_8 > 1, \quad x_i, y_i > 1 (i = 4, 5, 6, 7), \tag{32}$$

$$Area(P_1 P_2 \cdots P_9) \geq \frac{9}{2} + 2\sqrt{2} = 7.3284 \cdots, \tag{33}$$

*and*

$$Area(P_1 P_2 \cdots P_9) > \frac{1}{2}(x_3 y_5 + 5). \tag{34}$$

*Proof.* The inequalities $x_i > 1$ for $3 \leq i < 8$ and $y_j > 1$ for $3 < j \leq 8$ are derived from the convexity of $P_1 P_2 \cdots P_9$.

For (33), we have seen that there is an equi-peripheral octagon $P_1' P_2' \cdots P_8'$ in $P_1 P_2 \cdots P_8$, therefore,

$$
\begin{aligned}
\text{Area}(P_1 P_2 \cdots P_8 P_9) &= \text{Area}(P_1 P_2 \cdots P_8) + \text{Area}(P_8 P_9 P_1) \\
&\geq \text{Area}(P_1' P_2' \cdots P_8') + \frac{1}{2} \\
&\geq \frac{8}{4 \sin^2(\pi/8)} \cdot \frac{1}{2} + \frac{1}{2} = \frac{9}{2} + 2\sqrt{2}.
\end{aligned}
$$

For (34), we have

$$
\begin{aligned}
\frac{1}{2}(x_3 y_5 + 5) &= \frac{1}{2} \begin{vmatrix} x_3 & 1 & 1 \\ x_5 & y_5 & 1 \\ 0 & 0 & 1 \end{vmatrix} + \frac{1}{2} \cdot x_5 + \frac{3}{2} + 1 \\
&= \text{Area}(P_1 P_3 P_5) + \text{Area}(P_1 P_5 P_9) \\
&\quad + \text{Area}(P_1 P_2 P_3) + \text{Area}(P_3 P_4 P_5) + \text{Area}(P_5 P_6 P_7) + 1 \\
&= \text{Area}(P_1 P_2 \cdots P_9) - \text{Area}(P_5 P_7 P_8 P_9) + 1 < \text{Area}(P_1 P_2 \cdots P_9).
\end{aligned}
$$

$\square$

In view of Proposition 3, the equation system (29) can be split into the following two systems:

$$
\text{(Eqs I):} \begin{cases} h_{1,20}^{(1,3,2,2)} = 0, \\ h_{2,923}^{(5,10,6,8)} = 0, \\ h_{3,688}^{(4,10,6,7)} = 0, \\ h_{4,882}^{(5,11,6,7)} = 0. \end{cases} \quad \text{(Eqs II):} \begin{cases} h_{1,240}^{(3,7,4,5)} = 0, \\ h_{2,923}^{(5,10,6,8)} = 0, \\ h_{3,688}^{(4,10,6,7)} = 0, \\ h_{4,882}^{(5,11,6,7)} = 0. \end{cases}
$$

At first, we prove that the smallest equi-peripheral nonagon $P_1 P_2 \cdots P_9$ cannot be a solution to the system (Eqs I). Namely, we have

**Proposition 4.** *If $x_i (3 \leq i < 8)$, $y_j (3 < j \leq 8)$ are the coordinates of an equi-peripheral nonagon $P_1 P_2 \cdots P_9$ as defined in (28), and $S$ is the area of the nonagon, then they do not satisfy the equation system (Eqs I).*

*Proof.* Let $f, g \in \mathbb{Z}[u_1, \cdots, u_n, x]$, $h \in \mathbb{Z}[u_1, \cdots, u_n]$, and

$$
\begin{aligned}
\text{resultant}(f, g, x) &= c_1 \cdot r_1^{d_1} \cdots r_k^{d_k} \\
(c, d_1, \cdots, d_k &\in \mathbb{Z}_+, r_1, \cdots, r_k \in \mathbb{Z}[u_1, \cdots, u_n])
\end{aligned}
$$

be the irreducible factorization of the Sylvester resultant of $f, g$ that satisfies $\gcd(r_i, r_j) = 1$ for all $1 \leq i < j \leq k$. Define

$$
\text{resultant1}(f, g, x, h) = \prod_{\substack{1 \leq i \leq k \\ \gcd(r_i, h) = 1}} r_i,
$$

that is, resultant1$(f, g, x, h)$ is the square-free product of all factors of the Sylvester resultant which has no common divisor with $h$. Let

$$h_0 = y_5(x_3 - 1)(x_3 y_5 + 1)(2S - 3)(2S - x_3 y_5 - 3)(2S - x_3 - 3).$$

In view of (3), it is obvious that $h_0 > 0$ for all equi-peripheral nonagons defined by (28). Then we have

$$\texttt{resultant1}(h_{1,240}, h_{2,923}, x_5, h_0) = h_{120,72}^{(4,4,3)}(S, x_3, y_5),$$

$$\texttt{resultant1}(h_{1,240}, h_{3,688}, x_5, h_0) = h_{130,24}^{(2,2,2)}(S, x_3, x_5),$$

$$\texttt{resultant1}(h_{1,240}, h_{4,882}, x_5, h_0) = h_{140,33}^{(3,4,2)}(S, x_3, x_5),$$

where $h_{120,72}^{(4,4,3)}(S, x_3, y_5)$, $h_{130,24}^{(2,2,2)}(S, x_3, y_5)$, $h_{140,33}^{(3,4,2)}(S, x_3, y_5)$ are polynomials in $\mathbb{Z}[S, x_3, y_5]$, with $n = \texttt{nops}(h_{i,n}^{(d_1,d_2,d_3)})$, $d_1 = \texttt{degree}(h_{i,n}^{(d_1,d_2,d_3)}, S)$, and so on. Hence, any real solutions of the system (Eqs I) that forms a convex nonagon must satisfy

$$h_{120,72} = 0, \quad h_{130,24} = 0, \quad h_{140,33} = 0. \tag{35}$$

To complete the proof, observe that

$$\texttt{resultant}(h_{130,24}, h_{140,33}, x_5) = (2S - 3)^2(2Sx_3 - 2S - 2x_3 + 3)^8. \tag{36}$$

Since

$$2S - 3 > 0, \quad 2Sx_3 - 2S - 2x_3 + 3 = 2(S - 1)(x_3 - 1) + 1 > 0$$

hold for all convex nonagons, we conclude that $h_{130,24}$ and $h_{140,33}$ have no common zero, and therefore, the coordinates $(x_i, y_i)$ and the area $S$ of any equi-peripheral nonagon do not satisfy (Eqs I) as stated in the proposition.  □

To solve the equation system (Eqs II), we need the following lemma.

**Lemma 1.** *Let*

$$\begin{aligned}
q_0 &:= (2S - x_3 - 4) \cdot (2Sx_3 - 2S - 4x_3 + 3) \\
&\quad \cdot (2Sx_3 - x_3^2 - 4x_3 - 1) \cdot (2Sx_3 - 2S - x_3^2 - 2x_3 + 2) \\
&\quad \cdot (4S^2 x_3 - 2Sx_3^2 - 4S^2 - 10Sx_3 + 2x_3^2 + 12S + 6x_3 - 9), \tag{37} \\
q_2 &:= x_3 (x_3 - 1) x_5 (x_5 - 1)(2S - 3) \\
&\quad \cdot \big(4S^2 x_3 - 2S x_3^2 - 4S^2 - 8Sx_3 + 2x_3^2 + 10S + 4x_3 - 6 \\
&\quad\quad - 2Sx_3 x_5 + x_3^2 x_5 + 2x_3 x_5\big) \\
&\quad \cdot \big(4S^2 x_3 - 2S x_3^2 - 4S^2 - 10Sx_3 + 2x_3^2 + 12S + 6x_3 - 9 \\
&\quad\quad - 2Sx_3 x_5 + x_3^2 x_5 + 3x_3 x_5\big). \tag{38}
\end{aligned}$$

*Then*

$$(q_0 = 0 \vee q_2 = 0) \wedge (\text{Eqs II}) \Rightarrow S < 9/2 + 2\sqrt{2}.$$

*Proof.* The proof is essentially to add a relatively simple polynomial, for example, $2Sx_3 - 2S - 4x_3 + 3$, to the system (Eqs II), and eliminate the variables $x_3, x_5, y_5$ from the new system, using resultant computation. The computation is direct, for saving space we omit the details here.     □

**Proposition 5.** *Let* $P_1 P_2 \cdots P_9$ *the equi-peripheral nonagon of minimal area with the coordinates defined in (28), and* $S$ *the minimal area of the nonagon. Then*

$$8S^3 - 108S^2 + 324S - 243 = 0, \tag{39}$$

*Proof.* Observing that

$$q_1 := (-x_3 y_5 + 2S - 3)(2S - 3)(x_3 - 1)(2S - 2 - x_3)$$
$$\cdot (x_3 y_5 + 1)(-x_3 y_5 + 2S + y_5 - 5) y_5 > 0$$

for any convex equi-peripheral nonagon determined by (28), we have

$$\left.\begin{array}{l}
\texttt{resultant1}(h_{1,240}, h_{2,923}, x_5, q_1) = h_{120,1761}^{(15,14,11)}(S, x_3, y_5), \\[4pt]
\texttt{resultant1}(h_{1,240}, h_{3,688}, x_5, q_1) = h_{130,832}^{(10,10,8)}(S, x_3, y_5) \cdot h_{131,8}^{(2,2,1)}(S, x_3, y_5), \\[4pt]
\texttt{resultant1}(h_{1,240}, h_{4,882}, x_5, q_1) = h_{140,1257}^{(14,14,9)}(S, x_3, y_5),
\end{array}\right\} \tag{40}$$

where $h_{120,1761}^{(15,14,11)} \in \mathbb{Z}[S, x_3, y_5]$ is irreducible, with $\texttt{nops}(h_{120,1761}^{(15,14,11)}) = 1761$, and the degrees of $S, x_3, y_5$ are $15, 14, 11$, respectively, and so on. Using resultant to eliminate $y_5$ from the equation system (40), we obtain

$$\texttt{resultant1}(h_{120,1761}, h_{130,832} \cdot h_{131,8}, y_5, q_0)$$
$$= h_{12130,9}^{(3,2)} \cdot h_{12131,9498}^{(104,110)} \cdot h_{12132,147}^{(13,11)} \cdot h_{12133,432}^{(23,25)}, \tag{41}$$

$$\texttt{resultant1}(h_{140,1257}, h_{130,832} \cdot h_{131,8}, y_5, q_0)$$
$$= h_{13140,5}^{(1,2)} \cdot h_{13141,7571}^{(92,96)} \cdot h_{13142,306}^{(20,21)} \cdot h_{13143,58}^{(7,7)} \cdot h_{13144,3}^{(1,1)}, \tag{42}$$

where $q_0 \in \mathbb{Z}[S, x_3]$ is the polynomial defined in (37), and $h_{1213i,n_i}, h_{1314j,n_j} \in \mathbb{Z}[S, x_3]$ are irreducible polynomials. As we have proven that the equi-peripheral nonagon satisfies $q_0 \neq 0$, we assert that a necessary condition for minimal equi-peripheral nonagon can be given as follows:

$$\texttt{resultant1}\left( \prod_{0 \le i \le 3} h_{1213i,n_i}, \prod_{0 \le j \le 4} h_{1314j,n_j}, x_3, 1 \right) = r_1(S) r_2(S) \cdots r_{53}(S), \tag{43}$$

where $r_1, r_2, \cdots, r_{53}$ are irreducible polynomials in $\mathbb{Z}[S]$, and their degrees are given by the following list:

```
[1, 1, 1, 1, 1, 1, 1, 2, 2, 2, 2, 2, 2, 3, 3, 3, 3, 3, 3, 3, 3, 4,
 4, 4, 5, 5, 5, 6, 6, 6, 7, 10, 14, 17, 22, 26, 28, 42, 46, 75, 90,
 92, 120, 171, 209, 209, 270, 535, 640, 1220, 2045, 2110, 9210].
```

This elimination process can be done in different orders. For example, we can also eliminate $y_5$ at first, and then eliminate $x_5$, and $x_3$ at the final step as follows.

$$\left. \begin{array}{l} \texttt{resultant1}(h_{1,240}, h_{2,923}, y_5, q_2) = h_{12y,2651}^{(14,18,11)}(S, x_3, x_5), \\[6pt] \texttt{resultant1}(h_{1,240}, h_{3,688}, y_5, q_2) = h_{13y,1444}^{(13,16,8)}(S, x_3, x_5), \\[6pt] \texttt{resultant1}(h_{1,240}, h_{4,882}, y_5, q_2) = h_{14y,1273}^{(11,14,9)}(S, x_3, x_5), \end{array} \right\} \quad (44)$$

where $q_2$ is the polynomial defined by (38), and $h_{12y}, h_{13y}, h_{14y}$ are irreducible polynomials with $2651, 1444, 1273$ terms, respectively. Then eliminate $x_5$ as follows:
we have

$$\texttt{resultant1}(h_{12y,2651}, h_{13y,1444}, x_5, q_0)$$
$$= h_{1213y0,147}^{(13,11)} \cdot h_{1213y1,29107}^{(173,198)}, \quad (45)$$
$$\texttt{resultant1}(h_{13y,1444}, h_{14y,1273}, x_5, q_0)$$
$$= h_{1314y0,5}^{(1,2)} \cdot h_{1314y1,12180}^{(116,123)} \cdot h_{1314y2,3}^{(1,1)} \cdot h_{1314y3,58}^{(7,7)}, \quad (46)$$

where $q_0 \in \mathbb{Z}[S, x_3]$ is defined by (37), and $h_{1213y0}, h_{1213y1}, h_{1314y0}, \cdots, h_{1314y3}$ are irreducible polynomials in $\mathbb{Z}[S, x_3]$. Finally, eliminate $x_3$ via computing the resultant as follows:

$$\texttt{resultant1}\left(\prod_{i=0}^{1} h_{1213yi,n_i}, \prod_{j=0}^{3} h_{1314yj,n_j}, x_3, 1\right) = r_{y1}(S) r_{y2}(S) \cdots r_{y17}(S), \quad (47)$$

where $r_{y1}, r_{y2}, \cdots, r_{y16}$ are irreducible polynomials in $\mathbb{Z}[S]$, and $r_{y17} \in \mathbb{Z}[S]$ is not irreducible (MAPLE 18 was not able to complete the factorization of $r_{y17}$ in $332,498.40$ seconds in the machine mentioned in the end of Sect. 1). The degree information of $r_{yj}(j = 1, 2, \cdots, 17)$ is recorded in the following list:
[1, 1, 1, 1, 1, 2, 3, 10, 14, 90, 114, 156, 160, 160, 1234, 1595, 30402].

Consequently, we obtain a simple necessary condition for $S$ to be the minimal area of the optimal equi-peripheral nonagon as follows

$$\gcd\left(\prod_{i=1}^{53} r_i, \prod_{j=1}^{17} r_{yj}\right) = (S-1)(2S-3)(2S-5)(S-3)(2S-7)(8S^2 - 12S + 13)$$

$$\cdot (8S^3 - 108S^2 + 324S - 243) \cdot r_{8,11}^{(10)}(S) \cdot r_{9,15}^{(14)}(S). \quad (48)$$

Using the MAPLE built-in command $\texttt{sturm}(f, x, a, b)$ we can check that among all factors, only the cubic one has a real root in the interval $[6, 10]$. Therefore, the optimal equi-peripheral nonagon satisfies the cubic equation as claimed in the proposition. $\qquad \square$

*Remark 1.* The time for computing the resultant (43) was $46,433.95$ seconds, and for computing (47) was $127,865.93$ seconds on the machine we used.

Finally, we can give a proof to Theorem 2 as follows. Adding the equation $8S^3 - 108S^2 + 324S - 243 = 0$ to the equation system (Eqs II), we get the following equations: the univariate equation $x^3 - 3x^2 + 3 = 0$ (which has two positive roots: $1 + 2\cos(4\pi/9) = 1.3472\cdots$, $1 + 2\cos(2\pi/9) = 2.5320\cdots$) for $x_3, y_4, x_7, y_8$; the univariate equation $y^3 - 6y^2 + 9y - 3 = 0$ (which has two positive roots greater than 1: $2 + 2\cos(5\pi/9) = 1.6527\cdots$, $2 + 2\cos(\pi/9) = 3.8793\cdots$) for $x_4, y_5, x_6, y_7$; and the univariate equation $z^3 - 3z^2 - 6z - 1 = 0$ (which has one positive root $1 + 2\sqrt{3}\cos(\pi/18) = 4.4114\cdots$) for $x_5, y_6$. After checking the $2^4 \times 2^4 \times 1^2 = 256$ possible combinations we can see that there is only one combination

$$P_1 = (0,0), \quad P_2 = (1,0), \quad P_3 = (a,1), \quad P_4 = (b,a), \quad P_5 = (c,b),$$
$$P_6 = (b,c), \quad P_7 = (a,b), \quad P_8 = (1,a), \quad P_9 = (0,1),$$

where

$$a = 1 + 2\cos(2\pi/9), \quad b = 2 + 2\cos(\pi/9), \quad c = 1 + 2\sqrt{3}\cos(\pi/18)$$

that satisfies the equation system (21) and all *four-point-conditions* (14) for a simple polygon. It is also easy to check that the minimal area of the nonagon is the largest root ($S = 9.6172\cdots$) of the Eq. (39), which can also be represented as $9/(8\sin^2(\pi/9))$. This proves Theorem 2.

*Remark 2.* We have used Gröbner bases computation for solving the equation systems (Eqs I) and (Eqs II), neither were succeeded for insufficient memory space.

# References

1. Behrend, F.: Über die kleinste umbeschriebene und die größte einbeschriebene Ellipse eines konvexen. Bereichs. Math. Ann. **115**(1), 379–411 (1938)
2. Buitrago, A., Huylebrouck, D.: Nonagons in the Hagia Sophia and the Selimiye Mosque. Nexus Netw. J. **17**(1), 157–181 (2015)
3. Cantrell, D.: The Heilbronn problem for triangles. http://www2.stetson.edu/~efriedma/heiltri/. Accessed 7 Sept 2019
4. Chen, L., Zeng, Z., Zhou, W.: An upper bound of Heilbronn number for eight points in triangles. J. Comb. Optim. **28**(4), 854–874 (2014)
5. Chen, L., Xu, Y., Zeng, Z.: Searching approximate global optimal Heilbronn configurations of nine points in the unit square via GPGPU computing. J. Global Optim. **68**(1), 147–167 (2017)
6. Comellas, F., Yebra, J.L.A.: New lower bounds for Heilbronn numbers. Electron. J. Comb. **9**, #R6 (2002)
7. Dress, A., Yang, L., Zeng Z.: Heilbronn problem for six points in a planar convex body. In: Combinatorics and Graph Theory 1995, vol. 1 (Hefei), pp. 97–118. World Scientific Publishing, River Edge (1995)
8. Friedman, E.: The Heilbronn Problem. http://www.stetson.edu/efriedma/heilbronn. Accessed 14 Mar 2019
9. Goldberg, M.: Maximizing the smallest triangle made by $n$ points in a square. Math. Mag. **45**, 135–144 (1972)

10. Komlós, J., Pintz, J., Szemerédi, E.: On Heilbronn's triangle problem. J. London Math. Soc. **24**(3), 385–396 (1981)
11. Komlós, J., Pintz, J., Szemerédi, E.: A lower bound for Heilbronn's problem. J. London Math. Soc. **25**(1), 13–24 (1982)
12. Weisstein, E.W.: Heilbronn Triangle Problem. From MathWorld - A Wolfram Web Resource. http://mathworld.wolfram.com/HeilbronnTriangleProblem.html. Accessed 14 Mar 2019
13. Yang, L., Zeng, Z.: Heilbronn problem for seven points in a planar convex body. In: Du, D.-Z., Pardalos, P.M. (eds.) Minimax and Applications. Nonconvex Optimization and Its Applications, vol. 4, pp. 191–218. Springer, Boston (1995). https://doi.org/10.1007/978-1-4613-3557-3_14
14. Yang, L., Zhang, J., Zeng, Z.: Heilbronn problem for five points. Intl Centre Theoret. Physics preprint IC/91/252 (1991)
15. Yang, L., Zhang, J., Zeng, Z.: A conjecture on the first several Heilbronn numbers and a computation. Chinese Ann. Math. Ser. A **13**(2), 503–515 (1992). (in Chinese)
16. Yang, L., Zhang, J., Zeng, Z.: On the Heilbronn numbers of triangular regions. Acta Math. Sinica **37**(5), 678–689 (1994). (in Chinese)
17. Zeng, Z., Chen, L.: On the Heilbronn optimal configuration of seven points in the square. In: Sturm, T., Zengler, C. (eds.) ADG 2008. LNCS (LNAI), vol. 6301, pp. 196–224. Springer, Heidelberg (2011). https://doi.org/10.1007/978-3-642-21046-4_11
18. Zeng, Z., Chen, L.: Determining the Heilbronn configuration of seven points in triangles via symbolic computation. In: England, M., Koepf, W., Sadykov, T.M., Seiler, W.M., Vorozhtsov, E.V. (eds.) CASC 2019. LNCS, vol. 11661, pp. 458–477. Springer, Cham (2019). https://doi.org/10.1007/978-3-030-26831-2_30
19. Zeng, Z., Shan, M.: Semi-mechanization method for an unsolved optimization problem in combinatorial geometry. In: SAC 2007, pp. 762–766 (2007)

# Extended Abstracts – Research Stream

# PseudoLinearSystems – A Maple Package for Studying Systems of Pseudo-Linear Equations

Moulay Barkatou, Thomas Cluzeau, and Ali El Hajj[(✉)]

University of Limoges, CNRS, XLIM UMR 7252, MATHIS, Limoges, France
{moulay.barkatou,thomas.cluzeau,ali.el-hajj}@unilim.fr

**Abstract.** *Pseudo-linear systems* constitute a large class of linear functional systems including the usual differential, difference and $q$-difference systems. The Maple package PseudoLinearSystems is dedicated to the study of this class of linear systems. It contains a generic procedure for computing a so-called *simple form* of a pseudo-linear system as well as local data useful for the local analysis: $k$-simple forms, super-irreducible forms, integer slopes of the Newton polygon, indicial equations, etc. It is also devoted to the computation of rational solutions (using simple forms) of a single linear differential, difference or $q$-difference system, as well as rational solutions of a system of mixed linear partial differential, difference and $q$-difference equations. In this software presentation, we demonstrate the use of several procedures of the package that are all based on the simple form procedure.

**Keywords:** Pseudo-linear systems · Simple forms · Rational solutions

The package is freely available online at [1]. All notions and details about the algorithms leading to the procedures presented in what follows are explained in our paper [2] and the references therein. For space restrictions, examples of computations of all procedures presented below, and more, are given in separate files at [1].

## 1 Simple Forms and Local Data

Let $K = C((t))$ be the field of Laurent series in a variable $t$ over a constant field $C \subset \overline{\mathbb{Q}}$, and equipped with the $t$-adic valuation $\nu$. A *pseudo-linear system* can be written in the form

$$L(y) := A\,\delta(y) + B\,\phi(y) = 0, \quad L = A\,\delta + B\,\phi, \tag{1}$$

where $A$ and $B$ are square matrices in $\mathbb{M}_n(C[[t]])$, with $C[[t]]$ the ring of power series in $t$, $\det(A) \neq 0$, $\phi$ is an automorphism of $K$ preserving the valuation, i.e., $\nu(\phi(a)) = \nu(a)$, for all $a \in K$, and $\delta$ is a $\phi-$derivation, that is an additive map from $K$ to itself satisfying the *Leibniz rule* $\delta(ab) = \phi(a)\delta(b) + \delta(a)b$, for all $a, b \in K$.

© Springer Nature Switzerland AG 2020
J. Gerhard and I. Kotsireas (Eds.): MC 2019, CCIS 1125, pp. 327–329, 2020.
https://doi.org/10.1007/978-3-030-41258-6_24

**Definition 1.** *The* leading matrix pencil *of a pseudo-linear system (1) is defined as the matrix polynomial* $L_\lambda = A_0\,\lambda + B_0$ *in the indeterminate* $\lambda$, *where* $A_0$ *and* $B_0$ *are the constant terms in the* $t$–*adic expansion of* $A$ *and* $B$. *We say that System (1) is* simple, *or in* simple form, *if* $\det(L_\lambda) \neq 0$.

Simple forms are very useful for many algorithms handling pseudo-linear systems. The `PseudoLinearSystems` package contains a procedure **SimpleForm** that is generic enough to compute simple forms of any pseudo-linear system of the form (1). The user enters the matrices $A$ and $B$ with rational function entries admitting power series expansions, specifies the local parameter $t$ (for instance $t = x - x_0$ for the singularity $x_0$, and $t = 1/x$ for the singularity $\infty$), and provides the automorphism $\phi$ and the derivation $\delta$ as MAPLE procedures. For instance, for a $q$-difference system written in the form (1), with $\phi$ defined as $\phi(x) = q\,x$ and $\delta = \phi - \mathrm{id}_K$, at the singularity $x = \infty$ the user defines:

```
> PhiAction:= proc(M,x) return subs(x=q*x,M) end:
> DeltaAction:= proc(M,x) return PhiAction(M,x)-M end:
> t:=1/x;
```

The user hence runs

```
> SimpleForm(A, B, x, t, DeltaAction, PhiAction);
```

The output is a list containing respectively the matrices $\widehat{A}$ and $\widehat{B}$ of the equivalent simple system $\widehat{L}(y) = \widehat{A}\,\delta(y) + \widehat{B}\,\phi(y) = 0$, the two invertible matrices $S$ and $T$ such that $\widehat{L} = S\,L\,T$, and the determinant of the leading matrix pencil of $\widehat{L}$, that is thus not identically zero.

`PseudoLinearSystems` contains procedures based on **SimpleForm** that are useful for computing local data of pseudo-linear systems. A so called *super-irreducible* form is computed by iteratively computing a $k$-simple form for $k = p - 1, \ldots, 0$, where $p$ is the *Poincaré rank* (see [2, Section 4.2], [3]). This can be performed using our generic **SimpleForm** procedure just by altering at each step the derivation $\delta$ in its input. For a general pseudo-linear system of the form

$$\delta(y) = M\,\phi(y), \qquad M \in \mathbb{M}_n(K), \tag{2}$$

the command

```
> SuperReduced(M,x,t,DeltaAction,PhiAction);
```

returns the matrix $\widehat{M}$ of an equivalent super-irreducible system $\delta(y) = \widehat{M}\,\phi(y)$, the list of the *characteristic polynomials* of each of the $k$-simple systems for $k = m, \ldots, 0$, where $m$ is the minimal Poincaré rank, and finally the matrix $T \in \mathrm{GL}_n(K)$ such that $\widehat{M} = T^{-1}(M\phi(T) - \delta(T))$.

`PseudoLinearSystems` also contains a procedure to compute the minimal Poincaré rank of System (2). The command

```
> MinimalPoincareRank(M, x, t, DeltaAction, PhiAction);
```

returns the minimal Poincaré rank, the matrix $\widehat{M}$ of an equivalent Moser-reduced system $\delta(y) = \widehat{M}\,\phi(y)$, the characteristic polynomial, and finally the matrix $T \in \mathrm{GL}_n(K)$ such that $\widehat{M} = T^{-1}(M\phi(T) - \delta(T))$.

The integer slopes of the Newton polygon, associated with their corresponding Newton polynomials can be also computed for a system of the form (2):

```
> IntegerSlopesNewtonPolygon(M, x, t, DeltaAction, PhiAction);
```

## 2   Rational Solutions

The package contains the procedure **RationalSolutions_1System**, devoted to computing all rational solutions of one single first order differential, difference, or $q$-difference system. This procedure calls the **SimpleForm** procedure to compute the indicial polynomial at *fixed singularities*. This is a main difference compared to similar procedures for rational solutions included in ISOLDE[1], resp., LinearFunctionalSystems[2], which uses super-reduction, resp. EG-eliminations, instead of simple forms.

**RationalSolutions_1System** takes as an input the matrix defining the system, the variable, and the type of the system (e.g., differential, difference, qdifference). It returns a matrix whose columns form a basis of all rational solutions ({} if there are no non-trivial rational solutions).

We now consider a fully integrable system $\{L_1(y) = 0, \ldots, L_m(y) = 0\}$ composed of $m$ partial pseudo-linear systems, where each $L_i(y) = 0$ is either a partial differential, difference or $q$-difference system. The PseudoLinearSystems package also contains the procedure **RationalSolutions** for computing rational solutions of such partial pseudo-linear systems. For example, for a system

$$\left\{ y(x_1 + 1, x_2, x_3) = A(\mathbf{x})y(\mathbf{x}), \quad y(x_1, q\,x_2, x_3) = B(\mathbf{x})y(\mathbf{x}), \quad \frac{\partial y}{\partial x_3}(\mathbf{x}) = C(\mathbf{x})y(\mathbf{x}) \right\},$$

defined over $K = C(x_1, x_2, x_3)$, with $A$, $B \in \mathrm{GL}_n(K)$, and $C \in \mathbb{M}_n(K)$, we enter the following lists:

```
> L:=[A,B,C]: x:=[x[1],x[2],x[3]]:
> type:=['difference', 'qdifference', 'differential']:
```
If the integrability conditions are satisfied, then the command
```
> y:=RationalSolutions(L, x, type);
```
returns a matrix whose columns form a basis of all rational solutions and {} if there are no non-trivial rational solutions.

## References

1. Barkatou, M., Cluzeau, T., El Hajj, A.: **PseudoLinearSystems** - A Maple Package for Studying Systems of Pseudo-Linear Equations (2019). http://www.unilim.fr/pages_perso/ali.el-hajj/PseudoLinearSystems.html
2. Barkatou, M., Cluzeau, T., El Hajj, A.: Simple forms and rational solutions of pseudo-linear systems. In: Proceedings of the ISSAC 2019 (2019)
3. Barkatou, M., El Bacha, C.: On $k$-simple forms of first-order linear differential systems and their computation. J. Symb. Comput. **54**, 36–58 (2013)

---

[1] http://isolde.sourceforge.net.
[2] https://www.maplesoft.com/support/help/Maple/view.aspx?path=LinearFunctionalSystems.

# Machine Learning to Improve Cylindrical Algebraic Decomposition in Maple

Matthew England[(✉)] and Dorian Florescu

Faculty of Engineering, Environment and Computing, Coventry University,
Coventry CV1 5FB, UK
{Matthew.England,Dorian.Florescu}@coventry.ac.uk

**Abstract.** Many algorithms in computer algebra systems can have their performance improved through the careful selection of options that do not affect the correctness of the end result. Machine Learning (ML) is suited for making such choices: the challenge is to select an appropriate ML model, training dataset, and scheme to identify features of the input. In this extended abstract we survey our recent work to use ML to select the variable ordering for Cylindrical Algebraic Decomposition (CAD) in Maple: experimentation with a variety of models, and a new flexible framework for generating ML features from polynomial systems. We report that ML allows for significantly faster CAD than with the default Maple ordering, and discuss some initial results on adaptability.

## 1  Introduction

Machine learning (ML) is the branch of artificial intelligence where computers learn computational tasks without being explicitly instructed. It is most attractive when the underlying functional relationship modelled is not well understood.

Software for computational mathematics will usually come with a range of choices which, while having no effect on the correctness of the end result, could have a great effect on how that result is presented and the resources required to find it. These choices range from the low level (in what order to perform a search that may terminate early) to the high (which of a set of competing exact algorithms to use for this problem instance). See for example the survey [5]. The issues are even more pronounced for computer algebra systems where the objects of computation are symbolic with often multiple acceptable representations. In practice these choices are taken by human-made heuristics − our hypothesis is that many could be improved by allowing ML algorithms to analyse the data.

In this extended abstract we discuss our work on using ML to select the variable ordering for Cylindrical Algebraic Decomposition (CAD): an algorithm that decomposes real space relative to a polynomial system. CAD was developed to perform quantifier elimination over the reals − see the collection [2]. Our experiments have focussed on the CAD implementation in MAPLE, part of the RegularChains Library as described in [3].

The authors are supported by EPSRC Project EP/R019622/1: *Embedding Machine Learning within Quantifier Elimination Procedures.*

J. Gerhard and I. Kotsireas (Eds.): MC 2019, CCIS 1125, pp. 330–333, 2020.
https://doi.org/10.1007/978-3-030-41258-6_25

## 2    Machine Learning for CAD Variable Ordering

### 2.1    Results from CICM 2014

The first application of ML for choosing a CAD variable ordering was [9] which used a support vector machine to select which of three human-made heuristics to follow. The experiments there identified substantial subsets of examples for which each of the three heuristics outperformed the others, and demonstrated that the ML meta-heuristic could outperform any one individual heuristic.

### 2.2    Results from CICM 2019

Our current project has revisited these experiments with the nlsat dataset[1], using 6117 SAT problems with 3 variables, and thus six possible orderings. We divided into separate datasets for ML training (4612) and testing (1505).

Our first experiments (see [6] for full details) used the CAD routine in MAPLE's `RegularChains` Library. We experimented with four common ML classifiers: K−Nearest Neighbours (KNN); Multi-Layer Perceptron (MLP); Decision Tree (DT); Support Vector Machine (SVM) with RBF kernel. The target variable ordering for ML was defined as the one that minimises the computing time for a given problem. All models used the same 11 features from [9].

The performance of the models is summarised in Table 1 (top half). Accuracy is how often the model picked an optimal ordering, while Time is the total computation time for all problems using that model's choices. The table also gives the results for virtual solvers who always pick the best (VB)/worst (VW) ordering, a random choice, and three human-made heuristics: sotd [4], Brown's heuristic (Br) [1], and the `RegularChains:-SuggestVariableOrdering` function (SVO).

SVO was omitted from [6]. It is the default used by MAPLE when the user neglects to specify an ordering. It does produce the lowest computation times of the human made heuristics, but all four ML models significantly outperform it.

### 2.3    Results from SC-Square 2019

We next considered how to extract further information from the input data. The 11 features used in [6,9] were derived in turn from the human heuristics. They can all be cheaply extracted from polynomials (e.g. measures of variable degree and frequency of occurence). The only other ideas from the literature involved more costly operations from CAD projection operators [4].

In [8] a new feature generation procedure [8] was presented, based on the observation that the original features can be formalised mathematically using a small number of basic functions (average, sign, maximum) evaluated on the degrees of the variables in either one polynomial or the whole system. Considering all possible combinations of these functions (1728) led to 78 useful and independent features for our dataset. The experiments were repeated with these, and the results are presented in Table 1 (bottom half). All four models were improved with these additional features.

---

[1] Available at http://cs.nyu.edu/~dejan/nonlinear/.

**Table 1.** Performance of ML models and three human-made heuristics.

| | | DT | KNN | MLP | SVM | VB | VW | rand | sotd | Br | SVO |
|---|---|---|---|---|---|---|---|---|---|---|---|
| From [6] | Accuracy | 62.6% | 63.3% | 61.6% | 58.8% | 100% | 0% | 22.7% | 49.5% | 51 % | 50.6% |
| 11 Fts. | Time (s) | 9 994 | 10 105 | 9 822 | 10 725 | 8 623 | 64 534 | 30 235 | 11 938 | 10 951 | 10 821 |
| From [8] | Accuracy | 65.2% | 66.3% | 67% | 65% | | | | | | |
| 78 Fts. | Time (s) | 9 603 | 9 178 | 9 399 | 9 487 | | | | | | |

## 3    Further Experiments on Adaptability of Classifiers

We are now experimenting with how adaptable these classifiers are: whether they can be applied elsewhere without further training. We first considered using them to predict the orderings for an alternative MAPLE CAD implementation (without retraining). The CAD from the `ProjectionCAD` Library [7] uses the traditional projection and lifting algorithm [2] rather than working via complex space with regular chains theory. The ML models still made better choices than the human-made heuristics but the difference is not as significant.

We also investigated how the models perform when presented with problems from a different CAD dataset[2] (one not large enough for training with). In this case a human-made heuristic outperformed the ML models. Thus our next challenge is to identify a training dataset that better represents the wide range of CAD problems that a Maple user may be interested in.

## References

1. Brown, C.: ISSAC 2004 Tutorial Notes (2004). http://www.usna.edu/Users/cs/wcbrown/research/ISSAC04/handout.pdf
2. Caviness, B.F., Johnson, J.R.: Quantifier Elimination and Cylindrical Algebraic Decomposition. Texts and Monographs in Symbolic Computation. Springer, Cham (1998). https://doi.org/10.1007/978-3-7091-9459-1
3. Chen, C., Moreno Maza, M.: Quantifier elimination by cylindrical algebraic decomposition based on regular chains. In: Proceedings of the ISSAC 2014, pp. 91–98. ACM (2014). https://doi.org/10.1145/2608628.2608666
4. Dolzmann, A., Seidl, A., Sturm, T.: Efficient projection orders for CAD. In: Proceedings of the ISSAC 2004, pp. 111–118. ACM (2004). https://doi.org/10.1145/1005285.1005303
5. England, M.: Machine learning for mathematical software. In: Davenport, J.H., Kauers, M., Labahn, G., Urban, J. (eds.) ICMS 2018. LNCS, vol. 10931, pp. 165–174. Springer, Cham (2018). https://doi.org/10.1007/978-3-319-96418-8_20
6. England, M., Florescu, D.: Comparing machine learning models to choose the variable ordering for cylindrical algebraic decomposition. In: Kaliszyk, C., Brady, E., Kohlhase, A., Sacerdoti Coen, C. (eds.) CICM 2019. LNCS (LNAI), vol. 11617, pp. 93–108. Springer, Cham (2019). https://doi.org/10.1007/978-3-030-23250-4_7

---

[2] Available from http://dx.doi.org/10.15125/BATH-00069.

7. England, M., Wilson, D., Bradford, R., Davenport, J.H.: Using the regular chains library to build cylindrical algebraic decompositions by projecting and lifting. In: Hong, H., Yap, C. (eds.) ICMS 2014. LNCS, vol. 8592, pp. 458–465. Springer, Heidelberg (2014). https://doi.org/10.1007/978-3-662-44199-2_69
8. Florescu, D., England, M.: Algorithmically generating new algebraic features of polynomial systems for machine learning. In: Proceedings of the $SC^2$ 2019. CEUR Workshop Proceedings, vol. 2460, 12 p. (2019). http://ceur-ws.org/Vol-2460/
9. Huang, Z., England, M., Wilson, D., Davenport, J.H., Paulson, L.C., Bridge, J.: Applying machine learning to the problem of choosing a heuristic to select the variable ordering for cylindrical algebraic decomposition. In: Watt, S.M., Davenport, J.H., Sexton, A.P., Sojka, P., Urban, J. (eds.) CICM 2014. LNCS (LNAI), vol. 8543, pp. 92–107. Springer, Cham (2014). https://doi.org/10.1007/978-3-319-08434-3_8

# Ball Arithmetic as a Tool in Computer Algebra

Fredrik Johansson$^{(\boxtimes)}$

Inria Bordeaux, Talence, France
fredrik.johansson@gmail.com
http://fredrikj.net

**Abstract.** This presentation gives an overview of ball arithmetic as a tool for computing with real numbers in the context of computer algebra, and discusses recent development to the Arb library.

**Keywords:** Ball arithmetic · Real numbers · Symbolic-numeric algorithms

## 1 Introduction

Computing with real numbers involves tradeoffs in the algorithms or interfaces, often between speed, correctness and simplicity. Depending on the application, it is useful to consider three levels of abstraction for real numbers, sometimes in combination: nonrigorous (floating-point arithmetic), rigorous (interval arithmetic), and exact (symbolic or lazy).

*Ball arithmetic* is interval arithmetic with a midpoint-radius representation of real numbers [1], for example: $\pi \in [3.141592654 \pm 4.11 \cdot 10^{-10}]$. This format is particularly suitable for the high precision calculations often needed in computer algebra applications, for instance to evaluate inequalities, to evaluate symbolic expressions that may involve large cancellations, and to recover exact coefficients and formulas from numerical approximations.

The Arb library [2] has been designed to replace many previous uses of non-rigorous numerics with rigorous (and often more efficient) versions. There are now several successful projects building on the ball arithmetic in Arb, validating this approach. This includes libraries and research code written directly against the low-level C interface, as well as projects working with the high-level interfaces to Arb available in Nemo (Julia) and Sage (Python). A good example is Mezzarobba's numeric extension of the ore_algebra Sage package, which supports effective analytic continuation of D-finite (holonomic) functions defined by linear ODEs with polynomial coefficients [3]. This relies heavily on arithmetic with Arb polynomials. Another good example is the class group computation in Hecke [4] which depends on Arb matrices.

The experience suggests that ball arithmetic not only makes sense for backend library code, but that it is a useful abstraction to expose to the end users of computer algebra software.

© Springer Nature Switzerland AG 2020
J. Gerhard and I. Kotsireas (Eds.): MC 2019, CCIS 1125, pp. 334–336, 2020.
https://doi.org/10.1007/978-3-030-41258-6_26

## 1.1  Algebraic Computation

One of the central tasks in computer algebra is to manipulate polynomials and matrices with coefficients in various domains, including the real and complex numbers. Floating-point computations often run into problems with ill-conditioning or cancellation for high-degree or high-dimensional objects, and errors are not always easy to detect, in particular when they occur inside a complicated algorithm. Ball arithmetic solves this problem elegantly. Indeed, it is often straightforward to translate an algorithm designed for exact quantities to a rigorous numerical version using ball arithmetic; the main caveat is that comparisons can be uncertain (for example, the truth value of $x = y$ or $x < y$ is unknown if $x$ and $y$ are represented by overlapping balls of nonzero width), and conditional cases must be handled accordingly.

The polynomial and matrix arithmetic in Arb has recently been optimized significantly and now generally runs several times faster than other arbitrary-precision software [6]. For linear algebra (including linear solving and eigen-decomposition), Arb provides two levels of functionality: classical nonrigorous numerical algorithms, and rigorous algorithms. The rigorous functions now use nonrigorous approximation followed by *a posteriori* certification to avoid the pessimistic blow-up resulting from direct methods such as Gaussian elimination in ball arithmetic. This makes it possible to handle dense linear algebra problems with dimensions in the thousands, using only modest working precision as long as the systems are well-conditioned. Direct methods are also used when appropriate for small systems or for very high precision.

## 1.2  Analytic Computation

Diverse numerical tools are required for analytic operations, such as the computation of limits, infinite series, integrals, and evaluation of transcendental functions. The computation of transcendental functions is a task that translates rather well to rigorous numerics with the help of ball arithmetic. The library of special functions in Arb is continually being expanded: additions in the last two years include the complex branches of the Lambert $W$-function, Coulomb wave functions, Dirichlet characters and Dirichlet L-functions (contributed by P. Molin), and arbitrary-index zeros of the Riemann zeta function (contributed by D.H.J. Polymath).

One of the more important recent additions to Arb is support for rigorous numerical integration of arbitrary (piecewise meromorphic) user-defined functions, based on adaptive Gaussian quadrature with automatic error bounds computed from complex magnitudes. This approach performs on par with the best nonrigorous integration tools for a wide range of integrals [5].

The main limitation of rigorous numerical integration (or similar operations) based on "black-box" evaluation of the integrand in ball arithmetic is that it cannot converge for improper integrals. Such integrals require symbolic processing, either in the form of explicit truncation, a suitable change of variables, or use of an integration scheme designed for the particular singularity at hand. Designing new symbolic-numeric algorithms in this area is an interesting research topic.

## 1.3   Relation to Other Software

Classical interval arithmetic using endpoints $[a, b]$ is more well known than ball arithmetic and exists in many implementations. For example, Maple has a built-in "range arithmetic" (although this is not strictly rigorous) and the external Maple package intpakX provides a wide range of tools for interval analysis [7].

It is natural to think of classical interval arithmetic as a better tool for subdivision of space and ball arithmetic as a better tool for representation of numbers [1], though it should be stressed that the formats are interchangeable for many tasks. This difference in goals is reflected in the content of the Arb library: Arb provides few subdivision-related functions and presently offers no functionality for tasks such as multivariate optimization. On the other hand, it has excellent support for transcendental functions.

A competitor to Arb is the C library MPFI which implements arbitrary-precision interval arithmetic using MPFR endpoints [8]. MPFI is designed to guarantee tightest possible intervals, while Arb makes virtually no *a priori* guarantees about the tightness of the balls. In part due to this tradeoff, Arb is generally more efficient than MPFI and also offers a much larger set of functions.

For Maple users, Arb may be most directly useful as drop-in replacement for individual mathematical functions, though one can also imagine more elaborate symbolic-numeric applications involving ball arithmetic. At this time, the author is not aware of any published projects combining Arb and Maple, but there should not be any technical obstacles since Arb is easy to interface in any environment with a C foreign function interface (for example, Arb has been used in projects written in C++, Fortran, Java, JavaScript, Julia and Python).

# References

1. van der Hoeven, J.: Ball arithmetic (2009). http://hal.archives-ouvertes.fr/hal-00432152/fr/
2. Johansson, F.: Arb: efficient arbitrary-precision midpoint-radius interval arithmetic. IEEE Trans. Comput. **66**, 1281–1292 (2017). https://doi.org/10.1109/TC.2017.2690633
3. Mezzarobba, M.: Rigorous multiple-precision evaluation of D-finite functions in SageMath (2016). arxiv:1607.01967
4. Fieker, C., Hart, W., Hofmann, T., Johansson, F.: Nemo/Hecke: computer algebra and number theory packages for the Julia programming language. In: ISSAC 2017, pp. 57–164. ACM (2017). https://doi.org/10.1145/3087604.3087611
5. Johansson, F.: Numerical integration in arbitrary-precision ball arithmetic. In: Davenport, J.H., Kauers, M., Labahn, G., Urban, J. (eds.) ICMS 2018. LNCS, vol. 10931, pp. 255–263. Springer, Cham (2018). https://doi.org/10.1007/978-3-319-96418-8_30
6. Johansson, F.: Faster arbitrary-precision dot product and matrix multiplication. In: ARITH26 (2019, to appear). arxiv:1901.04289
7. Geulig, I., Krämer, W., Grimmer, M.: intpakX 1.2. http://www2.math.uni-wuppertal.de/org/WRST/software/intpakX/
8. Revol, N., Rouillier, F.: Motivations for an arbitrary precision interval arithmetic and the MPFI library. Reliab. Comput. **11**(4), 275–290 (2005). https://doi.org/10.1007/s11155-005-6891-y

# The Lie Algebra of Vector Fields Package with Applications to Mappings of Differential Equations

Zahra Mohammadi[1(✉)], Gregory J. Reid[1], and S.-L. Tracy Huang[2]

[1] Department of Applied Mathematics, University of Western Ontario,
London, Canada
{zmohamm5,Reid}@uwo.ca
[2] Data61, CSIRO, Canberra, ACT 2601, Australia
tracy.huang49@gmail.com

**Abstract.** Lie symmetry groups of transformations (mappings) of differential equations leave them invariant, and are most conveniently studied through their Lie algebra of vector fields (essentially the linearization of the mappings around the identity transformation). Maple makes powerful and frequent use of such Lie algebras, mostly through routines that are dependent on Maple's powerful exact integration routines, that essentially automate traditional hand-calculation strategies. However these routines are usually heuristic, and algorithmic approaches require a deeper integration of differential elimination (differential algebraic) approaches in applications to differential equations. This is the underlying motivation of the LieAlgebrasOfVectorFields (**LAVF**) package of Huang and Lisle. The **LAVF** package introduces a powerful algorithmic calculus for doing calculations with differential equations without the heuristics of integration to calculate efficiently many properties of such systems.

We use **LAVF** in the development of our **MapDE** package, which determines the existence of analytic invertible mappings of an input DE to target DE. Theory, algorithms, and examples of **MapDE** can be found in [5,6]. Here we present a brief summary, through examples, of the application of **LAVF** to **MapDE**.

**Keywords:** Symmetry · Lie algebra · Structure constants · Differential algebra

Differential Equations (DE) are the main tools to mathematically express governing laws of physics and models in biology, financial and other applications. Examining the solutions of related DE helps to gain insights into the phenomena described by the DE. However, finding exact solutions of DE can be extremely difficult and often impossible. Analyzing solutions of DE using their symmetries is one of the main approaches to this problem. An important application of symmetries of PDE is to determine if the PDE is linearizable by an invertible

© Springer Nature Switzerland AG 2020
J. Gerhard and I. Kotsireas (Eds.): MC 2019, CCIS 1125, pp. 337–340, 2020.
https://doi.org/10.1007/978-3-030-41258-6_27

mapping and construct the linearization in terms of exploiting the Lie symmetry invariance algebra of vector fields of the given PDE and construct a linear target and the mapping when the existence is established.

Here we describe the application of the LAVF package in our MapDE algorithm that determines the existence and construction of local mapping relating a given *Source* DE system to a more tractable *Target* DE. Please see [5,6] for additional details, examples, algorithms, references and theoretical results. MapDE will be integrated into the LAVF package of Huang and Lisle [2,3]. LAVF is a powerful object-oriented Maple package for determining the structure of symmetry of differential equation by extracting algebraic and geometric information about Lie algebras of vector fields (e.g. isomorphism invariants, diffeomorphism invariants, various sub-algebras, upper and lower series, etc). LAVF can automatically compute defining systems and Lie algebra structure for various objects including derived algebras which are core operations needed by MapDE.

*Example 1.* Consider a class of Schrödinger equations in $N$ space (for $N = 1, \cdots, 10$) with time varying harmonic oscillator potentials for $u(x_1, \cdots, x_N, t)$:

$$i \, u_t = \nabla^2 u + \sum_{j=1}^{N} t^2 x_j^2 . u$$

Application of pdsolve fails to find the symmetries or to solve any of the above equations (even for $N = 1$). However LAVF can efficiently compute the structure of Lie Symmetry algebras of the above equations for any $N = 1, \cdots, 10$. We note that the above equations with such harmonic oscillator potentials are fundamental in physics.

LAVF computes Lie Algebra of Vector fields for the determining equations of differential equation and Derived Algebras by using command lines of LAVF e.g. VectorField, SymmetryLAVF, DerivedAlgebra, etc. These commands improve the construction stage of MapDE algorithm and other commands such as SolutionDimension help us to build an efficient linearizations existence test.

**Table 1.** Table presents the CPU times for $\left(\frac{d}{dx}\right)^d (u(x)^2) + u(x)^2 = 0$. Timings correspond to the existence (3rd and 4rd row) and construction (5th row) of linearization by MapDE.

| Existence and construction of linearization ODE in secs using MapDE | | | | | | | | | | | | |
|---|---|---|---|---|---|---|---|---|---|---|---|---|
| Order ODE | 3 | 4 | 5 | 6 | 7 | 8 | 9 | 10 | 11 | 12 | 13 | 14 | 15 |
| ExistenceLGMTest | .406 | .266 | .515 | .609 | 1.000 | 1.329 | 1.734 | 2.500 | 2.953 | 4.204 | 6.000 | 7.922 | 10.359 |
| ExistenceHilbertTest | .484 | .329 | .578 | .687 | 1.094 | 1.454 | 1.844 | 2.625 | 3.078 | 4.360 | 6.203 | 8.109 | 10.609 |
| Existence and construction | .578 | .407 | .672 | .812 | 1.234 | 1.735 | 2.062 | 3.125 | 3.453 | 4.860 | 7.000 | 9.156 | 11.688 |

**Procedure 1** LGMLinearizationTest

IsLinearizable := proc (Q)

    local n, xi, eta, Y, L, m, DA, Linearizability;

    n := PDEtools[difforder] (Q);

    Linearizability := false;

    Y := VectorField ([[xi(x, u), x], [eta(x, u), u]]);

    L := SymmetryLAVF ([Q], Y);

    m := SolutionDimension (L);

    DA := DerivedAlgebra (L);

    **if** $2 < n$ and $m = n + 4$ **then** Linearizability := true;

        **return** Linearizability;

    **elif** $2 < n$ and ($m = n + 1$ or $m = n + 2$) **then**

        DA := DerivedAlgebra (L);

        **if** IsAbelian (DA) and n = SolutionDimension (DA) **then**

            Linearizability := true

        **end if**

    **end if**;

    **return** Linearizability;

end proc;

---

*Example 2* (Lyakhov, Gerdt and Michels ODE Test Set [4]). To illustrate the flexibility and power of LAVF we use it to implement algorithm (I) of [4] using LAVF commands for $\mathrm{ODE}[n] = \left(\frac{d}{dx}\right)^n (u(x)^2) + u(x)^2 = 0$ of order $n \geq 3$. See Procedure 1.

It admits the linearization:

$$\Psi = \{\hat{x} = \psi = x, \, \hat{u} = \phi = u^2\}$$

The times for detecting the existence of the linearization by Lyakhov et al. test in range from 0.2 s for $d = 3$ to about 150 s for $d = 15$. Their linearization construction method takes 7512.9 s for $n = 9$ and out of memory for $n \geq 10$. (See [4] for these results). Our runs of the same tests to determine the existence and construct the map are displayed in Table 1. We also report the time for our other linearization test, Hilbert test, see our paper [6] for more details.

# References

1. Bluman, G.W., Cheviakov, A.F., Anco, S.C.: Applications of Symmetry Methods to Partial Differential Equations. Springer, Heidelberg (2010). https://doi.org/10. 1007/978-0-387-68028-6
2. Huang, S.L.: Properties of lie algebras of vector fields from lie determining system. Ph.D. thesis, University of Canberra (2015)
3. Lisle, I.G., Huang, S.-L.T: Algorithmic calculus for Lie determining systems. J. Symbolic Comput. **79**(part 2), 482–498 (2017)
4. Lyakhov, D., Gerdt, V., Michels, D.: Algorithmic verification of linearizability for ordinary differential equations. In: Proceedings ISSAC 2017, pp. 285–292. ACM (2017)

5. Mohammadi, Z., Reid, G., Huang, S.-L.T.: Introduction of the MapDE algorithm for determination of mappings relating differential equations. arXiv:1903.02180v1 [math.AP] (2019). To appear in Proceedings of ISSAC 2019. ACM
6. Mohammadi, Z., Reid, G., Huang, S.-L.T.: Symmetry-based algorithms for invertible mappings of polynomially nonlinear PDE to linear PDE. Submitted to Mathematics of Computer Science (Revision requested 15 May 2019)

# Polynomial Factorization in Maple 2019

Michael Monagan$^{(\boxtimes)}$ and Baris Tuncer

Department of Mathematics, Simon Fraser University, Burnaby, Canada
mmonagan@cecm.sfu.ca

## Extended Abstract

Maple 2019 has a new multivariate polynomial factorization algorithm for factoring polynomials in $\mathbb{Z}[x_1, x_2, ..., x_n]$, that is, polynomials in $n$ variables with integer coefficients. The new algorithm, which we call MTSHL, was developed by the authors at Simon Fraser University. The algorithm and its sub-algorithms have been published in a sequence of papers [3–5]. It was integrated into the Maple library in early 2018 by Baris Tuncer under a MITACS internship with Maplesoft. MTSHL is now the default factoring algorithm in Maple 2019.

The multivariate factorization algorithm in all previous versions of Maple is based mainly on the work of Wang in [6,7]. Geddes is the main author of the Maple code. The algorithm and sub-algorithms are described in Chap. 6 of [1]. Wang's algorithm is still available in Maple 2019 with the `method="Wang"` option to the factor command.

Wang's method can be exponential in $n$ the number of variables. MTSHL is a random polynomial time algorithm. In [3] we found that it is faster than previous polynomial time methods of Kaltofen [2] and Zippel [8] and competitive with Wang's method in cases where Wang's method is not exponential in $n$.

Here we give an overview of the main idea in MTSHL. Let $a$ be the input polynomial to be factored. Suppose $a = fg$ for two irreducible factors $f, g \in \mathbb{Z}[x_1, \ldots, x_n]$. The multivariate polynomial factorization algorithm used in all computer algebra systems is based on Multivariate Hensel Lifting (MHL). For a description of MHL see Chap. 6 of [1]. MHL first chooses integers $\alpha_2, \alpha_3, \ldots, \alpha_n$ that satisfy certain conditions and factors the univariate image $a_1 = a(x_1, \alpha_2, \ldots, \alpha_n)$ in $\mathbb{Z}[x_1]$. Suppose $a_1(x_1) = f_1(x_1)g_1(x_1)$ and $f_1(x_1) = f(x_1, \alpha_2, \ldots, \alpha_n)$ and $g_1(x_1) = g(x_1, \alpha_2, \ldots, \alpha_n)$. Next MHL begins Hensel lifting. Wang's design of Hensel lifting recovers the variables $x_2, \ldots, x_n$ in the factors $f$ and $g$ one at a time in a loop. Let us use the notation

$$f_j = f(x_1, \ldots, x_j, \alpha_{j+1}, \ldots, \alpha_n) \text{ for } j \geq 1.$$

So at the $j$'th step of MHL we have the factorization $a_{j-1} = f_{j-1}g_{j-1}$ and we want to obtain the factorization $a_j = f_j g_j$. Consider the polynomials $f_j$ and $g_j$ expanded as a Taylor polynomial about $x_j = \alpha_j$

$$f_j = \sum_{i=0}^{\deg(f_j, x_j)} \sigma_i(x_j - \alpha_j)^i \text{ and } g_j = \sum_{i=0}^{\deg(g_j, x_j)} \tau_i(x_j - \alpha_j)^i$$

© Springer Nature Switzerland AG 2020
J. Gerhard and I. Kotsireas (Eds.): MC 2019, CCIS 1125, pp. 341–345, 2020.
https://doi.org/10.1007/978-3-030-41258-6_28

Here $\sigma_i, \tau_i \in \mathbb{Z}[x_1, \ldots, x_{j-1}]$ and $\sigma_0 = f_{j-1}$ and $\tau_0 = g_{j-1}$ are known. MHL recovers the coefficients $\sigma_i, \tau_i$ one at a time in a loop. Let $\mathrm{supp}(\sigma)$ denote the support of $\sigma$, that is, the set of monomials in $\sigma$. Before continuing we give an example to fix the ideas and notation presented so far. Let $f = x_1^3 - x_1 x_2 x_3^2 + x_2^3 x_3^2 + x_3^3 - 27$. For $\alpha_3 = 2$, we have $f_2 = f(x_1, x_2, 2) = x_1^3 + 4 x_2^3 - 4 x_1 x_2 - 19$. Expanding $f$ about $x_3 = 2$ we have

$$f = \underbrace{(x_1^3 + 4 x_2^3 - 4 x_1 x_2 - 19)}_{\sigma_0} + \underbrace{(4 x_2^3 - 4 x_1 x_2 + 12)}_{\sigma_1}(x_3 - 2)$$
$$+ \underbrace{(x_2^3 - x_1 x_2 + 6)}_{\sigma_2}(x_3 - 2)^2 + \underbrace{1}_{\sigma_3}(x_3 - 2)^4.$$

We have $\mathrm{supp}(\sigma_0) = \{x_1^3, x_2^3, x_1 x_2, 1\}$, $\mathrm{supp}(\sigma_1) = \mathrm{supp}(\sigma_2) = \{x_2^3, x_1 x_2, 1\}$, and $\mathrm{supp}(\sigma_3) = \{1\}$. Multivariate Hensel Lifting (MHL) computes $\sigma_i$ and $\tau_i$ by solving the multivariate polynomial diophantine (MDP) equation

$$\sigma_i g_{j-1} + \tau_i f_{j-1} = c_i \quad \text{in} \quad \mathbb{Z}_p[x_1, \ldots, x_{j-1}]$$

where the polynomial $c_i$ is the Taylor coefficient

$$\mathrm{coeff}\left( a_j - \left( \sum_{i=0}^{k-1} \sigma_i (x_j - \alpha_j)^i \right) \left( \sum_{i=0}^{k-1} \tau_i (x_j - \alpha_j)^i \right), \ (x_j - \alpha_j)^k \right).$$

Most of the time in MHL is solving these MDP equations. Wang's method for solving them is recursive. If the $\alpha_2, \ldots, \alpha_n$ are non-zero Wang's method is exponential in $n$. For many polynomials it is possible to use zero for some or all $\alpha_j$ and avoid this exponential behaviour. But this is not always possible as there are several conditions that $\alpha_2, \ldots, \alpha_n$ must satisfy. The sparse Hensel lifting methods of [2] and [8] were developed to solve this problem. It turns out that if the integer $\alpha_j$ is chosen randomly from a large set then

$$\mathrm{supp}(\sigma_i) \supseteq \mathrm{supp}(\sigma_{i+1}) \quad \text{for} \ \ 0 \le i < \deg(f_j, x_j) \tag{1}$$

with high probability. The reader may verify this support chain holds in Example 1 where $\alpha = 2$ but it does not hold if $\alpha = 0$. See Lemma 1 in [3] for a precise statement for the probability and proof. MTSHL exploits (1) by using $\mathrm{supp}(\sigma_{i-1})$ as the support for $\sigma_i$ to construct linear systems to solve for the coefficients of $\sigma_i$ in $x_1$. The linear systems are $t_j \times t_j$ transposed Vandermonde systems where $t_j = \#\mathrm{coeff}(\sigma_i, x_1^j)$. We use Zippel's method from [9] to solve them in $O(t_j^2)$ time and $O(t_j)$ space. Since the number of terms in $\sigma_i$ and $\tau_i$ is not more than those in $f$ and $g$ respectively, our algorithm takes advantage of sparse factors $f$ and $g$.

We present two benchmarks comparing the new algorithm MTSHL in Maple 2019 with Wang's method in Maple 2019. The following Maple code creates an input polynomial $a \in \mathbb{Z}[x_1, \ldots, x_n]$ which is a product of two factors $f \times g$. Each factor has $n$ variables, 100 terms, and degree at most $d$. The Maple command randpoly creates each term randomly to have degree at most $d$ with an integer coefficient chosen at random from $[-10^6, 10^6]$.

```
kernelopts(numcpus=1); t := 100; d := 15;
for n from 5 to 12 do
  X := [seq( x||i, i=1..n )];
  f := randpoly(X,coeffs=rand(-10^6..10^6),terms=100,degree=d);
  g := randpoly(X,coeffs=rand(-10^6..10^6),terms=100,degree=d);
  a := expand(f*g);
  h := CodeTools[Usage]( factor(a,method="Wang") );
  #h := CodeTools[Usage]( factor(a) ); # Uses MTSHL in Maple 2019
od:
```

Shown in column (MDP) in Table 1 is the percentage of time Wang's algorithm spent solving Multivariate Diophantine Equations. MTSHL is not impacted significantly by the number of variables. In theory the cost of MTSHL is linear in $n$ which is supported by this example.

**Table 1.** Factorization timings in CPU seconds

| $n$ | Wang (MDP) | MTSHL | $n$ | Wang (MDP) | MTSHL |
|---|---|---|---|---|---|
| 5 | 4.87 s (89.4%) | .509 s | 10 | 65.55 s (98.0%) | .911 s |
| 6 | 8.67 s (85.8%) | .589 s | 11 | 154.8 s (98.0%) | .989 s |
| 7 | 6.77 s (91.2%) | .616 s | 12 | 169.8 s (99.0%) | 1.78 s |
| 8 | 35.04 s (94.7%) | .718 s | 13 | 163.8 s (96.5%) | 1.16 s |
| 9 | 40.33 s (99.6%) | .788 s | 14 | 603.6 s (98.7%) | 2.37 s |

Let $C_n$ denote the $n \times n$ cyclic matrix. See Fig. 1. Observe that $\det C_n$ is a homogeneous polynomial in $\mathbb{Z}[x_1, \ldots, x_n]$. Because the factors of $\det C_n$ are dense, MTSHL has no inherent advantage over Wang's method and we expected it to be slower than Wang's method.

$$\begin{bmatrix} x_1 & x_2 & \ldots & x_{n-1} & x_n \\ x_n & x_1 & \ldots & x_{n-2} & x_{n-1} \\ \vdots & \vdots & \vdots & & \vdots \\ x_3 & x_4 & \ldots & x_1 & x_2 \\ x_2 & x_3 & \ldots & x_n & x_1 \end{bmatrix} \qquad \begin{array}{c} (x_1 + x_2 + x_3 + x_4) \\ (x_1 - x_2 + x_3 - x_4) \\ \left(x_1^2 - 2\,x_1\,x_3 + x_2^2 - 2\,x_2\,x_4 + x_3^2 + x_4^2\right) \end{array}$$

**Fig. 1.** The cyclic $n \times n$ matrix $C_n$ and the factors of $\det(C_4)$.

Maple code for computing $\det C_n$ and factoring $\det C_n$ is given below. Note, for a homogenous input polynomial $\det C_n$, the **factor** command evaluates one variable $x_i = 1$, factors $\det(C_n)(x_i = 1)$ then homogenizes the factors. To fix $i$ we compute and factor $\det(C_n(x_n = 1))$.

```
kernelopts(numcpus=1);
for n from 6 to 10 do
   Cn := Matrix(n,n,shape=Circulant[x]);
   Cn := eval(Cn,x[n]=1); # dehomogenize Cn
   d  := LinearAlgebra[Determinant](Cn,method=minor);
   F  := CodeTools[Usage](factor(d));
   #F := CodeTools[Usage](factor(d,method="Wang"));
od:
```

**Table 2.** Timings (CPU time seconds) for factoring $\det(C_n(x_n = 1))$

| $n$ | #det | $\deg(f_i)$ | max $\#f_i$ | MTSHL | Wang (MDP) | Magma |
|---|---|---|---|---|---|---|
| 8 | 810 | 1, 1, 2, 4 | 86 | 0.140 s | 0.096 s (52%) | 0.12 s |
| 9 | 2704 | 1, 2, 6 | 1005 | 0.465 s | 0.253 s (76%) | 1.02 s |
| 10 | 7492 | 1, 1, 4, 4 | 715 | 3.03 s | 1.020 s (49%) | 10.97 s |
| 11 | 32066 | 1, 10 | 184756 | 1.33 s | 12.43 s (88%) | 142.85 s |
| 12 | 86500 | 1, 1, 2, 2, 2, 4 | 621 | 4.97 s | 20.51 s (65%) | 7575.14 s |
| 13 | 400024 | 1, 12 | 2704156 | 10.24 s | 212.40 s (88%) | 30,871.9 s |
| 14 | 1366500 | 1, 1, 6, 6 | 27132 | 666.0 s | 1364.4 s (68%) | $>10^6$ s |

Table 2 contains data for $\det C_n$ and timing data for factoring $\det C_n(x_n = 1)$. Column 2 is the number of terms of $\det C_n$. Column 3 is the degrees of the factors of $C_n$. Column 4 is the number of terms of the largest factor. Columns 5–7 are the CPU time to factor $\det C_n$ using our new algorithm MTSHL in Maple 2019, Wang's algorithm in Maple 2019 and Wang's algorithm in the Magma computer algebra system.

# References

1. Geddes, K.O., Czapor, S.R., Labahn, G.: Algorithms for Computer Algebra. Kluwer Academic Publishers, Boston (1992)
2. Kaltofen, E.: Sparse hensel lifting. In: Caviness, B.F. (ed.) EUROCAL 1985. LNCS, vol. 204, pp. 4–17. Springer, Heidelberg (1985). https://doi.org/10.1007/3-540-15984-3_230
3. Monagan, M., Tuncer, B.: Using sparse interpolation in Hensel lifting. In: Gerdt, V.P., Koepf, W., Seiler, W.M., Vorozhtsov, E.V. (eds.) CASC 2016. LNCS, vol. 9890, pp. 381–400. Springer, Cham (2016). https://doi.org/10.1007/978-3-319-45641-6_25
4. Monagan, M., Tuncer, B.: Factoring multivariate polynomials with many factors and huge coefficients. In: Gerdt, V.P., Koepf, W., Seiler, W.M., Vorozhtsov, E.V. (eds.) CASC 2018. LNCS, vol. 11077, pp. 319–334. Springer, Cham (2018). https://doi.org/10.1007/978-3-319-99639-4_22
5. Monagan, M., Tuncer, B.: The complexity of sparse Hensel lifting and sparse polynomial factorization. Symb. Comput. **99**, 189–230 (2019)

6. Wang, P.S., Rothschild, L.P.: Factoring multivariate polynomials over the integers. Math. Comput. **29**(131), 935–950 (1975)
7. Wang, P.S.: An improved multivariate polynomial factoring algorithm. Math. Comput. **32**, 1215–1231 (1978)
8. Zippel, R.E.: Newton's iteration and the sparse Hensel algorithm. In: Proceedings SYMSAC 1981, pp. 68–72. ACM (1981)
9. Zippel, R.E.: Interpolating polynomials from their values. J. Symb. Comput. **9**(3), 375–403 (1990)

# Extended Abstracts –
# Education/Applications Stream

# Distributive Laws Between the Operads *Lie* and *Com*

Murray Bremner[1]([✉])(ID) and Vladimir Dotsenko[2](ID)

[1] Department of Mathematics and Statistics, University of Saskatchewan, Saskatoon, Canada
bremner@math.usask.ca
[2] Institut de Recherche Mathématique Avancée, UMR 7501, Université de Strasbourg et CNRS, 7 rue René-Descartes, 67000 Strasbourg CEDEX, France
vdotsenko@unistra.fr

**Abstract.** We apply computer algebra, especially linear algebra over polynomial rings and Gröbner bases, to classify inhomogeneous distributive laws between the operads for Lie algebras and commutative associative algebras.

**Keywords:** Computer algebra · Linear algebra over polynomial rings · Gröbner bases · Algebraic operads · Distributive laws · Lie and commutative algebras

## 1 Theory

We refer the reader to [1,5] for a systematic treatment of algebraic operads and Gröbner bases, and to [4] for specific information on weight graded operads. All objects in this paper are defined over an arbitrary field $\Bbbk$. We denote by $\circ$ the composition of (underlying symmetric) collections, and by $\circ'$ the infinitesimal composition.

Let $\mathscr{P} = \mathscr{T}(\mathscr{X})/(\mathscr{R})$ and $\mathscr{Q} = \mathscr{T}(\mathscr{Y})/(\mathscr{S})$ be two weight graded operads presented by generators and relations; we assume the standard weight grading for which the generators are of weight 1. We say that an operad $\mathscr{O}$ generated by $\mathscr{X} \oplus \mathscr{Y}$ is obtained from $\mathscr{P}$ and $\mathscr{Q}$ by an *inhomogeneous distributive rewriting rule* if the defining relations of $\mathscr{O}$ are $\widetilde{\mathscr{R}} \oplus \mathscr{D} \oplus \mathscr{S}$, with subcollections $\widetilde{\mathscr{R}}$ and $\mathscr{D}$ of the free operad $\mathscr{T}(\mathscr{X} \oplus \mathscr{Y})$ satisfying two constraints: (i) There should exist a map of weight graded collections $\rho \colon \mathscr{R} \to \mathscr{T}(\mathscr{X}) \circ \mathscr{T}(\mathscr{Y}) \subset \mathscr{T}(\mathscr{X} \oplus \mathscr{Y})$ such that the post-composition of $\rho$ with the projection $\mathscr{T}(\mathscr{X} \oplus \mathscr{Y}) \twoheadrightarrow \mathscr{T}(\mathscr{X})$ is zero, and the subcollection $\widetilde{\mathscr{R}}$ consists of all elements of the form $r - \rho(r)$ with $r \in \mathscr{R}$. (ii) There should exist a map of weight graded collections $\lambda \colon \mathscr{Y} \circ' \mathscr{X} \to \mathscr{T}(\mathscr{X} \oplus \mathscr{Y})_{(2)}$ such that the post-composition of $\lambda$ with the projection $\mathscr{T}(\mathscr{X} \oplus \mathscr{Y}) \twoheadrightarrow \mathscr{T}(\mathscr{X})$ is zero, and the subcollection $\mathscr{D}$ consists of all elements $v - \lambda(v)$ with $v \in \mathscr{Y} \circ' \mathscr{X}$.

The first author was supported by the Discovery Grant *Algebraic Operads* from NSERC. The authors thank the anonymous referees for useful comments.

© Springer Nature Switzerland AG 2020
J. Gerhard and I. Kotsireas (Eds.): MC 2019, CCIS 1125, pp. 349–352, 2020.
https://doi.org/10.1007/978-3-030-41258-6_29

These constraints imply $\mathcal{O}/(\mathscr{Y}) \cong \mathscr{P}$; we choose a splitting $\alpha \colon \mathscr{P} \to \mathcal{O}$ on the level of weight graded collections, allowing us to define the following maps:

$$\mathscr{P} \circ \mathscr{Q} \hookrightarrow \mathscr{T}(\mathscr{P} \oplus \mathscr{Q}) \to \mathscr{T}(\mathcal{O}) \to \mathcal{O}.$$

We say that an inhomogeneous distributive rewriting rule is an *inhomogeneous distributive law* if the composite map $\eta$ is an isomorphism on the level of collections.

## 2   Computations

The operad *Com* is generated by the symmetric sequence $\mathscr{X}$; a basis of $\mathscr{X}(2)$ is the commutative operation $a_1a_2$, and $\mathscr{X}(n) = 0$ for $n \neq 2$. The relations $\mathscr{R} \subset Com(3)$ are the $S_3$-module generated by associativity: $(a_1a_2)a_3 - (a_2a_3)a_1$; we have written this relation in commutative normal form. The operad *Lie* is generated by the symmetric sequence $\mathscr{Y}$; a basis of $\mathscr{Y}(2)$ is the anticommutative operation $[a_1, a_2]$, and $\mathscr{Y}(n) = 0$ for $n \neq 2$. The relations $\mathscr{S} \subset Lie(3)$ are the $S_3$-module spanned by the Jacobi identity: $[[a_1, a_2], a_3] - [[a_1, a_3], a_2] + [[a_2, a_3], a_1]$, written in anticommutative normal form.

If an operad $\mathcal{O}$ is obtained from *Com* and *Lie* by an inhomogeneous distributive law, then $\mathcal{O}$ must be a quotient of $\mathscr{T}(\mathscr{X} \oplus \mathscr{Y})$ by relations of arity 3. We determine the viable candidates for the relations of $\mathcal{O}$. They must contain $\mathscr{S}$, meaning in our case the Jacobi identity $[[a_1, a_2], a_3] - [[a_1, a_3], a_2] + [[a_2, a_3], a_1]$. The canonical projection $\mathscr{T}(\mathscr{X} \oplus \mathscr{Y}) \twoheadrightarrow \mathscr{T}(\mathscr{X})$ sends to zero the image of $\rho$; hence $\rho \colon \mathscr{R} \to (\mathscr{X} \circ' \mathscr{Y}) \oplus (\mathscr{Y} \circ' \mathscr{Y})$. The $S_3$-module $\mathscr{R}$ is generated by $(a_1a_2)a_3 - (a_2a_3)a_1$, which satisfies skew-symmetry under the transposition (13), and the sum over cyclic permutations is 0. We obtain

$$(a_1a_2)a_3 - (a_2a_3)a_1 - t_3[[a_1, a_3], a_2].$$

Finally, the canonical projection $\mathscr{T}(\mathscr{X}) \circ \mathscr{T}(\mathscr{Y}) \twoheadrightarrow \mathscr{T}(\mathscr{X})$ must send to zero the image of the map $\lambda$; hence $\lambda \colon \mathscr{Y} \circ' \mathscr{X} \to (\mathscr{X} \circ' \mathscr{Y}) \oplus (\mathscr{Y} \circ' \mathscr{Y})$. By symmetry of $[a_1a_2, a_3]$ under the transposition (12), we obtain

$$[a_1a_2, a_3] - t_1([a_1, a_3]a_2 + [a_2, a_3]a_1) - t_2([[a_1, a_3], a_2] + [[a_2, a_3], a_1]).$$

We focus on checking that the map $\eta$ is an isomorphism in arity 4. To that end, we impose the condition $\dim \mathcal{O}(4) = 24$, and study this condition using the methods we applied for classification of regular parametrized one-relation operads in [2]. This is done using Maple, especially the packages LinearAlgebra and Groebner.

First, we order a basis of $\mathscr{T}(\mathscr{X} \oplus \mathscr{Y})(3)$, and find a reduced row echelon matrix whose rows form a basis of relations of $\mathcal{O}$; those relations are

$$r_1 = (a_1a_2)a_3 - (a_2a_3)a_1 - t_3[a_1, a_3]a_2,$$
$$r_2 = (a_1a_3)a_2 - (a_2a_3)a_1 - t_3[[a_1, a_3], a_2] + t_3[[a_2, a_3], a_1],$$
$$r_3 = [a_1a_2, a_3] - t_1[a_1, a_3]a_2 - t_1[a_2, a_3]a_1 - t_2[[a_1, a_3], a_2] - t_2[[a_2, a_3], a_1],$$
$$r_4 = [a_1a_3, a_2] - t_1[a_1, a_2]a_3 + t_1[a_2, a_3]a_1 - t_2[[a_1, a_3], a_2] + 2t_2[[a_2, a_3], a_1],$$
$$r_5 = [a_2a_3, a_1] + t_1[a_1, a_2]a_3 + t_1[a_1, a_3]a_2 + 2t_2[[a_1, a_3], a_2] - t_2[[a_2, a_3], a_1],$$
$$r_6 = [[a_1, a_2], a_3] - [[a_1, a_3], a_2] + [[a_2, a_3], a_1].$$

Pre- and post-composing the relations $r_1, \ldots, r_6$ with the generators of $\mathscr{O}$, and applying all 24 permutations of the arguments, we obtain a spanning set of 1152 consequences of arity 4 of those relations. These consequences are linear combinations of the basis elements of the vector space $\mathscr{T}(\mathscr{X} \oplus \mathscr{Y})(4)$ of dimension 120. This gives us a $1152 \times 120$ matrix $M$ which has entries in the polynomial ring $\mathbb{Q}[t_1, t_2, t_3]$, which we equip with the *deglex* (*tdeg* in `Maple`) monomial order $t_1 \succ t_2 \succ t_3$.

Since $\mathbb{Q}[t_1, t_2, t_3]$ is not a PID, the matrix $M$ has no Smith form, but since many entries of $M$ are $\pm 1$, we can compute a partial Smith form; see [1, Chapter 8] and [2]. The result is a block matrix

$$\begin{pmatrix} I_{96} & 0_{96 \times 24} \\ 0_{1056 \times 96} & L' \end{pmatrix}$$

where the $1056 \times 24$ lower right block $L'$ has many zero rows. Deleting the zero rows, we obtain a $372 \times 24$ matrix $L$ which contains 126 distinct elements of $\mathbb{Q}[t_1, t_2, t_3]$. We replace each of these elements by its monic form and obtain a set $S$ of 56 distinct polynomials of degrees 2 and 3. Finally, we compute the reduced Gröbner basis for the ideal $I$ generated by $S$ and obtain the set

$$t_2, \qquad t_3(t_1 - 1), \qquad t_1(t_1 - 1).$$

Hence the zero set of $I$ consists of the point $(0, 0, 0)$ and the line $(1, 0, t_3)$. By the results of [3, 6], each of the corresponding operads is indeed obtained from *Com* and *Lie* by an inhomogeneous distributive law. Since any isomorphism between two different such operads $\mathscr{O}$ and $\mathscr{O}'$ must send the symmetric generator of $\mathscr{O}$ into a nonzero scalar multiple of the symmetric generator of $\mathscr{O}'$, and the anti-symmetric generator of $\mathscr{O}$ into a nonzero scalar multiple of the anti-symmetric generator of $\mathscr{O}'$, we immediately obtain a classification up to isomorphism, as follows.

**Theorem 1.** *The only operads obtained from the symmetric operads Com and Lie by an inhomogeneous distributive law are defined by the following relations:*

$$\begin{cases} (x_1 x_2) x_3 - x_1(x_2 x_3) = 0, \\ [x_1 x_2, x_3] = 0, \\ [[x_1, x_2], x_3] + [[x_2, x_3], x_1] + [[x_3, x_1], x_2] = 0. \end{cases} \tag{1}$$

$$\begin{cases} (x_1 x_2) x_3 - x_1(x_2 x_3) + q[[x_1, x_3], x_2] = 0 \quad (q \in \Bbbk), \\ [x_1 x_2, x_3] - [x_1, x_3] x_2 - x_1 [x_2, x_3] = 0, \\ [[x_1, x_2], x_3] + [[x_2, x_3], x_1] + [[x_3, x_1], x_2] = 0. \end{cases} \tag{2}$$

*To classify such operads up to isomorphism, replace $q \in \Bbbk$ by $q \in \Bbbk/(\Bbbk^\times)^2$ in the first relation of the operads (2). In particular, over $\mathbb{C}$, the operads (2) are all isomorphic for nonzero $q$; that is, the same operad admits many different presentations.*

# References

1. Bremner, M., Dotsenko, V.: Algebraic Operads: An Algorithmic Companion. Chapman and Hall/CRC, Boca Raton (2016)
2. Bremner, M., Dotsenko, V.: Classification of regular parametrised one-relation operads. Can. J. Math. **69**(5), 992–1035 (2017)
3. Dotsenko, V., Griffin, J.: Cacti and filtered distributive laws. Algebraic Geom. Topol. **14**(6), 3185–3225 (2014)
4. Dotsenko, V., Markl, M., Remm, E.: Non-Koszulness of operads and positivity of Poincaré series. Preprint arXiv:1604.08580
5. Loday, J.L., Vallette, B.: Algebraic Operads. Grundlehren der mathematischen Wissenschaften, vol. 346. Springer, Heidelberg (2012). https://doi.org/10.1007/978-3-642-30362-3
6. Markl, M., Remm, E.: Algebras with one operation including Poisson and other Lie-admissible algebras. J. Algebra **299**(1), 171–189 (2006)

# Classifying Discrete Structures by Their Stabilizers

Gilbert Labelle[(✉)] [ID]

LaCIM, Université du Québec à Montréal, Montréal, (QC), Canada
labelle.gilbert@uqam.ca
http://www.lacim.uqam.ca

**Abstract.** Combinatorial power series are formal power series of the form $\sum c_{n,H} X^n/H$ where, for each $n$, $H$ runs through subgroups of the symmetric group $S_n$ and the coefficients $c_{n,H}$ are complex numbers (or ordinary power series involving some "weight variables"). Such series conveniently encode species of combinatorial (possibly weighted) structures according to their stabilizers (up to conjugacy). We give general lines for expressing these kinds of series – as well as the main operations $(+, \cdot, \times, \circ, d/dX)$ between them – by making use of the `GroupTheory` package and give suggestions for possible extensions of that package and some other specific procedures such as `collect`, `expand`, `series`, etc. An analysis of multivariable combinatorial power series is also presented.

**Keywords:** Discrete structures · Stabilizers · Combinatorial operations

## 1  Encoding Species by Combinatorial Power Series

Any two rooted trees of Fig. 1 in a rectangle share the <u>same</u> stabilizer group but are <u>not</u> isomorphic. For example, in the last rectangle, this common stabilizer is the 2-element subgroup $\langle (1,2) \rangle$ of $S_5$ generated by the transposition $(1,2)$[1].

**Definition 1.** *Two discrete structures $s$ and $t$ are* similar *if they share the same stabilizer subgroup of $S_n$ after suitable relabelings of their underlying sets by $[n] = \{1, 2, \ldots, n\}$. Equivalently, $s$ and $t$ are similar if they have conjugate stabilizers in $S_n$ when their underlying sets are arbitrarily relabeled by $[n]$.*

In order to enumerate structures according to the nature of their stabilizers, fix, for each $n \geq 0$, a system $\mathcal{H}_n = \{ H_{n,1}, H_{n,2}, \cdots, H_{n,c_n} \}$ of representatives of the $c_n$ conjugacy classes of subgroups of the symmetric group $S_n$ and let $\mathcal{H} = \cup_{n=0}^{\infty} \mathcal{H}_n$. Then, any class $F$ of discrete structures on arbitrary finite sets

---

— See [1] and [4] for more references about combinatorial species.
[1] Two subgroups $G$ of $S_n$ and $H$ of $S_m$, with $n \neq m$, are always considered to be be different, even if they consist of the "same" permutations.

© Springer Nature Switzerland AG 2020
J. Gerhard and I. Kotsireas (Eds.): MC 2019, CCIS 1125, pp. 353–356, 2020.
https://doi.org/10.1007/978-3-030-41258-6_30

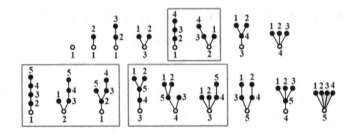

**Fig. 1.** The 17 non isomorphic rooted trees on $\leq 5$ elements grouped by similarity.

that is closed under arbitrary relabellings of their underlying sets[2] is encoded by a combinatorial power series[3] (CPS for short)

$$F(X) = \sum_{n \geq 0} \sum_{H \in \mathcal{H}_n} c_{n,H}(F) X^n / H \tag{1}$$

where $c_{n,H}(F) =$ the number of similar non isomorphic $F$-structures on $[n]$ whose stabilizers are conjugate to $H$. Of course, when $c_{n,H}(F) = 0$, the corresponding term does not appear in (1). Also, if $H = \{id_n\}$, the trivial identity subgroup of $S_n$, then $X^n / \{id_n\}$ is denoted by $X^n$. For example, let $L, C, E$ be respectively the species of finite *linear orders*, *cyclic permutations* and *sets* then it is easily checked that

$$L(X) = \sum_{n \geq 0} X^n, \quad C(X) = \sum_{n \geq 1} X^n / C_n, \quad E(X) = \sum_{n \geq 0} X^n / S_n,$$

where $C_n = \langle (1, 2, \ldots, n) \rangle$, $S_n = \langle (1, 2), (1, 2, \ldots, n) \rangle$. Also, Fig. 1 shows that the first terms, up to degree 5, of the CPS $T(X)$ of the species $T$ of rooted trees are given by

$$T(X) = X + X^2 + X^3 + \frac{X^3}{S_2} + 2X^4 + \frac{X^4}{S_2} + \frac{X^4}{S_3}$$

$$+ 3X^5 + 3\frac{X^5}{S_2} + \frac{X^5}{\langle (1,2)(3,4) \rangle} + \frac{X^5}{S_3} + \frac{X^5}{S_4} + \cdots.$$

## 2    CA Implementation of CPS Calculus

Various operations (including $+, \cdot, \times, \circ, d/dX$) between CPS's have been defined by Yeh to faithfuly reflect the corresponding combinatorial operations between

---

[2] Technically, such classes are *species* in the sense of Joyal [2]. A *species* is an endofunctor $F$ of the category of finite sets with bijections as morphisms. For each finite set $U$, each $s \in F[U]$ is called an $F$-structure on $U$ and for each bijection $\beta : U \to V$, the bijection $F[\beta] : F[U] \to F[V]$ is said to "relabel" (or "transport") each $F$-structure $s$ on $U$ to an isomorphic $F$-structure $t = F[\beta](s)$ on $V$..

[3] This kind of series was introduced by Yeh [5] to deal with species.

species defined by Joyal, by which species can be defined (explicitly or implicitly) in terms of simpler ones. Moreover, the CPS of a species $F$ is the most refined series associated to $F$. It contains, by specialization, all the underlying classical power series : $F(x)$ (which counts labelled $F$-structures), $\widetilde{F}(x)$ (which counts unlabelled $F$-structures), $F(x,q)$ (which $q$-counts $F$-structures), as well as the Pólya-Joyal cycle-index series $Z_F(x_1, x_2, x_3, \ldots)$.

In this presentation we shall deal with the implementation of CPS calculus using and suggesting extensions of the Maple `GroupTheory` package and other specific procedures. The following points will be discussed.

**1.** Construct convenient/canonical lists of generators for subgroups of $S_n$.
**2.** Efficiently decide whether or not two given subgroups of $S_n$ defined in terms of generators are conjugate (as subgroups).
**3.** Construct convenient/canonical (exhaustive or not) ordered lists of systems of representatives $\mathcal{H}_n = \{ H_{n,1}, H_{n,2}, \cdots, H_{n,c_n} \}$ of the conjugacy classes of subgroups of $S_n, n \geq 0$. Give suggestive combinatorial names to some $H_{n,i}$.
**4.** Efficiently decide (using Yeh's criteria [5], or otherwise) whether or not a given subgroup of $S_n$ is atomic. A subgroup $A$ of $S_n$ is said to be *atomic* if it cannot be expressed as an external product $H * K$, in the sense of [5], of two subgroups $H$ of $S_U$ and $K$ of $S_V$, $U \cup V = [n]$, $U \cap V = \emptyset$, $U \neq \emptyset \neq V$.
**5.** Construct convenient/canonical (exhaustive or not) ordered lists of systems of representatives $\mathcal{A}_n = \{ A_{n,1}, A_{n,2}, \cdots, A_{n,a_n} \}$ of the conjugacy classes of atomic subgroups of $S_n, n \geq 0$. A theorem of Yeh [5] asserts that every subgroup $H$ of $S_n$ can be expressed in a unique way (up to conjugacy) as an external product $A * B * \cdots$ of atomic subgroups.
**6.** Implement the operations $+, \cdot, \times, \circ, d/dX$ on CPS's making use of linearity, bilinearity, concatenation, wreath products, double cosets, etc.
**7.** Implement methods (see [3]) to compute the CPS of special classes of species to large degrees. For example, the class $\mathcal{E}$ of *set-like species*, defined as the smallest class of species containing $n$-sets, $n \geq 0$ that is closed under summability, $\cdot$, and $\circ$. Class $\mathcal{E}$ contains the species $T$ of rooted trees, and the CPS $T(X)$ can be expanded up to large degrees by such methods.
**8.** Explore the extensions of the above points to the analysis of CPS of weighted multisort species. These are CPS on several variables $X, Y, \ldots$, of the form

$$F(X, Y, \ldots) = \sum_{n,m,\cdots \geq 0} \sum_{H \in \mathcal{H}_{n,m,\ldots}} c_{n,m,\ldots,H}(F) X^n Y^m \cdots / H$$

where $\mathcal{H}_{n,m,\ldots}$ is a system of representatives of the conjugacy classes of subgroups of the Young subgroup $S_{n,m,\ldots}$ of $S_{n+m+\cdots}$ and each coefficient $c_{n,m,\ldots,H}(F)$ is a polynomial or power series in some *weight-variables* $u, v, \ldots$.

# References

1. Bergeron, F., Labelle, G., Leroux, P.: Combinatorial Species and Tree-like Structures. Ency. of Mathematics and Its Applications, vol. 67. Cambridge University Press, Cambridge (1998)
2. Joyal, A.: Une théorie combinatoire des séries formelles. Adv. Math. **42**, 1–82 (1981)
3. Labelle, G.: New combinatorial computational methods arising from pseudo singletons. In: Discrete Mathematics and Theoretical Computer Science, pp. 247–258 (2008)
4. Labelle, G.: Binomial species and combinatorial exponentiation. J. Électronique du Séminaire Lotharingien de Combinatoire **78**, B78a (2018)
5. Yeh, Y.-N.: The calculus of virtual species and K-species. In: Labelle, G., Leroux, P. (eds.) Combinatoire énumérative. LNM, vol. 1234, pp. 351–369. Springer, Heidelberg (1986). https://doi.org/10.1007/BFb0072525. ISBN 978-3-540-47402-9

# How Maple Has Improved Student Understanding in Differential Equations

Douglas B. Meade$^{(\boxtimes)}$

Department of Mathematics, College of Arts and Sciences,
University of South Carolina, Columbia, SC 29205, USA
meade@math.sc.edu

**Abstract.** In this talk I will provide a quick tour through some of the different ways in which I have used Maple to improve student understanding of traditional topics in an introductory differential equations course, such as direction fields, the phenomenon of beats, and developing an understanding of solutions to first-order systems in phase space.

**Keywords:** Ordinary differential equations · Conceptual understanding · Multiple representations · Symbolic and graphical · Maplet · Embedded component · Maple Cloud

The algebraic manipulations involved in finding an explicit solution to an ordinary differential equation can obscure the concepts and structure of the solution—particularly for students seeing the material for the first time. Maple's symbolic capabilities are one way to circumvent some of the algebraic complications, but students' abilities to see structure in mathematical expressions.

## 1 Visualizing Slope Fields for First-Order ODEs

The traditional introductory course in differential equations is an ideal course to utilize Maple's symbolic and graphical features to increase student understanding of fundamental concepts about slope fields. Figures 1 and 2 show two different maplets intended to help students develop their understanding of slope fields.

These maplets are part of the 201 maplets in the Maplets for Calculus [1] collection that the author co-wrote with Professor Philip Yasskin of Texas A&M University. While the maplet technology is now somewhat outdated, the Maplets for Calculus are still very effective at helping students develop good habits for solving calculus problems as well as improving their conceptual understanding of the calculus and differential equations.

Supported in part by NSF DUE grants 0737209 and 1123170, Michael Monagan, Maplesoft, and the University of South Carolina's College of Arts and Sciences.

J. Gerhard and I. Kotsireas (Eds.): MC 2019, CCIS 1125, pp. 357–361, 2020.
https://doi.org/10.1007/978-3-030-41258-6_31

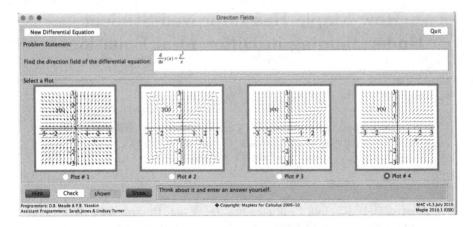

**Fig. 1.** Maplet for identifying slope field for a given ODE. [1]

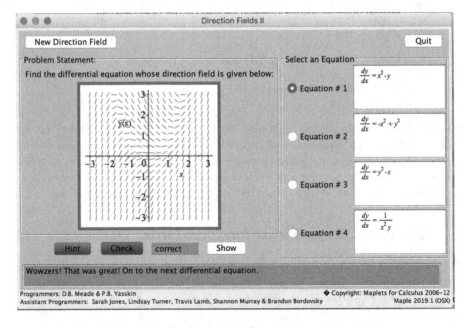

**Fig. 2.** Maplet for identifying ODE for a given slope field. [1]

## 2  Second-Order Linear ODEs

For second-order linear ordinary differential equations with constant coefficients, Maple is an ideal tool for reinforcing the process of finding the solution of the homogeneous equation (Fig. 3), developing an understanding of solutions in phase space (Fig. 4), and understanding beats and other phenomena that appear in solutions of an undamped spring-mass system with periodic external force (Fig. 5).

While Fig. 3 is another example from the Maplets for Calculus collection, Figs. 4 and 5 are built directly in a Maple worksheet using embedded components. These specific examples illustrate some of Maple's interactive features for both symbolic and graphical representations of solutions. I like to challenge students to see if they can find an example that shows a specific characteristic.

## 3   And More ..

In this talk I will share additional examples of interactive Maple-based resources developed utilizing combinations of symbolic and graphical representation. While the specifics are different for each topic, all have a common goal of communicating fundamental mathematics in a way that allows students to better understand the underlying mathematics.

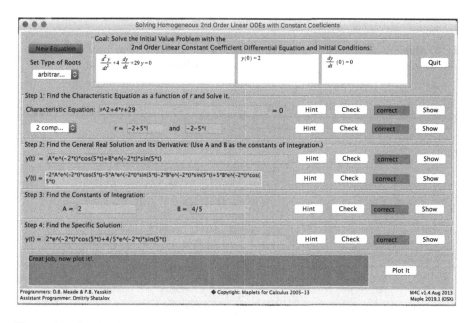

**Fig. 3.** Maplet for finding homogeneous solution of second-order constant coefficient linear ODE. [1]

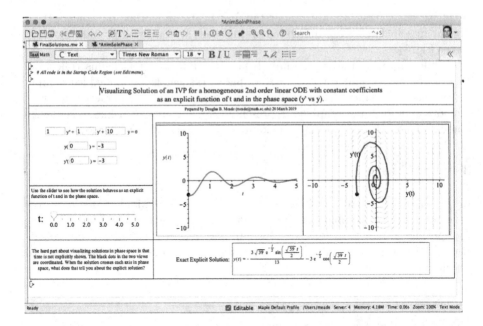

**Fig. 4.** Maple Cloud-based interactive worksheet for multiple visualizations of homogeneous solution of second-order constant coefficient linear ODE. [2]

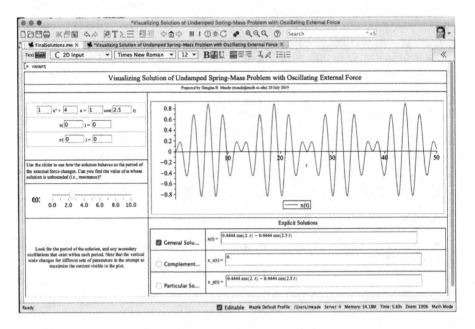

**Fig. 5.** Maple Cloud-based interactive worksheet for visualizing solutions of undamped spring-mass problem with oscillating external force. [3]

# References

1. Maplets for calculus, version 1.3.1. https://m4c.math.sc.edu/. Accessed 10 Sept 2019
2. Visualizing solution to second-order linear IVP: explicit and phase space, Maple Cloud. https://maple.cloud/app/5171234265366528/Visualizing+Soln+to+2nd+order+linear+IVP%3A+explicit+and+phase+space. Accessed 10 Sept 2019
3. Visualizing solution to undamped spring-mass problem with oscillating external force, Maple Cloud. https://maple.cloud/app/6287234297757696/Visualizing+Solution+of+Undamped+Spring-Mass+Problem+with+Oscillating+External+Force. Accessed 10 Sept 2019

# Author Index

Printed in the United States
By Bookmasters